LEIBNIZ IN PARIS

LEIBNIZ IN PARIS
1672-1676

HIS GROWTH TO
MATHEMATICAL MATURITY

JOSEPH E. HOFMANN

CAMBRIDGE UNIVERSITY PRESS

CAMBRIDGE UNIVERSITY PRESS
Cambridge, New York, Melbourne, Madrid, Cape Town, Singapore, São Paulo

Cambridge University Press
The Edinburgh Building, Cambridge CB2 8RU, UK

Published in the United States of America by Cambridge University Press, New York

www.cambridge.org
Information on this title: www.cambridge.org/9780521202589

*Die Entwicklungsgeschichte der Leibnizschen Mathematik während des Aufenthalts
in Paris (1672–1676)* was published by R. Oldenbourg Verlag, Munich,
in 1949. This revised translation first published 1974
This digitally printed version 2008

A catalogue record for this publication is available from the British Library

Library of Congress Catalogue Card Number: 73-80469

ISBN 978-0-521-20258-9 hardback
ISBN 978-0-521-08127-6 paperback

To my very dear wife and
ever constant helpmate in gratitude

CONTENTS

PREFACE

The original German edition (Munich 1949) of this book, completed in manuscript in 1946, was, unavoidably, to a large extent based on shorthand excerpts from sources that could not at that time, immediately after the end of the war, be checked and verified. When, therefore, twenty years later, Mr A. Prag of Oxford and Mr D. T. Whiteside of Cambridge approached me with a proposal for preparing its English translation, it soon became clear to me that the original text would require thorough revision, now that its documentation could be based directly on sources once again fully available, and correction following standard new editions which had appeared in the meantime.

The present revised translation may serve to supplement the first volume, now in the press, of Leibniz' *Mathematisch-naturwissenschaft-lich-technischer Briefwechsel* (in the third series of the current Berlin Academy edition of his *Sämtliche Schriften und Briefe*). This contains all texts, derived wherever possible from the Latin, English, French, Dutch and German primary manuscripts relating to Leibniz' formative stay in Paris during 1672-6, together with a general introduction, explanatory notes and detailed indexes. The present monograph seeks to make known to a wider circle of readers what seems to me the essence of this textual volume, stressing the complicated interconnections and developments of this period of Leibniz' life. To make clear to what extent Leibniz was indebted to others and how he himself reacted to such external stimulating influences, I discuss on the one hand the many communications sent to him or to which he had access: where it will become evident that his main advance was to synthesize their isolated results in an ingenious and perceptive manner; and on the other hand I draw for enlightenment on his surviving research papers and annotations, most as yet still unprinted. For the notes made during his stay in Paris we have Rivaud's brief indications of their content in his *Catalogue critique* of the Hanover manuscripts, but this can, because of its highly

ix

condensed form, serve merely for a first orientation. We may also there find detailed locations of the manuscripts, and in the present work therefore I add such information only when we have to do with papers not listed in the *Catalogue*. By such a policy I hope to render my presentation easier to follow, shortening the footnotes without, I trust, sacrificing their scholarly exactitude. With the same end in view I have condensed the references by employing codings whose denotations are given in the list of abbreviations. All dates of letters quoted are given in full with corresponding Gregorian (New Style) dates supplied in parentheses when appropriate – where only one form appears, the date is the New Style; moreover, since the British New Year then began on 25 March, it is convenient to add the equivalent Gregorian figure: thus for instance I write '25 January (4 February) 1675/6'. In citing printed works a main consideration has been the wish to refer to those which Leibniz himself saw or to which he might reasonably have had access. Where able to quote reliable modern editions, I refer the reader to the best – in the case of letters generally to one only, so as not to overburden the chronological register. Books are cited by standard short-titles by which they will be readily identifiable. Papers and treatises by the main characters in our story are listed not in the general chronological register but will be found under their author in the Index of names. As an exception, papers first published in contemporary periodicals are briefly cited in the chronological index, with cross-reference to the author index where the full titles appear; the periodicals themselves are listed separately. Books and papers by secondary authors are noted individually in the footnotes, usually with their date of publication, in an abbreviated form, keyed to the author index, thus avoiding a separate listing of modern literature.

For the translation itself and for their continuous assistance in reshaping my text I owe both A. Prag and D. T. Whiteside a deeply felt debt of gratitude; I am also grateful for their many services to the members of the Leibniz Archive in Hanover: K. Müller, A. Heinekamp, H. Immel and H.-J. Hess – the last of whom has also helped with the proof-reading; to C. J. Scriba and his Berlin colleagues M. Folkerts and E. Knobloch; to W. Totok, Director of the Niedersächsische Landesbibliothek in Hanover for his ready assistance; and above all to my wife Josepha for her unswerving support and quiet collaboration.

If any reader of the book concerned over its finer detail is thereby stimulated to undertake further research of his own into the origins and development of Leibniz' mathematical ideas, then I would feel that my primary purpose in making this modest contribution to its continuing story has been achieved.

Ichenhausen   J. E. HOFMANN
May 1972

Early in October 1972 Professor Hofmann sent a last addition to his notes for this book incorporating one of his most recent finds in the Leibniz manuscripts. On 9 November he was knocked down by a motor-car while on his early morning walk and some six months later he died – on 7 May 1973, two months after his 73rd birthday.

1
INTRODUCTION

Sunt utique viri magni, de quibus non est temere pronuntiandum.
Wallis to Collins, 16 September 1676

Leibniz came only relatively late to any real acquaintance with the central mathematical notions of his day. By their special nature his early studies had merely awakened in him an interest in the deductive method of mathematics and the formal aspects of computation, while his technical knowledge long remained limited and superficial. It was neither in adolescence – like Pascal and Huygens – nor, like Torricelli and Newton, in his student days at university, nor even – as John Wallis before him – in his first graduate years that he entered the mathematical area, but rather in full intellectual maturity, his doctorate gained and with a developed awareness of his abilities and creative potentialities. The very reverse of the usual French image of a Germanic scholar as a heavy moralizing pedant, he rapidly became on his arrival in Paris in 1672 the charming, irresistibly fascinating centre of everyone's admiring attention, sparkling in his conversation and brilliant in his quick wit – at once an accomplished diplomat subtly skilful in promoting his ambitious political aims, and a vivacious young man of the world with an engaging zest for life and an infinite capacity for hard work. He joins in disputation with the Cartesians one day, with the Jansenists or Jesuits the next, dispatches in a few hastily written words a difficult diplomatic commission, shows interest in the latest inventions, is the guest of craftsmen, magicians, scholars, courtiers and charlatans, writes sharply detailed political reports, dreams of reforming the law on a natural basis, even of radically transforming the existing order of society of his day – and in every such activity he is totally absorbed, knows everything, grasps it all in his mind, has a thousand threads in his hands. It is to be his fate never to find firm ground under his feet, never to be able to create freely but always to remain a tool in the hands of men in power who know how to use him for their purpose but who are quick to set insurmountable barriers to his activities when he threatens to become troublesome.

A mind of such riches, such depths of knowledge, ability and experience, impressed with the necessity of gathering, comprehending and unifying needs a system, a *filum cogitandi* – an intellectual framework into which everything he thinks or encounters can be fitted. How could he be satisfied with the rigid, purely classificatory method of the late scholastics which he had mastered well enough at the German universities? Not even the system of the sciences promulgated by Descartes and then influencing all seriously enquiring minds of the day can altogether please him; he looks further and he looks deeper. In his opinion great progress has been made by Galileo and Descartes, both taking their start in nature, both unveiling her secrets in new ways and by new means. But did they penetrate to the deepest reaches? And is not the mechanistic basis on which they built their world concept too narrow? Will not Cartesianism become calcified just as the now slowly retreating Aristotelianism had done? Leibniz wants to create not a rigid edifice but an elastic structure which shall stand on firmer ground and be more durably constructed. To accomplish this purpose our mode of reasoning must itself be simplified, a method must be found which can comprehend and typify the essential features of the thought process. To think clearly and correctly must no longer be the prerogative of a few chosen spirits but must become the common property of all educated men. And Leibniz believes he is called to lead this march forward like a new Prometheus – struggling in the cause of his great art, the *ars inveniendi*, and ready to yield it up for the benefit of all mankind.

A man who places such thoughts into the forefront of his mind has mathematics in his blood even if he is still ignorant of its detail. And indeed, the actual breadth of knowledge of this twenty-six-year-old is still deplorable. With geometry he evidently gained no real acquaintance either at school or at university. In later years he often recalls how little knowledge in this field he then acquired. No doubt the first book of Euclid's *Elements* (probably in one of the widely used shortened versions by Clavius) was discussed in his presence, but his lessons in it made no impression on him.[1] From his school days

[1] According to Guhrauer (1846) I: 26 Leibniz heard lectures on Euclid by Kühn probably at the start of his studies in the faculty of philosophy, that is in the early summer of 1661; but because of their utter obscurity Leibniz alone, we are told, was somehow able to follow their course. In his memorandum 'De constructione' (winter 1674–5) Leibniz records that he was now reading attentively in Euclid's *Elements* having rarely done so before (*LMG* VII: 254). In his 'Historia et origo

Leibniz can remember only the insignificant *Arithmetic* by Lanz and the good, but limited, *Arithmetic* by Clavius.[2] As an undergraduate the two volumes of the well-commented Latin edition of Descartes' *Géométrie* by Schooten seemed much too complicated to him.[3]

In the *Disputatio arithmetica de complexionibus*[4] which he delivered as a new Magister artium in March 1666 at Leipzig university and which appeared still in the same year in print in a much enlarged version as the *Dissertatio de arte combinatoria*,[5] he mentions, apart from Descartes' *Géométrie*,[6] also Schooten's *Principia matheseos universalis*[7] and in addition (because its algebraic notation differs from his own) Barrow's edition of Euclid.[8] In technical detailed questions on combinations Leibniz starts from Schwenter–Harsdörffer's *Erquickstunden*;[9] from it he borrows references to Cardan,[10] Butéon's *Logistica*, Tartaglia's *General trattato* and Clavius' *Sphæra Johannis de Sacrobosco*,[11] but he has obviously read more closely only Cardan's *Practica arithmeticæ* where, as he emphasizes, the details mentioned by Schwenter do not occur:[12] that all given there is actually all but a literal translation from Clavius (though recognizable as such only to the expert) is briefly indicated. In Clavius' book[13] Leibniz had found remarks on combinations of letters. The title-page[14] itself of the *Dissertatio* has its origin in the illustration at the beginning of Clavius' first chapter. Of more recent foreign literature – Hérigone, Tacquet, Pascal – Leibniz is ignorant. True, he has at this period looked also at certain specialist mathematical books, but for the most part only turned their leaves without real attention; for a

calculi differentialis' (1714) he remarks in retrospect that initially, much occupied with other studies, he had not given sufficient attention to Euclid (*LMG* v: 398). His essay 'In Euclidis πρῶτα' (c. 1696) shows that Leibniz adheres to the Latin word forms of Clavius' edition of the *Elements* (*LMG* v: 183–211).

2 Leibniz–Jakob Bernoulli, April 1703, draft postscript (*LMG* III: 72).
3 *Ibid.* 72.
4 Critical edition by Kabitz in *LSB* VI. 1: 170–75, 228–30.
5 *Ibid.* 165–228.
6 *Ibid.* 171.
7 *Ibid.* 171.
8 *Ibid.* 173.
9 *Ibid.* 169, 173, 202–4, 215–18.
10 *Ibid.* 173, 178, 229.
11 *Ibid.* 173.
12 The passage does in fact not occur in the *Practica arithmeticæ* but in the *Opus novum*; see also *Opera* (1663) IV: 558.
13 Ch. 1: 'De numero et ordine elementorum', quoted with page reference in *LSB* VI. 1: 215.
14 *Ibid.* 166. The blockmaker has replaced the Jesuits' IHS-emblem of the original by a rose and has made a few other changes of little significance in redrawing it. *See* Knobloch (1971).

0	1	1	1	1	1	...	1
1	0	1	2	3	4	...	12
2	0	0	1	3	6	...	66
3	0	0	0	1	4	...	220
4	0	0	0	0	1	...	495
·	·	·					·
·	·	·					·
·	·	·					·
12	0	0	0	0	0		1
*	0	1	3	7	15	...	4095
+	1	2	4	8	16	...	4096

prolonged study of a lengthy chain of deductive reasoning he lacked patience. In the content of his *Disputatio* Leibniz scarcely goes beyond his model Schwenter. He gives the combinations of n elements taken p at a time according to the table above[15] and knows the addition law by which it is built up.[16] The *Dissertatio de arte combinatoria* which grew out of the disputation further contains the multiplication law for n! permutations of n different elements[17] and a related table (up to $n = 24$). It continues with the relation

$$2(n+1)! - n.n! = (n+1)! + n!$$

and the enumeration of certain permutations with repeated elements[18] – not in a tight methodically arranged form, but expounded by means of examples from the doctrine of syllogisms,[19] from the combination of letters into words,[20] of musical notes into melodies,[21] of long and short syllables into lines of verse[22] and the like. At the end of the disputation, in a passage omitted in the revised dissertation, appears the metaphysical corollary drawn from Cardan:[23] *Infinitum aliud alio maius est.* In his *Arithmetica infinitorum*, Leibniz adds, 'Seth Ward' (read: Wallis) is said to have adopted a different standpoint. When an unauthorized reprint of the *Ars combinatoria* appeared in 1690 Leibniz had a notice inserted in the *Acta Eruditorum*[24] pointing out

[15] *LSB* VI. 1: 174.
[16] *Ibid.* 175.
[17] *Ibid.* 211–12.
[18] *Ibid.* 212–13.
[19] *Ibid.* 180–8, in the course of a critical discussion of J. Hospinianus' writings on Logic of 1560 and 1576.
[20] *Ibid.* 223.
[21] *Ibid.* 218. An error due to carelessness in this calculation has been pointed out by M. Cantor in his *Vorlesungen über Geschichte der Mathematik* ₂III: 44.
[22] *LSB* VI. 1: 225–7.
[23] *Ibid.* 229. Leibniz refers the reader to *Practica arithmeticæ*, ch. 66, Nos. 165 and 260.
[24] *AE* (February 1691): 63–4 = *LPG* IV: 103–4.

INTRODUCTION [1

the unfinished and juvenile character of the work in its treatment of many details – probably referring to his indiscriminate use of equality signs and similar faults. Leibniz' application for his doctor's degree in law in Leipzig was refused in 1666 on account of his youth. Accordingly he went to Altdorf where his promotion[25] took place in February 1667. He subsequently stayed on in Nuremberg for several months, where for a while he moved entirely within a circle of Rosicrucians, trying his hardest to find out their secrets. Leibniz remembers two mathematical works he had in his hands during this time at Nuremberg. One is Léotaud's *Examen circuli quadraturæ*, the other Cavalieri's *Geometria indivisibilibus continuorum nova quadam ratione promota*.[26] For serious study there was, of course, no opportunity even yet. The second part of Léotaud's book, in which the erroneous circle quadrature of Grégoire de Saint-Vincent is refuted, he could in any case not have understood without a proper knowledge of the *Opus geometricum*.[27] The matter is different in regard to the first part, which contains a youthful piece (written a generation previously) by Artus de Lionne. In this, the *Amœnior curvilineorum contemplatio*,[28] curvilinear areas whose boundaries are circular arcs are squared in an ingenious way. The starting point is the quadrable circular lunule of Hippocrates cut off between the semicircle ACB and the quadrant ADB (fig. 1). A first major result is the equality of the mixtilinear triangle ACD and the rectilinear triangle AFN; a second result the equality of the mixtilinear triangles ACD and AED, each of which equals half the isosceles rightangled triangle ACE (fig. 2); a third result is the equality of the two lunules cut off between the semicircular arcs ACB, BFC, CGA to the rightangled triangle ABC whose sides are the diameters of the semicircles (fig. 3). Leibniz could remember none of these details in later life, but memory of the book itself lingered.[29]

[25] The dissertation *De casibus perplexis in jure* (Nuremberg 1666) which Leibniz probably had already completed in Leipzig provided the basis for the disputation of 15 Nov. 1666; the degree ceremony took place at the university of Altdorf on 22 Feb. 1667 (Müller–Krönert: 9). The origin of the titlepage which has doubtless been adapted from another work (*LSB* VI. 1: 233) has not yet been established.
[26] Leibniz–Jakob Bernoulli, 1703, postscript (*LMG* III: 72); 'Hist. et origo' (*LMG* V: 398); Leibniz–Conti, 9 Apr. 1716 (*LBG*: 278).
[27] Concerning this work which Leibniz mentions already in his edition of Nizolius (1670) (*LSB* VI. 2: 432) and in a note of winter 1671–2 (*ibid.* 480), but which he apparently has not yet seen himself, *see* below ch. 2: note 9.
[28] For the contents of this treatise *see* Hofmann (1938).
[29] Leibniz–Jakob Bernoulli, 1703, postscript (*LMG* III: 72); Leibniz–Bodenhausen, 5 Oct. 1692 (*LMG* VII: 375); 'Hist. et origo' (*LMG* V: 398).

5

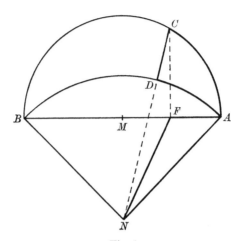

Fig. 1

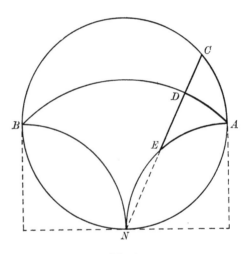

Fig. 2

In his 'proud ignorance' he believed himself capable of assimilating the contents of this and other mathematical books which he read 'like novels' by merely skimming through them.[30]

The ill effect of this skimping sort of study shows itself particularly in the case of Cavalieri's method of indivisibles which already contains within its compass a fair measure of vagueness and obscurity of its own. Leibniz did not at this time look too closely at the original

---

[30] Leibniz–Jakob Bernoulli, 1703, postscript (*LMG* III: 72).

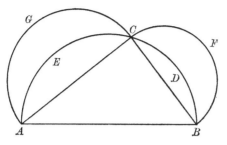

Fig. 3

but, we may add, effectively orientated his mathematical opinions in line with Hobbes' *Elementa philosophiæ*,[31] which, though a significant philosophical achievement, was nonetheless written by a man lacking proper mathematical expertise. The clearest indication of the level of Leibniz' mathematical knowledge of these days may be found in his *Hypothesis physica nova*[32] of 1671, where the method of indivisibles is briefly described as a *fundamentum prædemonstrabile*. Here two fundamental viewpoints, namely Cavalieri's concept of indivisibles and Archimedes' of finite increments, are everywhere confused – a feature of Cavalieri's book as well, though not so blatantly evident.[33] In an attempt to demonstrate the existence of an 'extensionless, indivisible' line-segment Leibniz constructs, quite perfunctorily, a continued bisection of a line.[34] He sets himself in conscious contrast to Euclid's well known definition: *Punctum est*

[31] Leibniz by 1670–1 knew the separate Latin parts of Hobbes' *Elementa philosophiæ* (1655, 1658, 1642) as well as the complete edition including the *Leviathan* (1651), the *Examinatio* (1660) and *De principiis* (1666). He refers to Hobbes' writings for the first time in his marginal notes (1663–4?) to Daniel Stahl's *Compendium* (1655) and Jakob Thomasius' *Philosophia practica* (1661) (*LSB* VI. 1: 22, 25, 60, 67). Several remarks in his letters of this time show that he was fascinated by many of Hobbes' ideas and greatly admired the subtleness of his thought. Hobbes' intention to represent all syllogistic conclusions by a symbolic calculus (*De corpore* I ch. 1, §2) touches closely on his own fundamental related ideas as does the expectation to find proofs for Euclid's axioms (*ibid.* ch. 6, §13); *see* below ch. 2: note 12. Leibniz however strongly opposes the suggestion that definitions might be arbitrarily fixed: *see* p. 21 and Leibniz–Tschirnhaus, early 1680 (*LBG*: 405, 411). Regarding the relations between Leibniz and Hobbes *see* Couturat (1901): 457–72.

[32] Part 1: *Theoria motus concreti*, part 2: *Theoria motus abstracti*; *see* also ch. 3: note 8.

[33] Without actually naming Cavalieri, Leibniz refers to the method of indivisibles in the preface to the *Theoria motus abstracti* (*LSB* VI. 2: 262). He had probably learnt of this method through his reading of Hobbes' *De corpore* I ch. 20, and at first appears to have informed himself about Cavalieri's procedure only at second hand – as he presumably also did with regard to Archimedes' writings.

[34] *LSB* VI. 2: 264.

7

cuius pars est nulla, and states that even though a point were an entity without an extension assignable in magnitude, or at all events less than any given quantity, it would still be possible to divide it; the parts indeed will be characterized as being without distance.[35] Other infinitesimal quantities can be regarded in a similar manner, for instance the indivisible of a circular arc.[36] This is certainly greater than its chord; while to a larger circle there pertains a larger arc-indivisible (on making the silent assumption of equal angle-elements in the two cases). In this way, Leibniz says, the puzzles of both the incommensurable and the angle of contact can be solved, particularly since every angle is an extensionless quantity.[37] The (regular) polygon of infinitely many sides indeed coincides, he says, with the corresponding circle, but only in extension and not in quantity, though the difference is less than any assignable quantity.[38] For an explanation of the measure of cylinder, cone and sphere the concept of movement is introduced;[39] but side-by-side with this idea of continuous quantities goes that of the discrete, as in the case of a point-by-point construction for the quadratrix or in Archimedes' measurement of the circle by polygons, where, as Leibniz says, it matters only to make the error immeasurably small.[40]

Leibniz is, as we see, completely under the spell of the concept of indivisibles, has no clear idea of the real nature of infinitesimal mathematics, and believes that for him, superficial conceptual allusions are sufficient. He further recounts that he had contrived during these early years a geometrical 'calculus' of his own, in which he operated with an unlimited number of squares and cubes,[41] unaware that it had all been done to far better purpose already by

[35] *Ibid.* 265 and 267. Here Leibniz refers to what he has read in Hobbes' *De corpore* (I ch. 15, §2), and *De principiis* (ch. 1). He sees in this new definition of a point the essential feature of Cavalieri's method.

[36] *Ibid.* 267.

[37] *Ibid.* 267; *see* also ch. 2: note 9.

[38] *Ibid.* 267.

[39] *Ibid.* 272.

[40] *Ibid.* 273. Leibniz is rather proud at having successfully worked out the essentials of Cavalieri's concept. He writes to this effect on 1 (11) Mar. 1671 to Oldenburg, early in May to Velthuysen and again on 22 June (?) and 17 Aug. of the same year to Carcavy (*LSB* II. 1: 90, 97, 126, 143).

[41] Leibniz–Jakob Bernoulli, 1703, postscript (*LMG* III: 72). Leibniz there uses the technical term *quadratillum*, to be found also in his *Theoria motus abstracti* (*LSB* VI. 2: 275). These earliest mathematical studies – of which no written notes have survived – are mentioned again in connection with Hobbes' rather absurd objections to the validity of the so-called Pythagorean theorem (*De corpore* I ch. 20, prop. 1; *Examinatio*, dial. 3; *De Principiis* ch. 23); *see LSB* VI. 2: 267.

Viète and Descartes.[42] But as an offshoot of these first trials there
developed a certain inward propensity towards mathematical lines of
reasoning. With the more primitive things – such as a typical proof
in elementary geometry or a lengthy transformation in algebra – he
never even in later years found it easy to cope, and errors in calculations are no rarity in his writings. In his case we see with all clarity
that purely formal schooling can only be accomplished during the
years of youth. His failing in this regard, however, obliged him ever
to think of new expedients and so to enrich science by new methods
in situations where a capable average brain would have achieved his
end without trouble. He was well aware of this and knew that his
strength lay not in elaborating the formal side, but in finding the
well-reasoned conclusion at the decisive point. Hence arises his
striving for a mechanization of purely technical aspects, in particular
the process of computation, hence his perennial struggle for a
calculating machine equally useful for all four species of computation,[43] the idea of which – so difficult to translate into practice – was
already in his mind before he came to Paris.

In the case of nearly all other important mathematicians the *grande
passion* is already recognizable during puberty and leads in the period
immediately following to decisive new ideas. In Leibniz' case this
biologically significant period passed without any specifically mathematical experience; that came to him only in full maturity; but then
indeed it took him by force, never to let him free again. During his
four years in Paris he pursued mathematics as an autodidact with a
strength and intensity known to few men. Chance has brought it
about that this man, forgetful to an unprecedented degree in his
individual mathematical results, carefully kept almost every sheet of
his notes – many with the exact date, others datable through watermarks, the quality of the paper, the character of their handwriting,
the notation employed (often changed in his early years) and technical terms used – a chaotic mass of scrap sheets from which Leibniz
in his old age in all seriousness wanted to compile a description of the
discoveries made or inaugurated by him.[44] Few of these notes have so

---

[42] Leibniz refers here to the progress made by Viète and Descartes in applying literal
calculation to geometry.
[43] For details *see* von Mackensen (1969).
[44] Leibniz talks of a planned treatise with the title 'Scientia infiniti' in his letters to
Johann Bernoulli of 31 Mar. 1694 (*LMG* III, 136); to Huygens of 22 June 1694
(*HO* 10: 640); to L'Hospital of 16 Aug. 1694 (conjectured from *LMG* II: 252); and
to Jakob Bernoulli of 12 Dec. 1695 (*LMG* III: 29). Johann Bernoulli mentions the
planned work in the *AE* of October 1694: 393–8 (*BJC* I: 119), so does L'Hospital

far been published, many have to be regarded as preliminary studies and are unfit to be printed, but it has all only been very partially sifted or drawn upon. Still, it is now possible from the letters exchanged during those years to gain an insight, clear in all essentials, into Leibniz' mathematical development during his Paris period, and so into the birth of the higher analysis – and this must rank as one of the greatest products of human intellect. We should not overlook the difficulties obtaining at that time in publishing books; nor that learned journals were only then just beginning to appear and that editors preferred, on the one hand encyclopedic descriptions of the everyday world, and on the other, the cleverly styled essay. The letter in its immediacy of expression of personal experience or its outline indication of discovery, was therefore substitute for, and precursor of today's periodical. Several generations of scholars have centred their research effort on these letters, but only now is the full text of the originals coming to be known, only now has the history of their genesis been clarified by careful study of their detail. This by no means involves Leibniz alone but an interwoven network of mutual relations which it is necessary to understand and survey completely before the deeper significance of the whole can be comprehended.

In all innocence, Leibniz ultimately became involved in the disputes between French and English scholars which were concerned less with matters of science and learning than with personal touchiness and questions of national vanity. In his boundless optimism, Leibniz believed he could catch hold of the threads of this fine-spun net in one bold grasp, but he overestimated his own valour and misjudged the situation. Indeed from his ignorance of the true circumstances arose those ever regrettable misunderstandings which a generation later brought down upon him the severest of reproaches and which finally reached their climax in the accusation made against him of intellectual theft. That this was baseless – that he had in fact, like his great adversary Newton, reached his grandiose results by his own unaided ingenuity can now accurately be demonstrated through the documents, by the exact agreement of his

in the preface to his *Analyse* (1696): e, fols. 2r–2v. Leibniz writes on 16 Dec. 1697 to Bodenhausen, that the work might well be expected to be completed if only he had someone to assist him (*LMG* VII: 392). If not even a preliminary draft was put together, we may see the reason in the great number of related publications by Leibniz beginning in 1694, full of results which were leading him further and made a concentrated treatise dealing with fundamental concepts hardly feasible.

private notes with the public statements in his letters and with other assertions in the writings of his contemporaries.[45] We must leave a detailed, more penetrating account of the topic to another occasion; here we will provide only a summary survey wherein we confine ourselves to a brief outline of the main issues involved.

[45] The publication (since 1959) of Newton's *Correspondence* edited first by H. W. Turnbull, then by J. F. Scott and now by A. R. Hall, together with the exemplary and rapidly progressing edition since 1967 of Newton's *Mathematical Papers* by D. T. Whiteside have brought us a wealth of new and reliably documented insights into Newton's mathematical activity.

2

THE 'ACCESSIO AD ARITHMETICAM INFINITORUM'

Leibniz arrived at Paris in March 1672 as the travelling companion of Melchior Friedrich von Schönborn (who, as a nephew of the Elector of Mainz and son-in-law of Boineburg,[1] had been sent on an important mission to France). His political duties[2] and the study of the French language did not allow him any personal freedom of movement for many months; only in the autumn of 1672 did he gain an opportunity to pay his respects to Huygens, then approaching the zenith of his fame and widely hailed as the leading mathematician and scientist of all Europe.[3] We learn details of his stay from letters exchanged at this time, from marginal notes inserted by Leibniz in his working copy of the *Commercium epistolicum* to which we shall later revert, and lastly from his autobiographical relation in 1714 of his discoveries in the realm of mathematics, the *Historia et origo calculi differentialis*.

Persuaded by reservations expressed by Hobbes,[4] Leibniz had taken a closer look at Euclid's axiom:[5] the whole is always greater than the part.[6] This seemed to lose its validity when applied to the angle of contact (that is the angle between a circle-arc and its tangent) – a difficulty over which, ever since Euclid's own days, dispute had raged with undiminished vigour showing no sign of coming to a halt. The basic difficulty concerned the definition of such an angle; this, in antiquity, and still in scholastic philosophy vaguely connected with areal concepts based on purely intuitive considera-

---

[1] M. F. Schönborn had in 1668 married Boineburg's daughter Sophie.

[2] For details of the political aspects of the time we refer the reader to Wiedeburg (1962, 1970).

[3] In the spring of 1666 Huygens had become a member of the Académie des Sciences recently founded by Colbert.

[4] *De corpore* i ch. 8, §25.

[5] *Elements* i, axiom 9 (according to the older style of numbering current at the time).

[6] Already in the *Disputatio de complexionibus* (1666) (*LSB* vi. 1: 171, No. 8 and 172, No. 10). Leibniz had pointed out that the whole, being the greater entity, had to be divided into (equal) lesser parts for the determination of *complexiones*.

tions,[7] was only clarified when the angle was finally declared to be the measure of a rotation and the angle of contact therefore necessarily of measure nil.[8] Leibniz did not yet employ this definition in terms of rotation. The failure of the axiom in the instance of the angle of contact[9] confirmed him in his suspicion that in fact we are here dealing with a provable theorem. He believed he found the crucial point to lie in his insight that the angle of contact lacks the quality of being a magnitude.[10] In principle he wished to admit only two types of unproved truths, namely definitions and identities.[11] The axiom in question is neither of these; hence it is required to find the constituent parts from which it can be built up. These are his conclusions:[12]

(1) If of two objects one is part of another, then the first is called smaller and the second larger, and this is a definition.

(2) Everything that is affected with magnitude is equal to itself, and this is a statement of identity.

(3) A magnitude which is equal to a part of another is smaller than this (by definition).

[7] For the Ancients the decisive question had been whether the angle should be understood as a quantity, a quality or a relation. It is thoroughly discussed by Proclus in his commentary to Euclid's *Elements* I when he expounds the definition of an angle (def. 8) basing himself on several earlier authors. Linked with it is the further problem whether the angle of contact partakes of the properties of a magnitude capable of being 'increased and diminished'. Many medieval philosophers recognized the angle of contact as a magnitude, although they were aware that it bore no assignable ratio to a rectilinear angle. Leibniz too speaks in this sense in the *Doctrina conditionum* (1669): (*LSB* VI. 1: 387).
[8] Möbius (1846) = *Werke* II (1885): 4–5.
[9] Grégoire de Saint-Vincent, *Opus geometricum* (1647): 871. The question is taken up in a note by Leibniz to his edition of Nizolius (1670) (*LSB* VI. 2: 432) and again later in a note of 1671-2 (*ibid.* 480). Probably these remarks are taken over from Hobbes' *De corpore* I ch. 8, §25 since Leibniz looked more closely at the *Opus geometricum* only towards the end of 1672, later also acquiring a copy of his own (*C* 1553) in which he made a number of interesting marginal notes. For these *see* Hofmann (1942): 10–15, 22, 39, 51, 53.
[10] In his 'In Euclidis πρῶτα' (c. 1696) (*LMG* V: 191), Leibniz follows J. Peletier's (1557) interpretation who resolves the difficulty by declaring the angle of contact actually to have zero magnitude – a remark which immediately roused considerable opposition.
[11] At a number of places Leibniz insists on the necessity of starting a logical sentence structure from suitably chosen definitions recognizing the axioms as in reality provable statements originating from the logical nexus of a chain of definitions. Basically, he asserts, there is but one axiom, that of identity. *See* further Couturat (1901): 186–8, 203–5.
[12] The following considerations occur first in a note of 1671-2 (*LSB* VI. 2: 482–3); later versions and variants are found in Leibniz for Gallois, late 1672 (*LSB* II. 1: 222–9): 'Specimen geometriæ luciferæ' (1687?) (*LMG* VII: 273); 'In Euclidis πρῶτα' (c. 1696) (*LMG* V: 207); Leibniz – Johann Bernoulli, 2 Sept. 1696 (*LMG* III: 321–2); 'Hist. et origo' (*LMG* V: 395–3); 'Initia rerum mathematicarum metaphysica' (1714) (*LMG* VII: 20); undated note: Couturat (1903): 518.

(4) The part is equal to a part of the whole.

(5) Therefore every part of a magnitude is smaller than the whole. By means of this axiom of identity the effectiveness of which his contemporaries failed to see,[13] Leibniz develops his main theorem on the summation of consecutive terms of a series of differences. Evidently[14]

$$a_0 - a_0 + a_1 - a_1 + a_2 - a_2 + \ldots + a_n - a_n = 0.$$

If now

$$0 < a_0 < a_1 < a_2 < \ldots < a_n \quad \text{and}$$
$$b_0 = a_1 - a_0, \quad b_1 = a_2 - a_1, \ldots, b_{n-1} = a_n - a_{n-1},$$

then

$$b_0 + b_1 + b_2 + \ldots + b_{n-1} = a_n - a_0.$$

That is, the sum of consecutive terms of a difference-series is equal to the difference of the two extreme terms of the related series. Thus we obtain from, for instance,

$$0 \quad 1 \quad 4 \quad 9 \quad 16 \quad 25\ldots$$
$$1 \quad 3 \quad 5 \quad 7 \quad 9 \ldots$$

the result that the sum of consecutive odd numbers can be expressed as the difference of two squares.[15] Considerations of this sort led him to the conviction that we should be able to derive the sum of any series whose terms are formed by some rule,[16] even where one has to deal with infinitely many terms – assuming only that the expected total sum approaches a finite limit.

This, his first major mathematical discovery in Paris, originated in thoughts strongly influenced by considerations of logic and philosophy – and, as so often with Leibniz, was not fully established, but

---

[13] Not one of the correspondents to whom Leibniz had sent his demonstration approved of it; in particular Johann Bernoulli made it clear in his reply of 22 Sept. 1696 (*LMG* III: 329–30) that he considered the conclusion to be circular.

[14] Referred to in 'Hist. et origo' (*LMG* v: 396).

[15] In 'Hist. et origo' (*LMG* v: 396) the scheme of differences of successive square numbers is declared to be one of the earliest results first mentioned in the 'Consilium de encyclopædia nova' (25 June 1679) (Couturat (1903): 30). Other places are enumerated in Couturat (1901): 262, quoting instances where Leibniz wishes to give a specially simple example for a difference scheme.

[16] The earliest relevant notes for a complete difference scheme have not been found. An attempt at a reconstruction, based on the inversion of the arithmetical triangle from its form in the *Disputatio de complexionibus* (1666) and employing number-pairs has been made in Hofmann (1970): 81–5. With letters in place of numbers the difference scheme appears in the letter to Johann Bernoulli of 16 Dec. 1694 (*LMG* III: 155); also in 'Hist. et origo' (*LMG* v: 397).

came as the fruit of a particular insight observed in simple examples and generalized by a stroke of genius. The result, that summation of the series can be carried out even over an infinite number of terms, was at once reported by him in conversation with Huygens.[17] Though he was not very familiar with topics in the general theory of series (then only in its infancy), the statement roused Huygens's interest, and he conceived it might be correct. Wishing to make an immediate test of the young man, he chose as an example a result which occurred to him in 1665 when involved in a discussion on questions of probability in games of chance with Hudde, namely to determine the sum of the infinite series of reciprocal triangular numbers[18]

$$\tfrac{1}{1} + \tfrac{1}{3} + \tfrac{1}{6} + \tfrac{1}{10} + \ldots$$

It was a fateful question; another example only slightly more difficult (and hence for Leibniz insoluble) would no doubt have quenched his enthusiasm for pursuing his new-found interest in mathematics; but with the one proposed here he could cope – moreover, by changing and extending the procedure of summation he arrived at new and beautiful results: thus he began to concern himself more deeply with particular mathematical problems. No doubt (as he usually did) during this conversation Huygens drew attention to the relevant literature, in particular to the summation of the geometrical progression in Grégoire de Saint-Vincent's *Opus geometricum*. Leibniz immediately borrowed this latter work from the Royal Library in Paris.[19] At the time when he still did not own the book he made excerpts of certain details in it.[20] The demonstrations concerning the geometrical progression seem to be the first he looked at more closely.

Geometrical progressions are treated fairly extensively in the second book of the *Opus geometricum*. To sum infinitely many terms Grégoire uses an intuitive geometrical procedure: in order to find the sum of a geometrical progression of line-segments with the initial terms AB, BC, he constructs the point sequence C, D, E, ... such that $AB:BC = BC:CD = CD:DE = \ldots$ (fig. 4); he further determines a point K from the proportion $AB:BC = AK:BK$ and shows that now $AB:BK = BC:CK = CD:DK = DE:EK = \ldots$; he then states that

---

17 The visit to Huygens is mentioned in the letters to Oldenburg of 16 (26) Apr. and 14 (24) May 1673 (*LBG*: 93 and 95); also in 'Hist. et origo' (*LMG* v: 404).
18 *See* below p. 19.
19 Leibniz–Jakob Bernoulli, 1703, postscript (*LMG* iii: 72).
20 These excerpts are no longer extant; references are to the sum of a geometric progression (*C* 510B, 794) and to its sum to infinity for the quadrature of the hyperbola (*C* 794, 820, 883, 1087).

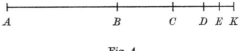

Fig. 4

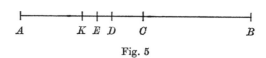

Fig. 5

AK is larger than the sum of every finite number of terms, hence certainly not less than the sum of the complete infinite series. There is left however a remainder which converges towards zero; therefore AK cannot be larger than the sum of the infinite geometrical progression either: hence AK is exactly equal to that sum. Grégoire gives several variants of this construction, the most important of which is indicated in fig. 5 where the consecutive line-segments BC, CD, DE ... are placed end to end from B towards A, and in this way the sums

$$\tfrac{1}{2}+\tfrac{1}{4}+\tfrac{1}{8}+\ldots = 1,$$
$$\tfrac{1}{3}+\tfrac{1}{9}+\tfrac{1}{27}+\ldots = \tfrac{1}{2},$$
$$\tfrac{1}{4}+\tfrac{1}{16}+\tfrac{1}{64}+\ldots = \tfrac{1}{3}$$

are constructed.

Leibniz perused the text of the *Opus geometricum* only very superficially. Grégoire's clumsy protracted manner of writing was far too heavy-handed for him.[21] With a sure eye he recognized the essential principle: the line-segments which are to represent the geometrical progression must not be placed end to end, but must all be made to start from one and the same point. When this is done then the differences of consecutive terms of the series turn out to be proportional to the original series. The significance of this becomes immediately clear from figs. 6 and 7. In these we can read off the sums, proceeding from right to left:

$$\tfrac{1}{2}+\tfrac{1}{4}+\tfrac{1}{8}+\ldots = 1, \quad \text{and} \quad \tfrac{2}{3}+\tfrac{2}{9}+\tfrac{2}{27}+\ldots = 1,$$

that is

$$\tfrac{1}{3}+\tfrac{1}{9}+\tfrac{1}{27}+\ldots = \tfrac{1}{2}$$

or generally

$$1/t + 1/t^2 + 1/t^3 + \ldots = 1/(t-1).$$

[21] Leibniz for Gallois, late 1672 (*LSB* II. 1: 223).

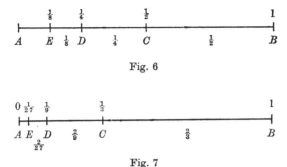

Fig. 6

Fig. 7

That from the geometrical progression $1, b, b^2, \ldots$ we may obtain by differencing (that is by $b^n - b^{n+1} = b^n(1-b)$) a new geometrical progression, proportional to the original one was a considerable insight.[22] Obviously this did not come as an immediate inspiration, but was suggested by the similar direction of Grégoire's considerations. Yet what Leibniz had derived from his predecessor's pattern of reasoning – namely, a general method – was something new and comprehensive. Here we see something which is typical of Leibniz's mode of thinking: his obtaining of a constructive general insight into relationships previously sketched but not yet finally clarified. The most interesting aspect of this thought process is his return to the purely conceptual and non-algorithmical: the reduction of a method hitherto difficult to a very simple basic idea which is then freed from its external cloak and is brought in its essential purity into the light: an 'explicatio' indeed, a development in the best sense of the word.

How the new idea can be applied we shall see at once when for instance we choose $AB = 1$, $AC = \frac{1}{2}$, $AD = \frac{1}{3}$, $AE = \frac{1}{4}$.... We now read off the relation

$$\frac{1}{1.2} + \frac{1}{2.3} + \frac{1}{3.4} + \frac{1}{4.5} + \ldots = 1,$$

and after multiplication by 2 we obtain the sum demanded by Huygens: $\frac{1}{1} + \frac{1}{3} + \frac{1}{6} + \frac{1}{10} + \ldots = 2$. This is the successful path which Leibniz took in his treatment of the problem proposed.[23] He also translated the topic into the realm of arithmetic proper[24] by writing out the two series

[22] 'Hist. et origo' (*LMG* v: 407). *See* also Leibniz, *AE* (February 1682): 45 (*LMG* v: 121).   [23] *See* Hofmann (1970): 87–8.
[24] Concerning Leibniz' first approaches *see* Mahnke's additions to Hofmann–Wieleitner (1931): 590–7. What follows is taken from *C* 1336 (February 1676).

17

$$A \quad \tfrac{1}{1} \quad \tfrac{1}{2} \quad \tfrac{1}{3} \quad \tfrac{1}{4} \quad \tfrac{1}{5} \dots$$
$$B \quad \tfrac{1}{1} \quad \tfrac{1}{3} \quad \tfrac{1}{6} \quad \tfrac{1}{10} \dots$$

and remarking that the product of consecutive denominators of terms in A is equal to twice the denominator of the intervening term of B; in other words, he has recognized $\tfrac{1}{2}B$ as the series of differences of A. A little later he states his result in the following form:

$$A - 1 = \tfrac{1}{2} + \tfrac{1}{3} + \tfrac{1}{4} + \tfrac{1}{5} + \dots$$
$$\tfrac{1}{2}B = \tfrac{1}{2} + \tfrac{1}{6} + \tfrac{1}{12} + \tfrac{1}{20} + \dots$$
$$\overline{A - 1 + \tfrac{1}{2}B = 1 + \tfrac{1}{2} + \tfrac{1}{3} + \tfrac{1}{4} + \dots = A,}$$

hence $B = 2$.

This sort of deduction would not satisfy us today, since the series A employed for the summation is divergent.[25] It is however quite easy to eliminate this difficulty, by combining only the finite partial series

$$A_n = 1 + \tfrac{1}{2} + \tfrac{1}{3} + \dots + 1/n$$

and

$$\tfrac{1}{2}B_n = \frac{1}{1.2} + \frac{1}{2.3} + \frac{1}{3.4} + \dots + \frac{1}{n(n+1)}$$

whence we now have

$$B_n = 2 - \frac{2}{n+1} \quad \text{and} \quad \lim_{n \to \infty} B_n = 2.$$

The genesis of this last sketched calculation shows us that Leibniz in fact meant to do exactly as much, but was still incapable of expressing himself in this way, lacking as he did any experience in algebraic transformations.

Leibniz did not stop with this success, indeed he at once tried how the choice of $AB = 1$, $AC = \tfrac{1}{3}$, $AD = \tfrac{1}{6}$, $AE = \tfrac{1}{10} \dots$ would work out. He now obtains

$$\tfrac{2}{3}(1 + \tfrac{1}{4} + \tfrac{1}{10} + \tfrac{1}{20} + \dots) = 1[= \tfrac{2}{3}C];$$

[25] Rather surprisingly the same fallacy occurs also in Jakob Bernoulli's first essay on series (1689), prop. 15 (*BKC*: 388–9). Newton employs a variant of the procedure in the 'Account' of the *CE* (*1712*) (*PT* 29 No. 342 for Jan./Feb. 1714/15: 183, reprinted under the title 'Recensio libri' as preface to *CE* (*1722*): 13).

that is the sum of the reciprocals of the pyramidal numbers. This too he translates into arithmetic by setting

$$
\begin{aligned}
B-1 &= \tfrac{1}{3}+\tfrac{1}{6}+\tfrac{1}{10}+\tfrac{1}{15}+\cdots \\
\tfrac{2}{3}C &= \tfrac{2}{3}+\tfrac{2}{12}+\tfrac{2}{30}+\tfrac{2}{60}+\cdots \\
\hline
B-1+\tfrac{2}{3}C &= \tfrac{3}{3}+\tfrac{4}{12}+\tfrac{5}{30}+\tfrac{6}{60}+\cdots = B,
\end{aligned}
$$

hence $C = \tfrac{3}{2}$.

In this case his deduction carries conviction for us too since the series B converges. Similarly he finds the series

$$
D = 1+\tfrac{1}{5}+\tfrac{1}{15}+\tfrac{1}{35}+\tfrac{1}{70}+\cdots = \tfrac{4}{3}
$$

of the reciprocal 'trigono-trigonal' numbers: and likewise in other cases. In his first flush of delight at his success Leibniz went at once to Huygens and explained his method in minute detail to him. Huygens was generous enough to share Leibniz's pleasure at his result and the method. He demonstrated his own method of summation (as we know accurately from his notes on the subject),[26] there making use of the formula $\Delta_n = \tfrac{1}{2}n(n+1)$ and from it deducing the relation

$$
1/\Delta_{2n}+1/\Delta_{2n+1} = 1/2\Delta_n,
$$

whence

$$
\tfrac{1}{1}+\underbrace{\tfrac{1}{3}+\tfrac{1}{6}}_{\tfrac{1}{2}}+\underbrace{\tfrac{1}{10}+\tfrac{1}{15}}_{\tfrac{1}{6}}+\underbrace{\tfrac{1}{21}+\tfrac{1}{28}}_{\tfrac{1}{12}}+\underbrace{\tfrac{1}{36}+\tfrac{1}{45}}_{\tfrac{1}{20}}+\underbrace{\tfrac{1}{55}+\tfrac{1}{66}}_{\tfrac{1}{30}}+\underbrace{\tfrac{1}{78}+\tfrac{1}{91}}_{\tfrac{1}{42}}+\underbrace{\tfrac{1}{105}+\tfrac{1}{120}}_{\tfrac{1}{56}}+\cdots
$$

$$
= 1+\tfrac{1}{2}+\tfrac{1}{4}+\tfrac{1}{8}+\cdots = 2.
$$

This is self-evident in Leibniz's way of deduction; it amounts only to omitting from the sequence $B, C, D, E, F, G, H, I, K, L, M\ldots$ first the intermediate point D, then the intermediate points F, G, H and so on.

Leibniz was very proud of this success and immediately began to write up the results of his investigation in a small treatise which he

[26] Addendum in an appendix to Schooten's *Exercitationes* of 1657: 521–34 (*HO 14*: 50–91 in Dutch with French translation on facing pages). The text of 1665 appears on *HO 14*: 144–50.

19

hoped to have inserted in the *Journal des Sçavans*.[27] His manner of expression is still awkward and heavy; Hobbes is still treated as a major mathematician, Grégoire as *geometra maximus*, Pascal's *Triangle arithmétique* is initially named only as a book not yet studied in depth, but Galileo's *Discorsi* seems to lie open on Leibniz' desk while he writes. He begins with a discussion of indivisibles; in non-technical language he refers to Cavalieri, Galileo, Wallis' *Arithmetica infinitorum*, James Gregory's *Vera circuli quadratura* and Archimedes, and then he attacks the summation of the infinite geometrical progression. Here he makes the remark, founded on a passage in the *Opus geometricum*,[28] misunderstood by him, that the general rule for the summation of the infinite series dates back to the Ancients – in reality nothing more than the summation of

$$1 + \tfrac{1}{4} + \tfrac{1}{16} + \cdots$$

(given in Archimedes' *Quadrature of the parabola*) has come down to modern times – and asserts, equally wrongly, that hitherto only an unproved rule for the sum was known. The unassailable proofs which Grégoire gave were evidently not appreciated by Leibniz. Next are given Huygens' problem and Leibniz' solution (agreeing with that found by Huygens) and then follows the generation of higher arithmetical progressions seen in the columns of the arithmetical triangle and therefrom he derives the sums of the reciprocal figurate numbers in the following tabular form

```
1
  1
1
  2   1
1
  3       4
  3
1
  4       10
  10
1
  5       20
  15
1
  6       35
  21
  7       56
  28
      84
```

	0	1	2	3	4	5 ...
	$\frac{0}{0}$	$\frac{1}{1}$	$\frac{1}{1}$	$\frac{1}{1}$	$\frac{1}{1}$	$\frac{1}{1}$ ...
	$\frac{0}{0}$	$\frac{1}{1}$	$\frac{1}{2}$	$\frac{1}{3}$	$\frac{1}{4}$	$\frac{1}{5}$ ...
	$\frac{0}{0}$	$\frac{1}{1}$	$\frac{1}{3}$	$\frac{1}{6}$	$\frac{1}{10}$	$\frac{1}{15}$ ...
	$\frac{0}{0}$	$\frac{1}{1}$	$\frac{1}{4}$	$\frac{1}{10}$	$\frac{1}{20}$	$\frac{1}{35}$ ...
	$\vdots$	$\vdots$	$\vdots$	$\vdots$	$\vdots$	$\vdots$ ...
sum	$\frac{0}{0}$	$\frac{0}{0}$	$\frac{0}{0}$	$\frac{2}{1}$	$\frac{3}{2}$	$\frac{4}{3}$ ...

[27] The *JS* contained apart from a few original contributions mainly book reviews which not always were to the authors' liking and created a good deal of hostility for the editors of the first volumes. Hence the issues came out rather irregularly and lengthy breaks in publication were not infrequent. From 1666 onwards the editor was Gallois, an influential founder-member of the Académie des Sciences and its secretary in 1668–9.

[28] Book II: 'De progressionibus geometricis', introduction.

The summation is done in columns. Leibniz takes care not to divulge his proof; he assures us only that the method of proof is difficult, requiring a number of lemmas, and that he wants to delay this to a later occasion.[29]

He has here written down the sum $\frac{1}{1} + \frac{1}{2} + \frac{1}{3} + \frac{1}{4} + \ldots = \frac{1}{0}$ of the harmonic series without giving any cogent reason for so doing, relying on analogy alone. He knows quite well that its sum becomes infinite, but how are we to understand that?[30] Already Galileo had concerned himself with the problem of the cardinal number 'infinity'.[31] From the fact that there are as many natural numbers as there are square numbers or cubes while there are evidently in the sequence of natural numbers more non-squares than squares and still more non-cubes than squares and yet more non-cubes than cubes, he had concluded that in the realm of the infinite the relations equal, more and less have no validity. This made him set the number 'infinity' equal to 1, since for this finite number only (as is true for infinity) can we have $x = x^2 = x^3$. To grasp this we ought to remember that 'number' at the time was understood properly to signify only a natural number and that 'one', according to the ancient Pythagorean conception, was not seen as a number at all but rather as the source and origin of all numbers. Leibniz does not approve of Galileo's conclusion since it fails at once for the ordinary multiples of natural numbers; hence he sets the number 'infinity' equal to zero, pointing out that the axiom that the whole is greater than its part has here lost its meaning and that the sum $1 + 1 + 1 + \ldots = \frac{1}{0}$ must be set equal to zero. There follows a sharp criticism of Hobbes' view that all valid propositions arose without exception from arbitrary statements,[32] and then a lengthy discussion regarding the utility of the conceptual script which Leibniz had previously strongly advocated in the *Ars combinatoria*;[33] and the tract ends with a detailed analysis of the axiom of the whole and its part and certain other axioms in the first book of the *Elements*.

[29] Mahnke (1931): 594 rightly concludes from this remark that Leibniz' original proof (now lost) must have been very elaborate.

[30] Leibniz for Gallois, late 1672 (*LSB* II. 1: 226).

[31] *Discorsi* (1639), Giornata prima. Leibniz read the second Italian edition (1656) of the *Opere*.

[32] *See* above ch. 1: note 31.

[33] In his *Ars combinatoria* (1666) (*LSB* VI. 1: 199–201) Leibniz coordinates numbers to certain geometrical concepts and statements and obtains through this not yet very usefully constructed conceptual script a number of geometrical definitions. They have become intelligible only since Couturat (1901): 554–61 decoded them into clear language.

As preserved, the complete treatise exists in two partially overlapping drafts which were subsequently combined in a fair copy by Leibniz, afterwards copied by a secretary and further revised and slightly altered by the author himself.[34] Leibniz intended to send the treatise to the *Journal des Sçavans*, but had to keep it by him because the periodical ceased to appear for a time after the issue for 12 December 1672 came out and its publication recommenced only at the beginning of 1674.[35] In the meantime Leibniz' paper had been superseded and the question of printing it no longer arose.

[34] LH 35. III. A 32 (*LSB* II. 1: 222–9).
[35] The editors now were Gallois and La Roque.

3

THE FIRST VISIT TO LONDON

The main theme of the paper just now briefly described, namely the summation of series, occupied Leibniz' attention at the time no less actively than his researches on his calculating machine of which a rough but nonetheless adequate working-model (regarded by Huygens as a promising project)[1] had been completed. Of greater (ultimate) significance, however, was the journey to London which Leibniz and Schönborn were preparing to undertake for urgent consultations with the English court. In the midst of the diplomatic preliminaries Boineburg died on 15 December 1672; in him Leibniz lost a congenial political patron and his main supporter at the court in Mainz. In the meantime the ever fluid political situation had changed again: the prospect of an early end to the war between France and the Low Countries had disappeared and with it the main purpose of the journey to London had become more or less illusory. It took place nevertheless. Schönborn and Leibniz arrived at Calais on 17 January 1673, but were prevented from sailing for England for a few days by stormy weather, finally reaching Dover on 21 January, and London probably on the 24th.[2] There, at the earliest opportunity, Leibniz visited Oldenburg, the permanent secretary of the Royal Society, with whom he had been in correspondence since 1670[3] and who had procured for him in 1671 the London edition of his *Hypothesis physica nova*.[4] On this topic there had been a debate at

[1] Huygens–Oldenburg, 4 (14) Jan. 1672/3 (*HO* 7: 244).
[2] Müller–Krönert: 31.
[3] The first letter to Oldenburg (*LSB* II. 1: 59–60) is dated 13 (23) July 1670.
[4] Leibniz had sent a copy of the *Hypothesis nova*, printed at Mainz, with an accompanying letter of 1 (11) Mar. 1670/1 to Oldenburg for consideration by the RS, hoping to have it reprinted in London (*LSB* II. 1: 91). Oldenburg had read Leibniz' letter at the meeting of 23 Mar. (2 Apr.): the book was to be examined by Boyle, Wallis and Hooke (*BH* II: 475) and first came to Wallis who wrote approvingly to Oldenburg (7 (17) Apr.); his letter was read at the meeting of the RS on 20 (30) Apr. and was to be collated with the opinion of the other two referees (*BH* II: 477). At the meeting of the RS on 4 (14) May (*BH* II: 479) the book was given to Hooke who at the next session on 11 (21) May voiced his disapproval. But since Hooke's

the time in the Royal Society: a copy of the Mainz printing had been presented by Oldenburg and a committee (of which Wallis and Hooke had been members) for examining its contents had been appointed. Wallis reported wholly favourably,[5] while Hooke indignantly rejected the little work.[6] At the time Oldenburg had sent on to Leibniz a collection of various opinions of Fellows of the Royal Society regarding it[7] and had had the *Hypothesis* reprinted under the Society's imprimatur.[8] So, then, Leibniz was not completely unknown in the circles of the Society, and from his compatriot Oldenburg, a man of broad views and noble thought, he could expect a friendly welcome, from Hooke and his friends at least no hindrance. Would the charm of his personality here too be victorious?

Leibniz came to Oldenburg with a number of commissions; among others with a request from Huet for help in obtaining transcripts of Greek manuscripts from the Bodleian Library to be used in a planned edition of the *Anthologiæ* of Vettius Valens.[9] Moreover, Leibniz wished to display the model of his calculating machine which he had brought over. The exhibition took place at the meeting of the Royal Society of 22 January (1 February) 1672/3.[10] Hooke, who could not easily tolerate being forestalled when others had made an invention and who usually declared afterwards that he himself had possessed

---

ever critical attitude was well known, Moray moved that the book should now go to Pell for a further opinion (*BH* II: 481). Nothing is known about Boyle's or Pell's reaction. Meanwhile Leibniz had also forwarded his *Theoria motus abstracti* with a further letter to Oldenburg on 29 Apr. (9 May) (*LSB* II. 1: 102). The letter was read at the RS meeting on 18 (28) May and the new book was at once handed to Hooke for his appraisal (*BH* II: 482) who at the next meeting on 25 May (4 June) again spoke adversely so that Oldenburg was charged to transmit this book also to Wallis (*BH* II: 482). On 2 (12) June Wallis sent an approving report to Oldenburg. Then on 17 (27) July the *PT* **6** No. 73: 2213–14 brought a review of the Mainz printing while Wallis' two letters were published in the *PT* **6** No. 74 (of 14 (24) Aug. 1671): 2227–31.

[5] Oldenburg forwarded copies of Wallis' two letters to Leibniz already on 12 (22) June 1671 (*LSB* II. 1: 131–4).

[6] Oldenburg did not tell Leibniz of Hooke's adverse comments.

[7] This collection sent early in August 1671 which Oldenburg mentions in his letter of 5 (15) Aug. (*LSB* II. 1: 142) has not been preserved: perhaps it never reached Leibniz.

[8] No more details are known. Leibniz later on spoke without enthusiasm about the scientific contents of the two treatises: see his letter to Foucher in the summer of 1693 (*LPG* I: 415).

[9] Huet had asked Leibniz for this early in 1673 (*LSB* II. 1: 229). Oldenburg wrote to Oxford about the manuscript and informed Huet in early June of the success of his application. Huet's own copy of the Greek text is now in the Bibliothèque Nationale in Paris (suppl. graec. 883). The edition he had planned did not materialize.

[10] Meeting of the RS on 22 Jan. (1 Feb.) 1672/3 (*BH* III: 72–3).

something similar or even better for quite a while previously, inspected the model minutely from all sides and evinced a strong desire to take it completely to pieces; the other Fellows present on this occasion also showed a lively interest.[11] At this same meeting Hooke read a long paper on the reflecting telescope, attacking Newton's design.

Leibniz was also present at the next meeting of the Society[12] on 29 January (8 February) 1672/3 when Oldenburg read the famous letter from Sluse on tangents[13] and Hooke reported on further improvements he wished to make to the reflecting telescope. More important for Leibniz was his meeting with Moray, one of the most respected and influential Fellows of the Society. He informed Leibniz that Samuel Morland, a man of much ingenuity with strong technical leanings, had also invented a calculating machine. Oldenburg took the matter in hand and arranged a meeting between Morland and Leibniz at which the two competitors should both show off their machines.[14] This took place not many days afterwards in Oldenburg's presence;[15] Leibniz later recorded in some detail that Morland's machine was in no way comparable with his own, since multiplication and division could not be carried out automatically[16] by it but only by using the so-called Napier's bones.[17]

A few days later Leibniz was a guest of Robert Boyle.[18] Ardently interested in his chemical experiments, he lost no time in approaching Boyle's laboratory assistant Schloer to learn, if only indirectly, any details that Boyle himself might not care to disclose.[19] During the

[11] This follows from the report in Leibniz–Oldenburg 26 Feb. (8 Mar.) 1672/3 (*LBG*: 82–3).
[12] Meeting of the RS on 29 Jan. (8 Feb.) 1672/3 (*BH* III: 73–4).
[13] Sluse–Oldenburg 17 Jan. 1673 (*SL*: 673–7). The letter was printed in the *PT* 7 No. 90 (of 20 (30) Jan. 1672/3): 5143–7. For the origin, contents and significance of this letter *see* below ch. 6: notes 43, 48, 50.
[14] Oldenburg–Leibniz, 30 Jan. (9 Feb.) 1672/3 (*LBG*: 73–4).
[15] The demonstration of both models probably took place on 31 Jan. (10 Feb.) 1672/3.
[16] Leibniz writes in this sense in letters to Jakob Bernoulli on 25 Mar. 1697 and in April 1703: (*LMG* III: 57, 68); to Johann Bernoulli on 25 June 1697 (*ibid.* 421); to L'Hospital on 26 Sept. 1701 (*LMG* II: 343) and to Hermann on 6 Sept. 1708 (*LMG* IV: 335).
[17] *Rabdologia* (1617).
[18] The invitation was for 2 (12) Feb. 1672/3. Concerning this date and the questions discussed at their meeting, *see* Leibniz for RS, 3 (13) Feb. 1672/3 (*LBG*: 74–5).
[19] Leibniz was especially interested in the solvent for tin which Boyle had discovered and about which Leibniz made enquiry from Oldenburg on 26 Feb. (8 Mar.) 1672/3 (*LBG*: 81). He also approached Briegel to obtain the recipe as appears from Briegel's letter of 14 (24) Mar. (*LSB* I. 1: 326–7), and again Schloer as we can see from his accounts of mid-April (*C* 366) and of 10 (20) Apr. (MS: Hanover LBr 870, 11). Oldenburg in fact knew the recipe recorded in a marginal note on the letter of

course of the evening Leibniz also met John Pell, who next to Wallis had the reputation of being the leading English mathematician. Pell, who suffered from a liver-complaint, was by nature a reserved character, always ready with a derogatory remark on other people's scientific achievements but not easily induced to talk about his own methods. He possessed a good library and had among other writings acquired several of Fermat's treatises[20] (few of which were yet available in print) and on which he made numerous annotations.[21] He worked especially on methods of approximating the numerical solutions of equations, and occasionally demonstrated a result, but always only by way of examples, never exposing his general ideas.[22] For the rest, he had a thorough knowledge of the relevant literature, but his health was poor, he was of a suspicious frame of mind, ever worried for his reputation as a scholar and obviously not endowed with the gift for easy expression. In the heart of this morose, rather melancholic sixty-year-old the sanguine young German did not awaken the slightest sympathy.

The conversation turned also to mathematical topics and Leibniz reported that he possessed a general method for the representation and interpolation of series by constructing a series of differences.[23] He had no idea of the progress achieved in this field in England, knowing neither Henry Briggs' *Arithmetica logarithmica* nor Pietro Mengoli's *Novœ quadraturœ arithmeticœ*, nor even Nicolaus Mercator's *Logarithmotechnia*, and he could not realize that Gregory had already in 1668 advanced to the refined theory of interpolation which he set out in his letters to Collins. This correspondence, of which Pell was only vaguely aware, was not mentioned at all on this occasion; on the other hand Pell was quick to observe in phrases in which, for all their ambiguity, Leibniz could not help noticing a veiled reproach of plagiarism, that all this could hardly be news even to a man who

8 Mar.: 2 parts of nitric acid to 1 part of hydrochloric acid. Presumably Leibniz' enquiry refers to experiments on the transmutation of tin into silver: perhaps he already knew of Schröter's assertions who had recently received money from Lord (William) Brereton for such trials, as Briegel reports on 14 (24) Mar. (*LSB* I. 1: 327).
[20] Pell had, for instance, while Fermat was still alive, received with Mersenne's and Carcavy's assistance copies of manuscript treatises of which no mention in print had been made as yet and of whose contents nothing was known in public. This may be gathered from Mersenne–Pell, 1 May 1641 (*MC* 10: 610), though we have no documentary evidence concerning Carcavy's intervention.
[21] *See* below p. 38.
[22] *See* below pp. 221–3.
[23] Leibniz for RS, 3 (13) Feb. 1672/3 (*LBG*: 74–5), *see* also note 26.

knew only what had been published in France: some years previously Gabriel Mouton had communicated similar methods in his book on the apparent diameters of the sun and the moon, as results found by Regnauld.

Leibniz was deeply embarrassed. The very next day he looked at Mouton's book in the library of the Royal Society and found confirmation of Pell's assertion regarding Regnauld. Perhaps Oldenburg thereupon advised him to deposit an explanation of the affair with the papers of the Royal Society.[24] The resulting document has manifestly been composed in great haste, but since Leibniz in his hurry could not trace the paper in question among his disordered notes,[25] it gives us a fairly good picture of the mathematical knowledge he could at the time immediately command without referring to books or notes. He points to the array of differences of a given series, which he visualizes to be written down quite generally (employing, rather inexpediently, the absolute values of the differences), and he gives as an example the series of cubes beginning from zero with first, second and third differences 1, 6, 6. He adjoins the arithmetical triangle, adapting it to determine binomial coefficients in the form here reproduced where the 'hooks' indicate the additive process of

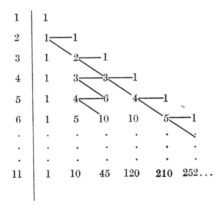

generating terms (for which he refers to his *Ars combinatoria*), and he now asserts that this is to be found neither in Mouton nor in Pascal's *Triangle arithmétique*. (In this he is, however, incorrect, for the

---

[24] This document for the RS would, as an official letter, be addressed to Oldenburg in his capacity as their secretary.

[25] *LBG*: 76.

generating rule is accurately explained in both these works.) By means of the initial term and the differences generated, and using the array of binomial coefficients, Leibniz imagines the terms of the original series (as becomes evident) to be built up by the formula[26]

$$0 \cdot \binom{n}{0} + 1 \cdot \binom{n}{1} + 6 \cdot \binom{n}{2} + 6 \cdot \binom{n}{3} = n^3.$$

This illustration is, however, only meant as an example and Leibniz indicates that he possesses the general rule, having attained it by means of the identity[27]

$$u^p - v^p = u^{p-1}(u-v) + (u^{p-1} - v^{p-1})v;$$

his procedure will not only, so he declares, give the individual terms of the series, but will also, for fractional n, solve the general problem of interpolation at equal intervals of argument. In conclusion he speaks of the summability of the infinite series formed from the reciprocal figurate numbers.

Later in the *Commercium epistolicum* Leibniz was specifically reproached for making no acknowledgment of the additive rule as it

[26] Leibniz at this time does not write the difference scheme from the top downward as we do now, but from below upward. How clumsily he expresses himself for

$$
\begin{array}{ccccccc}
 & & 0 & 0 & 0 & & \\
 & 6 & 6 & 6 & 6 & & \\
6 & 12 & 18 & 24 & 30 & & \\
1 & 7 & 19 & 37 & 61 & 91 & \\
0 & 1 & 8 & 27 & 64 & 125 & 216 \\
\end{array}
$$

the general case is evident from the array on *LBG*: 75. He gives the 'formula' expressing successive cubes in a mixture of words and numbers; though he has of course no symbols like $\binom{n}{k}$ at his disposal, the expressions could be formed purely numerically from the arithmetical triangle suitably combined with appropriate numerical factors to make up our formula for n^3: *see* Hofmann (1970): 87. The example appears to show that Leibniz was even at that time aware of the general interpolation formula for higher arithmetical progressions – contrary to Newton's allegation later in the 'Account' (1715): 215 = 'Recensio libri', *CE (1722)*: 49.

[27] This identity is not given as a formula but is expressed verbally. By repeated application it will yield the following result: where $P(x)$ is a polynomial of degree n in x and where x runs through the terms of an ordinary arithmetical progression, then the first differences of the progression will have the common divisor $(u-v)$. When this is removed they can be written in the form of a polynomial $Q(x)$ of degree $(n-1)$. Therefore in the array of the difference scheme of a higher arithmetical progression integers only will appear and the ultimate differences will be equal. Allusions to this occur in *C* 515. In 'Hist. et origo' (*LMG* v: 396) Leibniz remarks by way of supplement that the constant difference in the case of successive nth powers will be $n!$.

occurs in Pascal's *Triangle arithmétique*.[28] In his own marginal notes on this passage Leibniz rightly remarks that the whole paper shows rather clearly how little he understood of mathematics at the time[29] (and that he had merely skimmed through the pages of the *Triangle arithmétique*); the only point of any importance in the letter was his summation there of the reciprocal figurate numbers.[30]

With this explanation the contretemps with Pell was for the time being settled. At all events Oldenburg deemed it expedient not to invite his fellow countryman to the next meeting of the Royal Society on 5 (15) February 1672/3. During this session another attack on Leibniz took place: Hooke spoke in derogatory terms against the calculating machine and promised to produce a simple model using Napier's bones.[31] Oldenburg informed Leibniz of Hooke's remarks noting that he was universally regarded as a quarrelsome and cantankerous person, and advised his young friend urgently to speed up his technical improvements of the machine as much as possible. Oldenburg gladly fell in with Leibniz' wish to be admitted into the Royal Society.[32]

Having completed his business in London, Schönborn was impatient to depart.[33] Perhaps the web of intrigue spun by Schröter (a notorious person of doubtful repute) played its part in this.[34]

---

[28] Newton knew Leibniz' paper in a copy specially made for him (ULC. MS Add. 3971 fols. 27r–28v) and edited it in *CE* (*1712*): 32–6. On p. 37 he reprinted the French text from Pascal's *Traité* (1665): 2. In a marginal note about the arithmetical triangle (p. 35) Leibniz remarks that at the time he attached rather more importance to this really very simple connection – namely the additive structure as indicated by his 'hooks'.

[29] *CE* (*1712*): marginal note on p. 32.

[30] Leibniz has emphasized by underlining in his copy (*CE* (*1712*): 36) a brief explanatory passage where it was said that the sums to infinity of reciprocal figurate numbers can be found similarly.

[31] *BH* III: 75–6.

[32] Leibniz–RS, 10 (20) Feb. 1672/3 (*LBG*: 80). Leibniz had sent this letter with a (lost) personal greeting together with some books he had borrowed to Oldenburg through Schröter; *see* letter to Oldenburg, 26 Feb. (8 Mar.) (*LBG*: 81).

[33] Schönborn was to gain Charles II's assistance for the start of peace negotiations at a time when the initially successful military operations by French, British and allied troops had come to a standstill and when a continuation of hostilities was feared to lead to a wider European conflict: but the mission from Mainz had as little success in London as in Paris.

[34] Schröter obviously was an unstable personality, always short of money, with an exaggerated desire to be recognized, at the same time without moral restraint in his actions – he never reached the satisfying influential position he always wanted; in 1685 he published a work on political economy of considerable merit. He was employed incidentally to gather news for the Austrian government and had continually forwarded political reports from London. As a newly converted Catholic he was well received in Mainz, perhaps even given a counsellor's post there. Leibniz had first met him as a student in Jena, *see* Leibniz–Thomasius, 2 (12) Sept. 1663 (*LSB* II. 1: 3).

Making pretence of authority from Count Hanau,[35] Schröter had found means of becoming an intimate at court and duping the King with his alchemical performances; he now feared exposure by the embassy from Mainz and probably suspected that his secretary Briegel was secretly collaborating with Leibniz and supplying him with copies of Schröter's political correspondence – a state of affairs not at all atypical of the diplomacy of the time[36] and, indeed, a few weeks later, Schröter found himself discovered and had to leave London in disgrace.[37]

Preparations for the return of the Mainz legation to Paris went ahead so speedily[38] that Oldenburg found no opportunity personally to take leave of Leibniz, missing him when he paid a return call; all he could do was to send a brief note[39] enclosing a letter for Huygens[40] and presenting Leibniz with the latest number of the *Philosophical Transactions*[41] in which Sluse's tangent-method was published.

The affair of the finite-difference series was to have a lengthy sequel. In his very next letter from Paris[42] Leibniz sought Pell's opinion of his declaration for the Royal Society in particular as regarded Mengoli and received a brief reply from Oldenburg.[43] Later a more detailed answer arrived from John Collins, effectively Oldenburg's mathematical adviser. This man, a clearing house for much of the correspondence of London mathematicians with those outside London– in particular Wallis, Barrow, Gregory, Newton, Bertet and Borelli – had,

[35] Schröter was commissioned to prepare the transfer of Count Hanau's interests in the colony of Guyana in the West Indies to Britain; at the time it was leased from the Dutch.

[36] Briegel had enabled Leibniz to see copies of Schröter's correspondence, notably a letter from the Emperor's special envoy Meyersperg of 14 Jan. 1673 (*C* 294), also Schröter's reports to Vienna of 23 Jan. (2 Feb.), 1 (11) Feb. and mid-February 1672/3 (*C* 313, 319, 334). From these documents it becomes evident that Schröter was spreading the impression as if the delegation from Mainz was acting against the best interests of both the Emperor and individual German states and was really working on behalf of France.

[37] Schröter's game in London was up by February 1672/3 though he only left for Vienna in the summer. On the return journey he talked to Schönborn in Mainz alleging that he had bribed his secretary Briegel – a charge which Schönborn of course refuted: *see* his letter to Leibniz, July 1673 (*LSB* I. 1: 360).

[38] The chief reason for the hasty departure of the delegation on 10 (20) Feb. 1672/3 was doubtless the news of the serious illness of the archbishop-elector of Mainz: he died on 12 Feb. in Würzburg (*C* 326₁).

[39] Oldenburg–Leibniz, 9 (19) Feb. 1672/3 (*LBG*: 74).

[40] Oldenburg–Huygens, 9 (19) Feb. 1672/3 (*HO* 7: 256–7).

[41] *PT* 7 No. 90 of 20 (30) Jan. 1672/3.

[42] Leibniz–Oldenburg, 26 Feb. (8 Mar.) 1672/3 (*LBG*: 83).

[43] 6 (16) Mar. 1672/3, lost, recognizable from Oldenburg's annotation on Leibniz' letter 26 Feb. (8 Mar.) (*LBG*: 84).

as it happened, not met Leibniz during his visit to London. Collins had started as an accountant and arithmetical practitioner and though his understanding of the higher mathematics was never very deep he had come to be ardently interested in all new developments in it, especially at the hands of British mathematicians. Through his close contact with London booksellers he made vigorous efforts to get his friends' works quickly and adequately published. His contact, in person and by letter, with the most distinguished British scientists may have flattered his ambition no less than the position of trust he occupied as adviser to Oldenburg who had never himself done any serious mathematics and in his foreign mathematical correspondence with Sluse, Leibniz, Tschirnhaus and others had of necessity to rely to a large extent on Collins' prior drafts. Through an exaggerated national pride, unfortunately, Collins was incapable of an unprejudiced assessment of the achievement and character of foreign scholars; he was heavily biased against Frenchmen altogether and Descartes and the Cartesians in particular, and endeavoured in every way to extol the performance of his countrymen. The narrowness of his scientific horizon, in spite of his sincere wish to advance the truth and the new trends of thought, many times let him down badly. Indeed that more than a generation later there broke out a squabble over priority between Leibniz and Newton must in part be ascribed to Collins' intervention.

In Oldenburg's letter to Leibniz,[44] based on a draft by Collins, he tells briefly what he knows has been achieved by English mathematicians with regard to Regnauld's theory of interpolation.[45] Let $y = f(x)$ and set for $x = 1, 2, 3, \ldots$ values of $y = y_1, y_2, y_3, \ldots$ then the function can be determined by either of two methods. In the first when n values of y_k are known we try an expression for y as a polynomial in x of degree $n-1$. This method of general 'parabolic interpolation' was already widely used at the time; it is among others employed by Collins in his Decimal arithmetick;[46] in Oldenburg's letter it is represented by Mouton's two most difficult examples though in the second of these, given five values of y, there is deduced quite unnecessarily a function of degree five instead of four. For the second method certain multiplications are asserted to be required to

---

[44] Draft: Collins–Oldenburg for Leibniz, early April 1673; final version: Oldenburg–Leibniz, 6 (16) Apr. 1673 (LBG: 85-9).
[45] LBG: 85-6.
[46] Oldenburg refers to this treatise in (hitherto unpublished) addenda for Leibniz of 6 (16) Apr. 1673.

avoid the working out of the interpolating polynomial. This must be an allusion – so distorted as to become unrecognizable – to Gregory's method of interpolation[47] which would in the present case prove to be

$$y = y_1 + \binom{x-1}{1}\Delta y_1 + \binom{x-1}{2}\Delta^2 y_1 + \dots$$

and so essentially coincident with Leibniz' procedure.

As for summing the reciprocal figurate numbers[48]

$$\tfrac{1}{3}+\tfrac{1}{6}+\tfrac{1}{10}+\tfrac{1}{15}+\dots, \quad \tfrac{1}{4}+\tfrac{1}{10}+\tfrac{1}{20}+\tfrac{1}{35}+\dots, \quad \tfrac{1}{5}+\tfrac{1}{15}+\tfrac{1}{35}+\tfrac{1}{70}+\dots,$$

Oldenburg's letter reminds Leibniz that all these are discussed in Mengoli's *Quadraturæ arithmeticæ* though he could not find $\Sigma 1/k$, $\Sigma 1/k^2$ or $\Sigma 1/k^3$ which Collins can, so he says, perform by his method, and in particular he can find the sum to 100 terms of

$$\frac{10\,000}{106} + \frac{10\,000}{112} + \frac{10\,000}{118} + \dots,$$

a series (coming to very nearly 3200) of importance in repayment by simple interest. Collins had calculated this result by means of Newton's logarithmic approximation for the harmonic series. Newton's formula was derived in the following way:[49] let

$$S = a/b + a/(b+d) + a/(b+2d) + \dots + a/(c-d) + a/c,$$

with $m = b - \tfrac{1}{2}d$, $n = c + \tfrac{1}{2}d$ consider the case where

$$2mn/(b+c) \approx \sqrt{(mn)},$$

and then for a certain mean value e between $2mn/(b+c)$ and $\sqrt{(mn)}$ we may set

$$S = \frac{a}{e}\log\left(\frac{n}{m}\right)\Big/\log\left(\frac{2e+d}{2e-d}\right).$$

When the interval between the two bounds is sufficiently small, e can be taken as their arithmetic mean.

In his reply Leibniz[50] does not concern himself again with the interpolation problem. Writing in haste he was unable to inspect Mengoli's work, and so gained the impression from the Oldenburg–Collins letter that only a finite number of reciprocal figurate numbers had been summed in the *Quadraturæ arithmeticæ*, but not the related infinite series. He believed he could see confirmation of this assump-

[47] Gregory–Collins, 23 Nov. (3 Dec.) 1670 (*GT*: 119–21).
[48] Oldenburg–Leibniz, 6 (16) Apr. 1673 (*LBG*: 86).
[49] Newton–Collins, 20 (30) July 1671 (*NC* 1: 68–9); *see* Hofmann (1943): 41, 44.
[50] Leibniz–Oldenburg, 16 (26) Apr. 1673 (*LBG*: 90–3).

tion in the circumstance that Huygens himself expressly drew his attention to the novelty of the problem in asking for the sum of the reciprocal triangular numbers.[51] Leibniz reacted similarly in a later letter to Oldenburg,[52] but learnt in reply that Mengoli had indeed also found the sums of the infinite series;[53] he did not return to this topic in his further correspondence with Oldenburg. To be sure, Leibniz' result is wholly contained in Mengoli's far more penetrating analysis; the sum of the reciprocal triangular numbers is there achieved according to the Huygensian procedure, but the two are undoubtedly independent of each other. Leibniz did not know of his predecessor; the essence of his deduction is not the result but the method by which he obtained it.

With the other series proposed by Collins, $\Sigma 1/k$, $\Sigma 1/k^2$ and $\Sigma 1/k^3$ Leibniz never achieved any result. True, he discovered that the harmonic series diverges and hence cannot have a sum[54] – only to be informed by Collins[55] that this too had been demonstrated before by Mengoli[56] – though he had despaired of finding the sums of finite part-sums of the harmonic series, which the English again had accomplished.[57] All Leibniz' efforts to effect this summation were in vain – he could not guess that what he was told about concerned only an approximation to the sum – so that he ultimately offered to divulge his arithmetical quadrature of the circle with its proof if Collins would explain his method.[58] The only result Leibniz had found[59] was the sum

$$\sum_{n=2}^{\infty}\left(\frac{1}{2^n}+\frac{1}{3^n}+\frac{1}{4^n}+\ldots\right) = 1,$$

but Collins countered at once that this too was nothing new,[60] since all that was necessary was to rearrange it into a sequence of sums of

[51] *LBG*: 93.
[52] Leibniz–Oldenburg, 14 (24) May 1673 (*LBG*: 95).
[53] Oldenburg–Leibniz, 26 May (5 June) 1673 (*LBG*: 98).
[54] Leibniz–Oldenburg, 14 (24) May 1673 (*LBG*: 95).
[55] Oldenburg–Leibniz, 26 May (5 June) 1673 (*LBG*: 98).
[56] *Novæ quadraturæ* (1650): since

$$\frac{1}{3n-1}+\frac{1}{3n}+\frac{1}{3n+1} > \frac{1}{n},$$

Mengoli concludes

$$1 < \tfrac{1}{2}+\tfrac{1}{3}+\tfrac{1}{4} < \tfrac{1}{5}+\ldots+\tfrac{1}{13} < \tfrac{1}{14}+\ldots+\tfrac{1}{40} < \tfrac{1}{41}+\ldots+\tfrac{1}{121} < \ldots$$

[57] Further allusion to Newton's result, *see* pp. 32, 40.
[58] Leibniz–Oldenburg, 30 Mar. (9 Apr.) 1675 (*LBG*: 112); *see* further p. 127.
[59] Leibniz–Oldenburg, 14 (24) May 1673 (*LBG*: 96); the result occurs on *C* 279.
[60] Oldenburg–Leibniz, 26 May (5 June) 1673 (*LBG*: 98).

geometric progressions. The determination of $\Sigma 1/k^2$ occupied Leibniz all his life, but the solution never came within his grasp.[61] The affair with the calculating machine was to turn to Leibniz' disadvantage in the end. He was himself soon no longer satisfied with the working of his model and decided on far-reaching alterations,[62] but to his correspondents he continued to write over-optimistic reports which were not justified by the circumstances.[63] Hooke had already shown a model of his own machine to a meeting of the Royal Society on 5 (15) March 1672/3[64] and a few weeks later he handed in the detailed description which the Fellows had demanded of him.[65] Oldenburg continued to press Leibniz, reminding him of his promise to complete the machine as soon as possible,[66] but the only result of all this was that Leibniz forebore to send any answer to his latest

---

[61] Without knowing of Mengoli's relevant remark Jakob Bernoulli declares towards the end of his first dissertation on series (1689) that though he had not yet succeeded in finding $s = \sum_1^\infty 1/k^2$, he now knew that

$$s < 1 + \int_1^\infty \mathrm{d}x/x^2 = 2$$

is finite (BKC: 398–9): this amounts to employing the bounding series

$$\sum_2^\infty 1/(k-1)k > \sum_2^\infty 1/k^2.$$

In propositions 22 and 24 of his second dissertation on series (1692) Bernoulli arrives at $t = \sum_0^\infty 1/(2k+1)^2 = \frac{3}{4}s$ (BKC: 526–7, 530), while in proposition 44 of the third dissertation he observes, in the context of the quadrature of the hyperbola, that

$$s = \int_0^1 \log\,(1/[1-x]).\,\mathrm{d}x/x$$

and

$$t = \tfrac{1}{2}\int_0^1 \log\,[(1+x)/(1-x)].\,\mathrm{d}x/x$$

(BKC: 759–60). In the subsequent discussion – traceable by other letters, as Leibniz–Johann Bernoulli, 16 and 19 Nov. 1696; Johann Bernoulli–Leibniz, 11 Dec. 1696; Leibniz–Jakob Bernoulli, 25 Mar. 1697, and a later remark in Jakob Bernoulli–Leibniz, 15 Nov. 1702 (LMG iii: 336, 337, 339, 61, 64) no new insight emerged. When Johann Bernoulli declared in his letter to Jakob Bernoulli on 1 June 1691 (BJS i: 104): that he now saw his way to the solution, he obviously was deceiving himself. On Euler's solution (1736) and Johann Bernoulli's attempt to appropriate this result for his own credit, see Spiess (1945).

[62] Leibniz–Oldenburg, 26 Feb. (8 Mar.) 1672/3; 16 (26) Apr., 14 (24) May, 1673 (LBG: 81–2, 92, 96–7).

[63] Letters to M.Fr. Schönborn, 10 Mar. 1673; Duke Johann Friedrich, 26 Mar. 1673; D. C. Fürstenberg, April (?) 1673 (LSB i. 1: 316, 488–90, 346).

[64] RS meeting 5 (15) Mar. 1672/3 (BH iii: 77–8).

[65] RS meeting 7 (17) May 1673 (BH iii: 85–8; RS Register Book iv: 197).

[66] Oldenburg–Leibniz, 8 (18) May and 26 May (5 June) 1673 (LBG: 94, 97).

letter. Hooke's arrangement did not differ much from Morland's and was of course in no way to be compared with Leibniz's own machine, capable of performing multiplication and division wholly mechanically as well – but that was only to become evident much later on when Leibniz had put his model into working order. For the time being he had to bear it that the members of the Society began to lower their earlier opinion of him, perhaps regretting their rashness in unanimously electing him to their fellowship.[67] Furthermore Leibniz, possibly in ignorance of the customary formalities, had reacted to Oldenburg's notification of his election[68] only with a few hasty words of thanks[69] and had to be reminded by Oldenburg that he was expected to signify his acceptance formally by letter.[70] The surprisingly forced style of his letter of acceptance shows how hard Leibniz – normally a fluent writer – found it to pen this document.[71] He was rather annoyed that he had thus exposed himself, yet he could not work up greater enthusiasm for the calculating machine because suddenly he found himself seized by a multitude of mathematical ideas which obsessed his creative mind for ever.

[67] RS meeting, 9 (19) Apr. 1673 (*BH* iii: 82–3).
[68] Oldenburg–Leibniz, 10 (20) and 14 (24) Apr. 1673 (*LBG*: 89, 94).
[69] Leibniz–Oldenburg, 16 (26) Apr. 1673 (*LBG*: 92).
[70] Oldenburg–Leibniz, 8 (18) May 1673 (*LBG*: 94).
[71] Leibniz–RS, 22 May (1 June) 1673 (*LBG*: 99).

4

OLDENBURG'S COMMUNICATION
OF 6 (16) APRIL 1673

We have still to relate that immediately after his return to Paris Leibniz met Jacques Ozanam (originally from Lyons and known for his work in algebra and number theory) and discussed with him[1] the solution of that class of higher equations that can be reduced to pure equations by the transform $x = \bar{x} + a$. When Leibniz remarked[2] upon this Oldenburg replied in a letter now lost of 6 (16) March 1672/3; later on Collins pointed out that where a quartic has rational quadratic factors the auxiliary cubic equation can be dispensed with.[3] The crux of the matter is to determine when the auxiliary cubic can be avoided and when not. Rather more important are several problems in number theory which Ozanam formulated in sequel to Jacques Billy's *Diophantus redivivus* and *Inventum novum* and which for many years remained unsolved challenging the mathematicians of Europe.[4] The most interesting of these questions seeks to determine three numbers whose sums and differences in pairs will be squares. Leibniz records that Ozanam gave him a very simple algebraic solution and he wished to know whether anything of that sort were known in England.[5] This too received an answer in Oldenburg's lost letter[6] of 6 (16) March 1672/3.

Oldenburg's detailed reply of 6 (16) April 1673 consists of three parts: first, the previously mentioned English draft by Collins[7] is rendered in Latin, together with some personal remarks;[8] next follows a Latin postscript which in previous printed versions has been placed out of context;[9] and lastly there is a supplement in English, so far

[1] Ozanam for Leibniz, early March 1673 (*C* 782).
[2] Leibniz–Oldenburg, 26 Feb. (8 Mar.) 1673 (*LBG*: 83).
[3] Oldenburg–Leibniz, 6 (16) Apr. 1673, postscript (*LBG*: 239–40).
[4] Ozanam for Leibniz, early March 1673 (*C* 781).
[5] Leibniz–Oldenburg, 26 Feb. (8 Mar.) 1673 (*LBG*: 83); *see* also ch. 7: note 41.
[6] Oldenburg's annotation on Leibniz' letter 16 Feb. (8 Mar.) 1673 (*LBG*: 84).
[7] *See* ch. 3: note 44.
[8] Oldenburg–Leibniz, 6 (16) Apr. 1673 (*LBG*: 85–9).
[9] *LBG*: 239–40.

unprinted, on the contemporary situation in mathematics, particularly in England.[10] At the time he received this letter Leibniz was too inexperienced to appreciate the contents fully; later, however, probably in April 1675, he made excerpts from all three documents[11] (which confirm the sequence now proposed, supporting the cross-references within the individual portions of Oldenburg's missive). From Leibniz' conversation with Collins during his second London visit we have supplementary notes upon it.[12]

In Huret's Perspective (which had been reviewed in the Philosophical Transactions[13]) Collins had read a fairly thorough and extremely derogatory critique of Desargues' theory of conics in his Brouillon project of 1639, one to which he could not at all assent. He had himself published in the Philosophical Transactions[14] an article on the solution of numerical equations in which he linked the solution of equations of third and fourth degree with the intersection of two appropriate conics. He considered the tangent plane to a sphere at its zenith and stereographically projected circles in the tangent plane from the nadir into circles on the surface of the sphere; these circles he transformed by optical projection from the centre of the sphere into conics in the tangential plane. This is elaborated in more detail in the letter to Leibniz,[15] but not without some mistakes. In particular it is not made clear how Collins thought he could establish the main issue, namely the proper relation of the curves so that the method aimed at, that is, the use of spherical trigonometry, could be applied. Collins conjectured that similar ideas might have been contained in the treatise of Desargues, which he had never seen, and in its continuation, the Essay on conics by Pascal. All he knew of Pascal's discoveries came from Mersenne's Cogitata;[16] the Essay of 1640 to which Mersenne there refers he had not seen. Since he had learnt that some of Pascal's posthumous papers were in the hands of a brother of the Paris bookseller Desprez in the Auvergne, he wished to induce Leibniz to concern himself with the manuscript. He was also to urge Picard to edit a tract by Florimond De Beaune on the solid angle (likewise never subsequently printed) which Bartholin had received after De Beaune's death and to which he had added a commentary.

---

[10] C 409A. Compare OC 9: 556–62.
[11] C 935.
[12] See below p. 291.
[13] PT 7 No. 86 of 19 (29) Aug. 1672: 5048–9.
[14] PT 4 No. 46 of 12 (22) Apr. 1669: 929–34.
[15] Oldenburg–Leibniz, 6 (16) Apr. 1673 (LBG: 87–8).
[16] Second treatise: 'Hydraulica pneumatica', Preface, 12th section.

Nobody wanted to print it since the diagrams belonging to it were complicated.[17]

Leibniz' communication of Ozanam's problems in number theory caused Collins to transmit a list of the contents of the second book of Kersey's *Algebra* which was just then being printed off. Already at the end of January 1673 a preliminary notice of its first book had appeared;[18] Collins had edited it for press and exerted himself to promote its wide distribution. Kersey was a mathematical practitioner and popular teacher who endeavoured to give his countrymen, chiefly by well chosen examples, a competent survey of the contemporary (almost exclusively Latin) literature in algebra and number theory, and, in particular, of course, of the publications inspired by Descartes' *Géométrie*. Collins names a number of Kersey's authorities and sources: in determining the limits of roots, for instance, and depressing the degree of the equation De Beaune and Johann Hudde with the related commentaries by Thomas Merry, and also Bartholin, Dulaurens, Jakob Brasser, Jan Jakob Ferguson, Rahn and Brancker. Kersey's *Algebra* would not appear for another six months at least, Collins warns, and perhaps by then Frénicle's previously announced book on topics in number theory, wherein essential supplements to Billy's *Inventum novum*[19] were expected, would be in print; Pell, incidentally, had raised hopes of a similar publication of his own.[20] Fermat, Wallis and Kersey had independently of each other attained the same rule for decomposing the sum of two cubes into two other cubes,[21] Collins went on, but nobody had been able to clarify Frénicle's decompositions in small numbers.[22] This touches upon a topic that had played a major part in the famous correspondence on number theory during 1657–8 between Fermat, Frénicle and Wallis. The problem originated with Fermat[23] following

[17] Schooten–Huygens, 10 (20) Nov. 1656 (*HO* 1: 512). Concerning the manuscript's rediscovery see Costabel (1968).
[18] *PT* 7 No. 90 of 20 (30) Jan. 1672/3: 5152–3.
[19] It appears as a first part (with independent pagination) in Samuel Fermat's new edition of Diophantus' *Arithmetica* (1670) in which he included his father's marginal notes to Bachet's original edition of 1621.   [20] *See* p. 26.
[21] The English mathematicians and, later, Leibniz too had informed themselves about these number theory questions through the Diophantus edition of 1670 on the one hand, and on the other through Wallis' own edition of the correspondence with Fermat about his challenge problems in number theory, the *Commercium epistolicum* of 1658 (*WCE*) – to which reference will be made in these notes.
[22] Brouncker–Wallis, 3 (13) Oct. 1657 (*WCE*: 19); addenda in Frénicle–Digby, enclosed with Digby–Wallis 20 (?) Feb. 1658 (*WCE*: 120–1).
[23] Fermat–Digby, 15 Aug. 1657 (*WCE*: 22 and 23–4): to find two positive rational numbers x, y so that $x^3 + y^3 = a^3 + b^3$ where a, b are positive whole numbers;

a lost porisma of Diophantus'[24] that had been restored by Bombelli,[25] Viète[26] and Bachet[27] and which stated that the difference of two cubes is also the sum of two cubes. By setting

$$a^3 - b^3 = (a - bt/a)^3 + (at/b - b)^3$$

they had obtained

$$t = 3a^2b^2/(a^3 + b^3),$$

but in every particular case it had to be decided from the ratio a/b whether the result was a sum or a difference. Fermat had discovered that by repetition of the process it was possible always again to proceed from a sum of cubes to another sum.[28] Wallis could not cope with the general problem,[29] and so Fermat proposed the example[30] $9 = 2^3 + 1^3$ for which Wallis gave the solution (easily found from the rule given above)[31]

$$2^3 + 1^3 = (\tfrac{20}{7})^3 - (\tfrac{17}{7})^3 = (\tfrac{20}{7})^3 + (-\tfrac{17}{7})^3,$$

and when this was not accepted, being after all only a difference of cubes, he defended his solution by claiming that $-\tfrac{17}{7}$ is also a 'number'. Billy elaborated Fermat's suggestions in greater detail in his *Inventum novum*[32] and after two intermediate steps he reached a solution in fractions of numbers of 21 and 22 decimal places. Frénicle had supplied a series of solutions in very small numbers, for instance,[33]

example: $x^3 + y^3 = 3^3 + 1^3 = 28$. The challenge was destined for Wallis and Frénicle, but Fermat's phrasing had been so imprecise that Frénicle read it to mean the determination (after clearing of fractions) of any four natural numbers.

[24] *Arithmetica* v: 19 [16 in the modern numbering].

[25] *Algebra* (1572), book III, probl. 233. Bombelli gives of course only a numerical example (where he has overlooked a possibility of cancelling in his algebraical approach).

[26] *Zetetica* (1595): IV, props. 18–20 = *Opera* (1646): 74–5. Viète gives general algebraic expressions with numerical examples.

[27] Addition by Bachet (1621) to *Arithmetica* IV: 2, treated purely numerically.

[28] Marginal note to *Arithmetica* II: 10 [9 in the modern numbering] p. 65 in the edition of 1670 says that 'this is a difficult problem'; note to IV: 2, *ibid.* pp. 133–4: attempted example $5^3 - 4^3 = (\tfrac{248}{63})^3 - (\tfrac{5}{63})^3$; *ibid.* p. 135: attempted example $2^3 + 1^3$ requiring two intermediate steps.

[29] In his letter to Digby of 21 Nov. (1 Dec.) 1657 (*WCE*: 41) Wallis gives only examples he derives from those of Frénicle by multiplication, adding that this type of problem does not appear to him to be very significant since it was too special.

[30] Fermat refutes Wallis' remark in a letter to Digby of 7 Apr. 1658 (*WCE*: 159) and declares that his example $x^3 + y^3 = 28$ is comparatively simple requiring only one intermediate step. A new example is quoted *WCE*: 160.

[31] Wallis–Digby, 20 (30) June 1658 (*WCE*: 180–1).

[32] (1670) I: 35 (p. 10).

[33] *See* note 22; for the following *see* Hofmann (1961).

$1^3 + 12^3 = 9^3 + 10^3$, which can be reduced to relatively simple equations of the type $X^2 - pY^2 = q$ by the substitutions

$$(x+u)^3 - x^3 = (y+v)^3 - y^3$$

and

$$u(2x+u)^2 - v(2y+v)^2 = \tfrac{1}{3}(v^3 - u^3)$$

with a suitable choice of u and v. His examples are the cases $p = 3$, $p = 7$, and $p = 13$.

Collins further gives a survey of the most important mathematical discoveries of recent times. He points to Hudde's rule for finding extreme values of a function of the form

$$y = a_0 x^n + a_1 x^{n-1} + \ldots + a_n,$$

which consists in setting

$$n a_0 x^{n-1} + (n-1) a_1 x^{n-2} + \ldots + a_{n-1} = 0;$$

this, he claims, goes far beyond the rule found by Bartholin. Again, using Sluse's method of tangents we could narrow down the limits for the roots of an equation by approximations from above and below and then further approximate logarithmically, or we could systematically develop approximations by constructing series expansions (the first allusion to Newton's method,[34] though with Sluse's name – instead of Newton's – wrongly attached); he reports moreover that one now could solve the problem of interpolation[35] and approximate the partial sums of the harmonic progression by logarithms;[36] the problem indicated by Wallis in his *Commercium epistolicum* (perhaps a reference to Wallis' interpolations[37]) could be tackled, awkward decimals could be approximated by continued fractions,[38] and spherical trigonometry[39] was being improved. For the

[34] The allusion is to Newton–Collins, 10 (20) Dec. 1672 (*NC* 1: 247–8), the 'tangent-letter', and to the logarithmic treatment of the annuity problem in Newton–Collins, 6 (16) Feb. 1669/70 (*NC* 1: 24–5), and also to Newton's series expansion of which Collins knew since he had received (and taken a copy of) the *De Analysi* in the summer of 1669 (*NP* 2: 206–47).

[35] This refers to the discovery of the binomial expansion in 1664–5 (*NP* 1: 104–11), of which Collins may have been informed in conversation. In his second letter to Oldenburg for Leibniz, 24 Oct. (3 Nov.) 1676 (*LBG*: 204–6) Newton gives a detailed account of the procedure he has now developed.

[36] *See* ch. 3: note 49.

[37] *WCE*: 51; *compare* ch. 17: note 73.

[38] Wallis–Digby, 6 (16) June 1657 (*WCE*: 9).

[39] Some misunderstanding appears to have crept in with regard to spherical trigonometry.

two last results the main credit belonged to Wallis, for they were included in his (still unprinted) *Angular sections*,[40] though Rahn had achieved similar results, and in the development of this topic Pell too had had a share. In Fermat's literary remains, copies of whose treatises Pell possessed and carefully annotated,[41] we might expect something on Euclid's porisms, on tangencies of spheres, on plane (straight line and circle), solid (conic) and linear (higher curve) loci, on surface loci (skew surfaces of second order). Collins then lists a number of studies by contemporary authors on Euclid's *Elements*, in particular of the celebrated book x (on types of irrationalities of the form $\sqrt{(a+\sqrt{b})}$), but maintains that the greatest progress had been made, not in this field but in the extension of algebraic method to geometry, a field in which Bartholin had done much detailed work and to which Newton promised to make a contribution.[42]

On conics, Collins gives credit to Strode for having gathered all he could find in the best authors, in particular Kinckhuysen. Collins possessed tracts by Barrow[43] and Newton[44] on the solution of higher equations by means of conics or mechanical instruments: Newton was able to construct any conic through five given points (any parabola through four points) by means of moving angles,[45] and could solve higher equations up to the ninth degree by means of two cubical parabolas, those up to the sixteenth degree by a pair of quartic parabolas;[46] for these curves, too, he had described a mechanical generation.[47] All these had arisen in elaboration of the graphical

[40] Concerning Wallis' treatise on *Angular sections*, largely completed in manuscript by 1648 but printed only in 1684/5, *see* Scriba (1966): 17–33.
[41] *See* ch. 3: note 20. Collins' information about Fermat's writings presumably stemmed from Carcavy's obituary notice in *JS* No. 6 of 9 Feb. 1665: 79–82 which was taken over in an abbreviated form in *PT* 1 No. 1 of 6 (16) Mar. 1664/5: 15–16.
[42] On Collins' suggestion Mercator had in 1670 translated Gerard Kinckhuysen's Dutch *Algebra* (1661) into Latin: the manuscript was not published at the time and had long been lost sight of until it was located by Scriba (1964b) and printed in *NP* 2: 295–364, together with Newton's 'Observationes' upon it (*NP* 2: 364–445). It contains (*NP* 2: 354–64) a section 'Quomodo quæstio aliqua ad æquationem redigatur'. Here the reference is to Newton's notes and additions to this section (*NP* 2: 422–45).
[43] Lost manuscript on solution of equations, probably cognate to what is said in *Lectiones geometricæ* (1670): 131–47.
[44] Newton's treatise (1670), previously unpublished, will be found on *NP* 2: 450–517.
[45] Newton–Collins, 20 (30) Aug. 1672: *NC* 1: 230–1 = *NP* 2: 156–8, reproduced in Oldenburg–Leibniz, 12 (22) Apr. 1675 (*LBG*: 119–20); *see* also ch. 10: note 76.
[46] These details are taken from the 1670 treatise (*see* note 44) (*NP* 2: 504–7). *See* also Oldenburg–Leibniz, 12 (22) Apr. 1675 (*LBG*: 119); and further ch. 10: note 74.
[47] Examples of geometrical and mechanical ('organic') constructions of curves are given in Newton's *Methodus* (1670–1) (*NP* 3: 266–9 and 272–5).

solution of quartic equations by means of a fixed parabola and a circle given by Descartes;[48] we might also, however, following Pappus, perform the construction by inserting a line of given length between two adjacent sides of a rectangle in such a way that it is inclined through the fourth corner of the rectangle.[49] What Descartes himself had to say on this question[50] was, in Pell's opinion, unnecessarily complicated.

In this context Collins maintains that a report recently transmitted by Leibniz on Pardies' use of the logarithmic curve in the solution of equations was felt to be not quite clear in this form;[51] in England one was able to deal with trinomial equations alone by the aid of logarithms.[52] (We have here an allusion to one of Newton's numerical methods which approximates the first terms in the series-expansion of a root by transition to the logarithmic series.) Should, however, Dulaurens be proved correct in his assertion that an intermediate term could be eliminated in every equation,[53] then indeed the application of logarithms would be self-explanatory.

Going on to the field of higher geometry Collins relates that Newton had developed, well before the publication of Mercator's *Logarithmotechnia*, a general method for squaring curvilinear areas, finding lengths of arcs, centres of gravity and volumes of solids of revolution and of their second segments, together with their surface areas. It was, further, now possible to calculate the natural trigonometric functions and their logarithms directly from the argument, and vice versa; moreover, affected equations could be solved and that in such a manner as to obtain a particular series for all cubics, another for all quartics and so on.[54] He hoped that Newton would publish his method together with his observations on Kinckhuysen's *Algebra*.[55]

---

[48] *Geometria* (1659): 85–95.
[49] We may assume that Collins refers to Manolessi's revised and augmented Pappus translation (1660): book VII, prop. 72; p. 311.
[50] *Geometria* (1659): 82–4, Schooten's commentary 315–17.
[51] Leibniz–Oldenburg 26 Feb. (8 Mar.) 1672/3 (*LBG*: 83–4).
[52] See note 34.
[53] *Specimina* (1667), preface, fol. b1r. *Compare* Collins–Gregory 1 (11) Nov. 1670 (*GT*: 111) and 15 (25) Dec. 1670 (*GT*: 143) where the text is reproduced.
[54] This report is similar to earlier accounts in Collins' letters, reproduced in English in *NP* **3**: 21–3 and in Latin in *CE* (*1712*): 26–8; in particular to Bertet, 21 Feb. (3 Mar.) 1670/1; to Borelli, December 1671; to Vernon, 26 Dec. 1671 (5 Jan. 1672). Here Leibniz learned for the first time of Newton's discoveries in the realm of higher analysis. The next indication came in Oldenburg–Leibniz, 8 (18) Dec. 1674 (*LBG*: 109). *See* ch. 7: note 91.
[55] For the unsuccessful endeavour to get the translation, together with Newton's 'Observationes', published *see* *NP* **2**: 279–91.

Gregory, too, had hit upon the same method[56] but had applied it especially to the solution of such mechanical problems as Kepler's[57] of dividing a semicircle in given proportion and the finding of tangents to non-geometrical curves.[58] To this should be added a practical method of approximation by Dary, deriving from his method of gauging the volume of wine barrels.[59] Several geometrical problems might be posed in regard to questions of dioptrics.[60]

Entirely without connection to the rest of the letter comes a short account of recent research into Apollonius' tangency problem, though Collins appears to have been unaware of Pascal's important extension of the problem to constructing circles which touch other circles and intersect straight lines and circles at given angles.[61]

To sum up, we can say that Leibniz had been given a survey, amply furnished with valuable references to current literature, of all that Collins considered most interesting, and that he was willing to divulge, in the recent achievements of English mathematicians. Occasionally, it is true, precise details are given, but on the whole his account remains vague to such a degree that even a specialist working in similar fields could not then readily have fathomed the nature of these new methods. Doubtless Collins intended here, as in similar reports to other foreign scholars, to register the achievements of his English friends and thus to secure their literary property. He saw himself driven to this course since English booksellers, heavily suffering from the losses of stock in the great fire of London in 1666, were all but wholly unable to publish any mathematical books and accordingly the researches of the London group of mathematicians could be made known only with the greatest difficulty. The booksellers in Oxford and Cambridge were in a similar position; for Barrow, Newton, Gregory, even for Pell and Wallis, there remained hardly any possibility of getting anything published; in particular Barrow's *Lectiones opticæ* and *Lectiones geometricæ*, despite their subsequent fame, initially bankrupted the several printers for whom

[56] *See* also Gregory–Collins, 19 (29) Dec. 1670 (*GT*: 148).
[57] First indication in Gregory–Collins, 23 Nov. (3 Dec.) 1670 (*GT*: 120); elaborated more fully in Gregory–Collins, 9 (19) Apr. 1672 (*GT*: 227–8). *See* also p. 218.
[58] Collins here refers in particular to Gregory's determination of the tangent to the *spiralis arcuum rectificatrix* in an addendum to Gregory's letter of 23 Nov. (3 Dec.) 1670 (*GT*: 134–5); *see* also ch. **15**: note 54.
[59] Presumably this is a reference to Dary's *Miscellanies* (1669).
[60] Though it is not clear why no author is named we may assume that an allusion to Newton's researches of 1670–1 (*NP* **3**: 514–42) is intended.
[61] Pascal–Fermat, 29 July 1654 (*FO* II: 298).

the separate titlepages under which the early editions (1669–74) of these works are now found were produced.

Collins' stand on behalf of his countrymen and the manner in which he made it will always deserve our gratitude. In his relations with Leibniz he always acted properly and correctly although he saw in him a Francophile and somewhat mistrusted him because of his close friendship with Huygens. He took him to be (like Oldenburg himself) a diplomat with scientific interests and an inventor who delighted in constructing instruments like his calculating machine. Of his mathematical ability he did not at that time gain any specific impression – and who could blame him for that? – yet by the synopsis he offered him he wished to facilitate his study of the relevant literature and to demonstrate how much had already been achieved in this field. Leibniz, who had by then, quite independently, made a series of elegant mathematical discoveries, was a novice in the field of mathematical literature. Almost everything he had discovered so far and believed to be unknown existed already in one publication or another; the novelty was his method, but in what respects it differed from that of his predecessors could not be recognized at the time. Collins' acquaintance was deeply significant for Leibniz; it made it very clear to him that in mathematics, too, nothing can be accomplished without painstaking study of detail; it dampened his completely unjustified optimism in this direction and spurred him on seriously to peruse the relevant literature.

It was not only the rarer books, like Mengoli's *Via regia* or his *Speculazioni di musica* or the writings of Grienberger, to which his attention was called, but he was equally strongly directed to such treatises circulating in manuscript only, as those by Fermat, Pascal, Roberval, Frénicle and others, and in particular to Pascal's posthumous papers. More and more he began to interest himself in the origin and the development of the new ideas to whose further formulation he consciously from now on made his contribution. What Oldenburg sent him was, in its subject matter, far beyond him at the time, but it set the right tone and gave him a strong stimulus to continue his own studies. With his subtle sensitivity Leibniz divined something of the mathematical activity that throbbed in the island of Britain, of the silent contest between Newton and Gregory regarding the new methods. He withdrew from Oldenburg's acquaintance in order to learn for himself and fill the gaps in his knowledge of which he was now painfully aware. He renewed the relationship only again in the summer of 1674, a different person, one now truly knowledgeable in this field.

Collins had then still not met Leibniz personally and it was only later that he came under the charm of his personality; for him the Leibniz of the summer of 1674 was the same man he had dismissed in 1673 as an interesting braggart with whom it was best to be brief and who was no one to worry about. This misjudgment and the lack of insight into Leibniz' change from the 1673 beginner to the near mature mathematician of 1674 help to explain in some measure the incomprehensible surprises which the later correspondence of Leibniz with his London contemporaries will hold out to us.

5

THE GREAT DISCOVERIES
OF THE YEAR 1673

In the midst of the London negotiations, on 12 February 1673, the Elector Johann Philipp of Mainz died. His successor Lothar Friedrich was related with the Schönborn family but no one could know with certainty what effect the change of ruler would have. M. F. Schönborn at once left Paris and succeeded in being confirmed in his office of 'Obermarschall' by the new prince; for Leibniz, Lothar Friedrich had no further mission in Paris, but he was permitted to continue his stay in France without danger of losing his post.[1] The Boineburg family hoped that Leibniz would undertake the education of the young Philipp Wilhelm Boineburg.[2] Leibniz had several times advised the old Freiherr against sending his son, as previously planned, to one of the Paris Collèges and he had submitted to him,[3] and later to his widow, a carefully drawn up plan of study[4] which the Boineburg family approved. The seventeen-year-old Philipp Wilhelm took up his quarters with Leibniz[5] but found himself too confined and too closely supervised in his new lodging – he had little inclination for serious studies and intended to enjoy his time in Paris to the full.[6] Unpleasantnesses and complaints of many sorts were the result. The widow Boineburg was short of money and reduced the honorarium agreed upon with Leibniz; in the end he found himself rather coolly dismissed from the family service.[7]

[1] Schönborn–Leibniz, 5 May 1673 (*LSB* I. 1: 349).
[2] Ph. W. Boineburg had begun his studies in Strasbourg with the historian and jurist J. H. Böckler and after his death had gone to Paris under the care of his tutor-equerry J. F. Sinold together with his brother-in-law M. F. Schönborn, arriving on 16 Nov. 1672: *see* Wiedeburg II. 1 (1970): 141.
[3] In letters to J. C. Boineburg, 25 and 26 Nov. 1672 (*LSB* I. 1: 288, 291).
[4] Leibniz for Ph. W. Boineburg, *c.* March 1673 and for Baroness von Boineburg, *c.* May 1673 (*LSB* I. 1: 332–3 and 349–53). Sinold's letter to Leibniz, January 1673 (*LSB* I. 1: 311) shows that the instruction in mathematics was entrusted to O. L. Mathion one of the many private teachers of the subject in Paris.
[5] Consenting letter from the Boineburg family secretary J. J. Münch, 10 Apr. 1673 (*LSB* I. 1: 338).   [6] Leibniz–Münch, autumn 1673 (*LSB* I. 1: 370–3).
[7] Baroness von Boineburg–Leibniz, 13 Sept. 1674 (*LSB* I. 1: 396). Young Boineburg appears to have left Paris early in 1676; M. F. Schönborn–Leibniz, 11 Feb. 1676 (*LSB* I. 1: 401).

He had long foreseen this turn of events. Already in the spring of 1673 the Duke of Hanover had offered him a post as counsellor at a salary of 400 Thaler[8] but Leibniz would not bring himself to accept it at once. He loved the marvellous city on the river Seine and believed he could not live without close personal communication with the choicest spirits of Europe that had there come together in Paris; he was further tempted by the game of diplomacy whose threads ran all together in the hands of *le roi soleil* and his entourage. Again and again he sought to be employed on a mission by one of his aristocratic correspondents so that he could stay on in the capital city, but one indispensable prerequisite – that of noble birth – was always lacking and without it there could be no chance of work or success in the diplomatic life of the time. He, one of the greatest intellects in Europe, remained an insignificant tool in the hands of the mighty. So far the chain that was later to clamp him fettered only lightly, he still had his hopes for the future and believed himself to be free – and what tempted him most of all at this moment was the freedom to learn all that was there to be learnt.

The letter which Oldenburg had given him for Huygens[9] provided Leibniz with a welcome opportunity to call again on Huygens, then in the midst of publishing his *Horologium oscillatorium*. Huygens came to like the studious and intelligent young German more and more, gave him a copy of the *Horologium* as a present[10] and talked to him about this latest work of his, the fruit of ten years of study, of the deep theoretical research to which he had been led in connection with the problem of pendular motion, and how eventually everything went back to Archimedes' methods for centres of gravity.[11] Leibniz listened intently; at the close he felt he had to say something, but what he brought up was clumsy to a degree; surely a straight line drawn through the centroid of a plane (convex) area will always bisect the area, will it not? This was nearly too much: if it had been one of his mathematical rivals like Gregory or Newton then Huygens would probably never have condoned such a remark, but what this innocent young German had to say one could not really take amiss; good-humouredly, Huygens corrected his error and advised him to seek out further details from the relevant works of Pascal, Fabri, Gregory, Grégoire de Saint-Vincent, Descartes and Sluse. Since no

[8] Duke Johann Friedrich–Leibniz, 25 Apr. 1673 (*LSB* I. 1: 490–1).
[9] *See* ch. **3**: note 40.
[10] Leibniz–Huygens, Oct. 1690 (not sent): *HO* **9**: 522, 'Hist. et origo' (*LMG* v: 398).
[11] Leibniz–Tschirnhaus, early 1680 (*LBG*: 407); to Jakob Bernoulli, 1703, postscript (*LMG* III: 72); 'Hist. et origo' (*LMG* v: 398).

official task detained him any longer and as life with young Boineburg had become increasingly unpleasant, Leibniz very readily and willingly took refuge in science. He procured the books named by Huygens and a few more from the Royal Library, made excerpt after excerpt and went really deeply into mathematics. As he learnt, his personality rapidly matured, digesting what he read and systematically penetrating its essence; he was concerned to acquire not facility in calculation or a mere catalogue of results, but basic insights and methods, and what he took in inspired continually in turn a surge of creative activity within himself. The whole was neither a simple reworking and recapitulation nor mere undirected thought, but an ingenious regrouping of certain long-known results according to general viewpoints hitherto unperceived – it was, to begin with, a diversion for a mind deprived of its customary field of action, but soon became a unified passion for knowledge.

The first penetrating insight dawned on Leibniz through his study of Pascal's method for finding the moment $\int_0^{\frac{1}{2}\pi a} y \,.\, \mathrm{d}s$ of a quadrant of a circle about the x-axis, itself based on Archimedes' determination of the surface of a sphere.[12] Pascal observes (fig. 8) the similarity of the triangle δx, δy, δs to the triangle y, $a-x$, a and obtains

$$\delta s : a = \delta x : y$$

and hence

$$\int_0^{\frac{1}{2}\pi a} y \,.\, \mathrm{d}s = \int_0^a a \,.\, \mathrm{d}x = a^2.$$

Leibniz noticed that this procedure is in no way confined to the circle but is quite general, if only a is replaced by the normal to the curve.[13] He realized at once that he had discovered something fundamental. When in conversation he told Huygens about it, his ardour was well received; Huygens confessed to having found by the same method his famous determination of the surface of a paraboloid of revolution[14] and of other solids of revolution of the second degree (quoted without proof in the *Horologium oscillatorium* III, 9)[15] which had ever remained

---

[12] *Lettres* (1659); 'Traité des sinus du quart de cercle': lemma and prop. 1. *See* also the references in Leibniz–Tschirnhaus, early 1680 (*LBG*: 407–8); to L'Hospital, 27 Dec. 1694 (*LMG* II: 259); to Jakob Bernoulli, 1703, postscript (*LMG* III: 72–3); *see* further Mahnke (1926): 15–16.

[13] That is, to spell it out, here is the discovery of the 'characteristic triangle'.

[14] Found on 27 Oct. 1657 (*HO* **14**: 234).

[15] Found in November 1657 (*HO* **14**: 314–46).

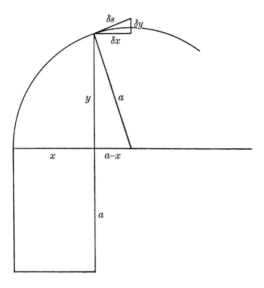

Fig. 8

a puzzle to Roberval and Boulliau.[16] This, his first geometrical discovery, Leibniz seems to have already come upon in the spring of 1673; perhaps he alludes to it in a remark to Oldenburg where he promises for later a more detailed description of what he has found.[17] At this same period Leibniz began serious study of Schooten's big two-volume edition of Descartes' *Geometria* then considered the best standard text in mathematics.[18] Here he learnt the representation of curves by coordinate geometry and the most important theorems on equations. He was particularly interested in the arithmetical and graphical treatment of cubic and biquadratic equations[19] and the 'fundamental theorem' of algebra on which Descartes had only lightly touched.[20] The rather singular Cartesian treatment of the problem of finding the normal at a point on a curve demanding the determination, on the axis, of the centre of a circle which touches the

[16] Leibniz–Jakob Bernoulli, 1703, postscript (*LMG* III: 73). Concerning Boulliau's lack of success there is a note in *C* 773; *see* also Leibniz–Tschirnhaus, early 1680 (*LBG*: 405), and Leibniz–Johann Bernoulli, 28 July 1705 (*LMG* III: 772).

[17] Leibniz–Oldenburg, 14 (24) May 1673 (*LBG*: 96). Accordingly the conversation with Huygens must have taken place in April or May 1673.

[18] Since the *Geometria* (1659/61) is frequently cited but no lengthy excerpts from it exist we may conclude that Leibniz had already bought his own copy of the book in Paris.

[19] Related remarks occur on *C* 769, 857, 865, 937, 1004 and 1168.

[20] *C* 834.

curve at the general point under consideration, appeared to him right from the start to be too contrived.[21] In the supplementary tracts in the *Geometria* he came in particular upon Heuraet's method of rectification; his knowledge of Huygens' determination of surfaces of revolution must[22] have come from Huygens himself. That Descartes wished to confine attention to 'geometrical' problems and deliberately cast aside all 'mechanical' ones always appeared to Leibniz as a highly rewarding narrowing of the field which he however tried to reverse as soon as he was able. He learnt a great deal from the *Geometria*, above all from the additional studies by Hudde on solving the cubic equation, on elimination, on reducing the degree of an equation, finding the highest common factor of polynomials and on the determination of extreme values, but also from De Witt's treatment of conics.

Pascal's *Lettres* had as their main theme the famous cycloid-problems regarding which a violent dispute had arisen some years before.[23] Leibniz also looked up what Pascal's opponents had produced, in the first place Fabri's tract on the sine-curve and cycloid[24] (reprinted in his *Synopsis*). Here he finds the rather pretty determination of the ordinate η of the centroid of the sine-area which, on taking

$$x = a \cos \theta, \quad y = a \sin \theta, \quad s = a\theta,$$

follows very easily from

$$\eta \int_0^s y \, . \, ds = \tfrac{1}{2} \int_0^s y^2 . \, ds, \quad \text{or} \quad \eta \int_0^x dx = \tfrac{1}{2} \int_0^x y . \, dx.$$

Probably under Fabri's influence[25] Leibniz adopted the habit in his diagrams of drawing the x-axis vertically downwards and the y-axis

[21] Later comments are to be found, for instance, on *C* 1125.
[22] Heuraet–Schooten, 18 Jan. 1658 (*HO 2*: 131). *See* also Mahnke (1926): 10, note 1; 29.
[23] Pascal issued his challenge in 1658; replies, printed only subsequently, were sent by Wallis (1659), Fabri (1659) and Lalouvère (1660); Pascal critically justified his refusal to award a prize in 1658 and gave his own assessment in the *Lettres* of 1659 consisting of his letters addressed to Carcavy, Huygens, Sluse and an unknown 'A.D.D.S.'; these letters had also been printed separately in 1658. In 1663 C. Dati also wrote on the subject. The questions asked for quadratures and centroids connected with the common cycloid, further for volume and centroid of the solid generated by rotating the cycloid about its base. Leibniz probably had seen, apart from the *Lettres*, only Pascal's challenge and *Historia*, but he had doubtless gathered more information in conversation since the whole affair was still being widely discussed.
[24] For an appreciation of Fabri's little known work *see* Fellmann (1959): 26–47.
[25] Mahnke (1926): 26.

horizontally to the right, a perfectly sensible arrangement, in agreement with our modern custom. Since the present-day convention follows from Leibniz' by a simple rotation to the left, we have preferred in the following diagrams to use always the modern one. From Fabri (who himself is in turn dependent on Cavalieri) Leibniz probably learnt the concept of static moment which he continually uses but which was rather thoughtlessly neglected by his contemporaries. Later he expressly mentioned Fabri's quadrature of the cycloid.[26] For both these topics Leibniz further refers to Lalouvère's *Geometria promota*,[27] but though he owned a copy himself he never seems to have more thoroughly perused this interesting and still only partly analysed work. Leibniz also saw Wallis' tract on the cycloid and from it he took a reference to Torricelli's achievements; yet he does not appear to have taken notice of Torricelli's *Opera geometrica* during his time at Paris.[28]

Of all that Leibniz learnt during these months the most essential technique was the theory of infinitesimals which went beyond Cavalieri's method of indivisibles by considering an area not as the totality of ordinates without breadth but as being generated from a very large number of narrow rectangles of equal width. Of its own accord there now appears the concept of an infinitesimal unit (the

[26] A lengthy excerpt from the *Synopsis* (1669) is found on *C* 500, more under *C* 612 indicating Leibniz's intensive study of this work; *C* 635 has the reference to the quadrature of the parabola (prop. 5 of part II: see Fellmann (1959): 17). In letters to Tschirnhaus, early March and late May 1678 (*LBG*: 353, 375) Leibniz speaks altogether favourably of Fabri on account of his methods of quadrature, while in a letter to L'Hospital, 27 Dec. 1694 (*LMG* II: 259), he names him as one of his original sources – as he also does in 'Hist. et origo' (*LMG* v: 399). When he published his own elementary quadrature of the cycloid in *JS* No. 18 of 23 May 1678 (*LMG* v: 116–17) – *see* below p. 58 – Leibniz refers incidentally to Fabri's quadrature (1659: props. 23–4; Fellmann (1959): 40–2). The high esteem of Fabri's work will explain Leibniz' remark *AE* January (1705): 30 (*CE* (*1712*): 108–9) in his review of Newton's *Quadratura curvarum* (printed 1704) – the remark which sparked off the priority dispute with Newton – where Leibniz asserts that Newton had everywhere used fluxions instead of the Leibnizian differentials just as Fabri had used Cavalieri's method in a variant form. Newton for whom Fabri was but a second-rate mathematician felt himself insulted by Leibniz' choice of phrase (note in *CE* (*1712*)); he refused to accept Leibniz' explanation in the letter to Conti, 9 Apr. 1716 (*LBG*: 274–5) – *see* his remarks of 19 (29) May 1716 (*LBG*: 285–6) and 'Account', *PT* 29 No. 342, January–February 1714/15: 202 = 'Recensio libri': *CE* (*1722*): 34).

[27] Later on the work is only briefly mentioned in the article on the elementary quadrature of the cycloid in 1678.

[28] Neither the summary reference to the parabola quadrature (*C* 635) nor the later citations indicate a thorough reading of the *Opera* (1644) during Leibniz' stay in Paris; Torricelli's best results indeed had not been printed at the time and were largely unknown; they were finally published in the *Opere* (1919) I. 2; *see* note 40.

width of the rectangle) which Wallis in particular employed in his *Arithmetica infinitorum*. Leibniz worked through this book only some while later.[29] The quadrature of the parabola in Wallis' arithmetical manner did not satisfy him, nor did the method of interpolation[30] by which Wallis had reached his infinite product for $4/\pi$. Wallis had, in modern terms, introduced a function

$$f(p, q) = 1 \bigg/ \int_0^1 (1 - t^{2/q})^{p/2} . \mathrm{d}t = 1 \bigg/ q \int_0^1 x^p y^{q-1} . \mathrm{d}y$$

where $x^2 + y^2 = 1$. On integrating by parts one can easily verify the symmetrical relation

$$f(p, q) = f(q, p)$$

and the recursion

$$pf(p, q) = (p + q) f(p - 2, q).$$

From this one obtains the values of $f(p, q)$ for all integral, non-negative p, q when $f(0, q)$ is set equal to 1. For

$$f(1, 1) = 1 \bigg/ \int_0^1 y . \mathrm{d}x = 4/\pi$$

Wallis puts the symbol $\square$; $f(0, 0)$ is not defined. He now step-by-step builds up a table of individual values of the function. For even p and q we get the ordinary binomial coefficients, for the other permissible values of p, q we obtain fractional multiples of 1 and $\square$. The law of formation typical of binomial coefficients,

$$f(p, q) = f(p - 2, q) + f(p, q - 2),$$

remains valid even for odd values of p, q. It is further true that

$$f(p - 2, q) < f(p, q) < f(p + 2, q),$$

with the additional inequality

$$f(p - 2, q) . f(p + 2, q) < f^2(p, q).$$

None of this, however, is rigorously proved by Wallis, but is all simply found heuristically by incomplete induction. Now Wallis further assumes that the last mentioned inequality holds not only for the arguments $p - 2$, p, $p + 2$ but also for $p - 1$, p, $p + 1$. By

[29] The first fuller reference to Wallis' product is found in a (still unpublished) draft of March 1675 (*C* 922 B) for Leibniz' letter to Oldenburg of 20 (30) Mar. 1674/5.
[30] *Arithmetica infinitorum* (1656), prop. 191; for the details *see* Whiteside (1961): 238–41.

systematic comparison of the functions $f(p, 1)$ he obtains from this successively

$$\sqrt{\frac{3}{2}} < \frac{3.3}{2.4}\sqrt{\frac{5}{4}} < \frac{3.3.5.5}{2.4.4.6}\sqrt{\frac{7}{6}} < \ldots < \square,$$

$$\frac{3}{2}\sqrt{\frac{3}{4}} > \frac{3.3.5}{2.4.4}\sqrt{\frac{5}{6}} > \frac{3.3.5.5.7}{2.4.4.6.6}\sqrt{\frac{7}{8}} > \ldots > \square$$

and hence

$$\square = \frac{3.3.5.5.7.7.9\ldots}{2.4.4.6.6.8.8\ldots}.$$

Leibniz accepted this product as correct – presumably he had talked it over with Huygens who could have told him that Brouncker had confirmed the result by a direct numerical evaluation,[31] perhaps based on the inequalities in the form

$$\tfrac{2}{3}(1+\tfrac{1}{15})(1+\tfrac{1}{35})(1+\tfrac{1}{63})\ldots < \pi/4 < (1-\tfrac{1}{9})(1-\tfrac{1}{25})(1-\tfrac{1}{49})\ldots$$

The conclusion by induction was not accepted by Leibniz; just like Fermat, Huygens and other mathematical critics of the time, he demanded a strict geometrical deduction. It may astonish us that nobody at the time seems to have noticed that the infinite product for $4/\pi$ could be deduced from

$$1/f(p-1, q) - 1/f(p, q) = q\int_0^1 (1-x)x^{p-1}\,y^{q-1}.\mathrm{d}y > 0.$$

As is well known, Brouncker at once transformed Wallis' infinite product for $\square$ into the continued fraction

$$\square = 1 + 1 \atop \displaystyle 2 + 9 \atop \displaystyle 2 + 25 \atop \displaystyle 2 + 49 \atop \displaystyle 2 + \ldots$$

and Wallis gives a very artificial proof of this in his *Arithmetica infinitorum*. It is not known how Brouncker actually proceeded;[32] as Euler later showed,[33] the reciprocal partial fractions of the continued fraction turn out to be the partial sums of the series

$$\pi/4 = 1 - \tfrac{1}{3} + \tfrac{1}{5} - \tfrac{1}{7} + \ldots,$$

[31] Wallis–Huygens, 12 (22) Aug. 1656 (*HO* 1: 476–7); see further Wallis–Digby, 6 (16) June 1657 (*WCE*: 9).

[32] See the reconstructions in Hofmann (1960) and Whiteside (1961): 211.

[33] Submitted to the Petersburg Academy on 22 Feb. 1776, printed in 1788.

so that Brouncker came very close to Leibniz' result. But no-one at the time recognized the close connection between the main topic of the *Arithmetica infinitorum* and Pascal's researches relating to quadratures, the determination of surfaces, volumes and centroids for the circle $x^2 + y^2 = a^2$, and certain solids connected with it (cylinder, sphere, circular ungula[34] and others), and questions pertaining to his challenge problems on the cycloid.[35] The wholly different forms of nomenclature and notation veiled the true equivalence both to the two rivals themselves and to their immediate successors. Leibniz did not recognize them either.[36]

Detailed excerpts made by Leibniz from the *Lettres* have been preserved; in these, moments, cylinder ungulae, solids of revolution and the related centres of gravity are singled out for consideration. On this occasion Leibniz hit upon the idea that an area may be approximated not merely by numerous narrow rectangles but equally well by numerous triangles concurrent in a small angle at a common point. In the beginning he several times went astray through not grasping the point at issue clearly and distinctly but he eventually broke through to a fundamental 'transmutation theorem' by means of which he could deduce briefly and simply all results so far gained in the field of geometrical quadratures.

Leibniz thinks of a (smooth convex) arc AB (fig. 9), suitably subdivided in selected points on it; he joins these points to the origin O, and through them draws a set of parallels given in direction to their intersection with the x-axis. Let now P, Q be two neighbouring points on the curve, R, S the end-points of the parallels on the x-axis, and let PQ produced intersect at T the line drawn through O in the given direction of the parallels. The parallel to the x-axis through T meets PR in U, QS in V, then the triangle

$$OPQ = \tfrac{1}{2}(RSVU).$$

To prove this Leibniz draws OW so that the angles $O\widehat{W}T$, $R\widehat{S}Q$ are equal; now from the similarity of the triangles PNQ, OWT (where PN is parallel to the x-axis, meeting QS in N) we obtain

$$\delta x : h = \delta s : z, \quad \text{hence} \quad h\,\delta s = z\,\delta x,$$

---

[34] The current designation for an obliquely cut right circular cylinder.
[35] Pascal's challenge of 1658 had aroused considerable interest and by the ensuing discussions the theory of quadratures was greatly advanced.
[36] Details are found in connection with C 541, 544, 545AB, 546, 548, 549, 575, 617, 696, 697 in Mahnke (1926): 31–62.

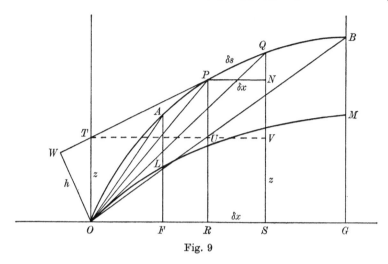

Fig. 9

and so on for the rest. Originally Leibniz established this proposition only for the case where PR, QS are perpendicular to the axis, but he later extended it to the case of parallels at any angle. We see clearly that we have to do with a product of affine transformations, by which in the limit the sector OBA of the curve is determined as one half of the area $FGML$. The idea of constructing a quadratrix[37] such as LM is by no means new; it occurs in particular in Lalouvère's writings. Only the transition from the concurrent slices of the area of a sector to the ordinary quadrature in strips is surprising.

What results is imposing enough: since $z = y - x\,\mathrm{d}y/\mathrm{d}x$ is linked to the theory of tangents, he has here established a highly important reciprocal relationship with the theory of quadratures by

$$\int_0^x y \,.\, \mathrm{d}x = \tfrac{1}{2}\left(xy + \int_0^x z \,.\, \mathrm{d}x\right).$$

Take, for instance, the problem of squaring the higher parabolas $(y/b)^q = (x/a)^p$ where $q > p > 0$: it follows that $q\,\mathrm{d}y/y = p\,\mathrm{d}x/x$ hence $z = (q-p)y/q$, and so

$$\int_0^x y \,.\, \mathrm{d}x = qxy/(p+q);$$

and in a corresponding manner the quadrature of the hyperbolas $(x/a)^q \,.\, (y/b)^p = 1$ is achieved.[38] This, of course, had long been known

[37] The name of quadratrix was at the time given to any curve $y(x)$ arising from representing an integral in the form $\int_a^t y(t) \,.\, \mathrm{d}t$.

[38] Leibniz–La Roque, late 1675 (C 1228).

before Leibniz; Fermat had first proved it rigorously using a step-diagram built up from rectangles whose bases are in geometrical progression,[39] and Torricelli had reached the same result by a similar reasoning,[40] but neither had made his method public. The general proposition appears in print for the first time in Wallis' *Arithmetica infinitorum*, but his approach by incomplete induction yields plausibility rather than a proof.[41] Huygens knew nothing of the method of Fermat and Torricelli. He constructed a subtle method of proof

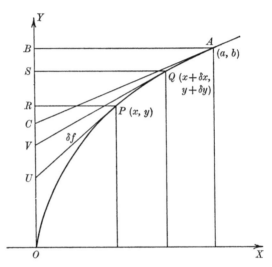

Fig. 10. $SR = \delta y$, $RU = py/q$; $SV = p(y+\delta y)/q$, $VU = (q-p)\delta y/q$.

which depends on comparing (fig. 10) the parabolic area $\int_0^a y \cdot dx$ with the area bounded by the parabola-arc, the tangent in the parabola's point (a, b) and the extended y-axis.[42] This 'tangent-area' is divided into a set of effectively triangular pieces by drawing the tangents at suitably chosen intermediate points on the parabola. If the abscissae of two such points differ by δx, then the subtangent for the point (x, y) is py/q and so the base of one of the triangular pieces proves to be $(q-p)\delta y/q$ while the related area is

$$\delta f = (q-p)x\,\delta y/2q = (q-p)y\,\delta x/2p;$$

[39] This occurs in the first part (c. 1629) of Fermat's treatise *De æquationum localium transmutatione et emendatione*, completed only in 1658 but previously circulating in manuscript (*FO* i: 255–67).

[40] *De infinitis parabolis* (1644–6), *De infinitis hyperbolis* (1647).

[41] Whiteside (1961): 237–9.

[42] Written down in a note in November 1657 (*HO* **14**: 273–93); reproduced in Hofmann (1953): 76–7.

the ratio of the parabola-area to the tangent-area is therefore $2p:(q-p)$ and to their triangular sum $\frac{1}{2}y^2\mathrm{d}x/\mathrm{d}y$ is $2p:(p+q)$, and to complete the rectangle ab a triangle of area $pab/2q$ is required and so the magnitude of the parabola area can be calculated. The result, rigorously demonstrated, as always with Huygens, by an indirect proof, is referred to briefly in his *Horologium oscillatorium*.[43] We might note in passing that we can deduce from Torricelli's approach the simple method founded on deriving the ratio

$$\int_0^a y.\mathrm{d}x:\int_0^b x.\mathrm{d}y = q:p.$$

The methods indicated so far are especially contrived for evaluating the areas under parabolas or hyperbolas; somewhat more general is

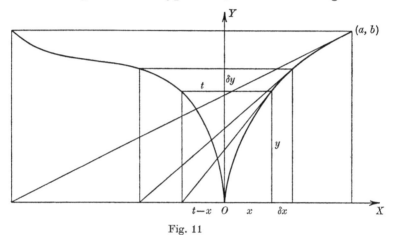

Fig. 11

the procedure developed by James Gregory in proposition 54 of his *Geometriæ pars universalis* (fig. 11). Here the subtangent $t = qx/p$ is drawn horizontally to the left of the point (x, y) and the ratio of the area of the corresponding trilineum $\int_0^b t.\mathrm{d}y = \int_0^a y.\mathrm{d}x$ to the difference $\int_0^b x.\mathrm{d}y = ab - \int_0^a y.\mathrm{d}x$ between the areas of the parabola and its circumscribing rectangle is found to be $q:p$; from this there then follows $\int_0^a y.\mathrm{d}x = abq/(p+q)$. Perhaps Gregory was led to this idea by his teacher Stefano degli Angeli, who was acquainted with the circle

43 *Horologium* (1673): 89–90.

round Torricelli and might also have known that Roberval had contrived a similar method.[44] We see that the quadrature of the parabola lies at the heart of the technical discussion; none of the participants however had reached Leibniz' comprehensive viewpoint.

Leibniz' superior vision was to show itself at once when he applied

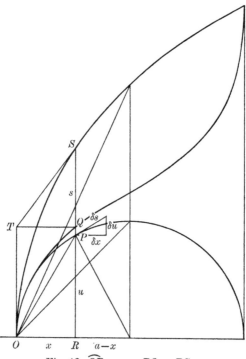

Fig. 12. $\widehat{OP} = s = RQ = PS$.

his method to the segment of the cycloid (fig. 12).[45] He starts from the circle $u^2 = 2ax - x^2$, finds

$$\mathrm{d}x/u = \mathrm{d}u/(a-x) = \mathrm{d}s/a = (\mathrm{d}u + \mathrm{d}s)/(2a - x)$$

and generates the cycloid by setting its ordinate $y = u + s$. Hence he obtains

$$z = y - x\,\mathrm{d}y/\mathrm{d}x = s,$$

and so after some simple transformation deduces that the area of its segment equals $\frac{1}{2}(au - s[a - x])$. When $x = a$, the corresponding area is rational, namely $\frac{1}{2}a^2$. Leibniz consciously connected his investi-

[44] Roberval–Torricelli, 1 Jan. 1646 (TO III: 355); Sluse–Huygens, 12 Jan. 1663 (HO 4: 292).   [45] C 545 B.

gation with an interesting result that Huygens had found in his *Horologium oscillatorium* (III, 7): for $x = \frac{1}{2}a$ the area bounded by the axis, ordinate and cycloid-arc is equal to one-quarter of the hexagon inscribed in the generating circle. Huygens had found his proposition in sequel to Pascal's challenge problems on the cycloid, but in a quite different way, namely from the relation between the area of the cycloid and the area of a circular ungula.[46]

The finest result that accrued to Leibniz during the year 1673 was his arithmetical quadrature of the circle. It follows immediately from his general transmutation theorem since, on setting $y^2 = 2ax - x^2$ we obtain

$$z/a = y/(2a - x) = x/y = \sqrt{(x/[2a - x])},$$

and so

$$x = 2az^2/(a^2 + z^2).$$

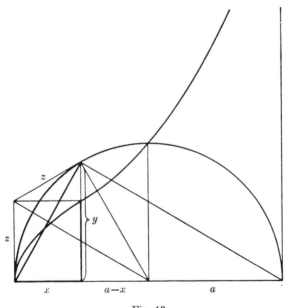

Fig. 13

This curve, the Versiera (fig. 13), has an asymptote $x = 2a$ and, says Leibniz, a point of inflexion at $x = a$, $z = a$ (in fact $x = \frac{1}{2}a$, $z = \frac{1}{3}a\sqrt{3}$) and its subtangent $t = z\,dx/dz = y^2/a$ is easily constructible.

[46] Note of July 1658 (*HO* **14**: 349–51).

The circle-sector can now, Leibniz continued, be simply calculated from

$$\tfrac{1}{2}(ay + \int_0^x z \,.\, \mathrm{d}x) = \tfrac{1}{2}(z[2a - x] + xz - \int_0^z x \,.\, \mathrm{d}z)$$

$$= az - \int_0^z az^2 \,.\, \mathrm{d}z/(a^2 + z^2).$$

In further treatment of this integral Leibniz closely follows what he had learnt meanwhile from Mercator's *Logarithmotechnia* and especially Wallis' improvements of its arguments. There the quadrature $\int_0^x v \,.\, \mathrm{d}u$ of the equilateral hyperbola $(b + u)v = b^2$ has to be effected, and is performed by long division and a subsequent integration term by term. Leibniz proceeds correspondingly in his example and finally arrives at the expression:

circle-sector $= az - z^3/3a + z^5/5a^3 - z^7/7a^5 + \dots .$

Regarding the limits between which this expansion is valid he could find information in Wallis' paper.[47] For $z = a$ he finds the famous series that bears his name

$$\frac{\pi}{4} = 1 - \frac{1}{3} + \frac{1}{5} - \frac{1}{7} + \dots \quad \text{or}$$

$$\frac{\pi}{8} = \frac{1}{1 \,.\, 3} + \frac{1}{5 \,.\, 7} + \frac{1}{9 \,.\, 11} + \dots = \frac{1}{2^2 - 1} + \frac{1}{6^2 - 1} + \frac{1}{10^2 - 1} + \dots .$$

He does not, however, stop at this result but goes on at once to consider the augmented series

$$\frac{1}{1 \,.\, 3} + \frac{1}{2 \,.\, 4} + \frac{1}{3 \,.\, 5} + \frac{1}{4 \,.\, 6} + \frac{1}{5 \,.\, 7} + \dots = \frac{3}{4}$$

from which he gets by omitting alternate terms the two series

$$\frac{1}{1 \,.\, 3} + \frac{1}{3 \,.\, 5} + \frac{1}{5 \,.\, 7} + \dots = \frac{1}{2}\left(1 - \frac{1}{3} + \frac{1}{3} - \frac{1}{5} + \dots\right) = \frac{1}{2},$$

$$\frac{1}{2 \,.\, 4} + \frac{1}{4 \,.\, 6} + \frac{1}{6 \,.\, 8} + \dots = \frac{1}{2}\left(\frac{1}{2} - \frac{1}{4} + \frac{1}{4} - \frac{1}{6} + \dots\right) = \frac{1}{4}.$$

By omitting groups of the three terms at a time in the original series Leibniz regains the previous series for $\pi/8$ on beginning with $1/1 \,.\, 3$, but on beginning with $1/2 \,.\, 4$ he derives a series

$$\frac{1}{2 \,.\, 4} + \frac{1}{6 \,.\, 8} + \frac{1}{10 \,.\, 12} + \dots .$$

[47] Wallis–Brouncker, 8 (18) July and 5 (15) Aug. 1668 (*PT* **3** No. 38 of 7 (17) Aug. 1668: 753–6, 756–9).

Now this series is the case of the Mercator series

$$bx - x^2/2 + x^3/3b - \ldots = \int_0^x b^2 \, . \, \mathrm{d}u/(b+u) = b^2 \log (1 + x/b)$$

for $b = x = \frac{1}{2}$ and thus it has the sum $\frac{1}{4} \log 2$. Hence the part-series

$$\frac{1}{1.3} + \frac{1}{5.7} + \frac{1}{9.11} + \ldots = \frac{\pi}{8}$$

depends on the quadrature of the circle, and the part-series

$$\frac{1}{2.4} + \frac{1}{6.8} + \frac{1}{10.12} + \ldots = \frac{1}{4} \log 2$$

on that of the hyperbola – a remarkable result which Leibniz

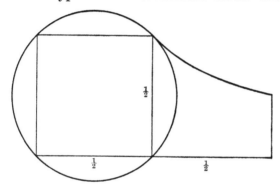

Fig. 14

illustrates on a diagram (fig. 14) where the area of the hyperbola of diameter $\sqrt{2}$ is $\frac{1}{4} \log 2$, that of the circle circumscribed to the square of side $\frac{1}{2}$ is $\pi/8$. Leibniz began to suspect a deeper connection between these two basic transcendental problems, the quadrature of the circle and that of the hyperbola.

How thoroughly Leibniz has made himself familiar with the essential background of quadratures and determinations of centres of gravity is evinced by his remark that the centre of gravity (ξ, ζ) of the area $\int_0^x z \, . \, \mathrm{d}x$ is dependent on the quadrature of the circle and the hyperbola, and that by rotating the area $\int_0^{2a} z \, . \, \mathrm{d}x$ about the asymptote a solid of finite volume is generated.[48]

[48] *C* 560–1.

In fact we obtain

$$(2a - \xi) \int_0^x z \, . \, \mathrm{d}x = \int_0^x (2a - x) z \, . \, \mathrm{d}x = a \int_0^x y \, . \, \mathrm{d}x;$$

$$\zeta \int_0^x z \, . \, \mathrm{d}x = \int_0^x \tfrac{1}{2} z^2 \, . \, \mathrm{d}x = \tfrac{1}{2} a^2 \int_0^x x \, . \, \mathrm{d}x / (2a - x) = a^3 \int_0^x \mathrm{d}x / (2a - x) - \tfrac{1}{2} a^2 x;$$

and

$$\int_0^x z \, . \, \mathrm{d}x = 2 \int_0^x y \, . \, \mathrm{d}x - xy.$$

From the first of these formulae follows that the distance of the centroid of the area $\int_0^{2a} z \, . \, \mathrm{d}x$ from the asymptote equals $\tfrac{1}{2} a$ and therefore (by Guldin's rule) the volume of the solid of revolution will be $\pi^2 a^3$.

6
READINGS IN CONTEMPORARY
MATHEMATICAL LITERATURE

That which had so far been gained in stormy sequence more by intuition than by systematic thought now had to be methodically clarified and rigorously demonstrated. To this was added the intention of drawing on the whole of accessible literature and throwing light on the essential ingredient in discovery, the *ars inveniendi* – a difficult undertaking for a scholar of such varied interests, in whose hands alien ideas sprang at once into new life and became a lure to further studies. Surely a mighty success had been scored by the arithmetical turn of the quadrature of the circle, but it remained uncertain how far the new method would carry. A conversation on this topic with Huygens seems to have taken place, during which Leibniz reported, in general terms only, on his researches to date.[1] On this occasion, we gather, the important question came to the fore whether one might by this new method be able to reach a decision on the possibility or impossibility of squaring the circle with ruler and compasses only by a finite number of steps, or at least to give a rule for the practical calculation of the circle area. Both these questions were much in Huygens' thoughts: all his life he believed in the possibility of squaring the circle by a finite construction. In his *De circuli magnitudine inventa* he had aimed at shortening the process of calculation, but above all at achieving a construction by this means. Amongst English mathematicians the opposite conviction had gained ground; Wallis in his *Arithmetica infinitorum*,[2] and Gregory in his *Vera circuli et hyperbolæ quadratura* each believed he had demonstrated the 'analytical impossibility' of circle quadrature. The latter had provoked an embittered clash with Huygens who would not accept Gregory's mode of deduction as cogent.[3] No clearcut decision had been reached on this highly controversial topic and Huygens hoped

[1] The visit to Huygens appears to have taken place on 30 Dec. 1673; *see* note 4.
[2] *Arithmetica infinitorum* (1656) fol. Bb1v (*WO* I: 359).
[3] On this debate *see* Dijksterhuis in *GT*: 478–86.

for a fresh contribution to the knotty question from Leibniz. He therefore handed to him the work in question[4] as the most important document to study requesting his judgment on this debate. That Wallis had worked with inadequate tools and hence needed not to be further considered was obvious, but Gregory's book was of a different order of difficulty.

For a long time Leibniz wrestled in vain with a concept which continued to elude him.[5] He was firmly convinced of the 'analytical impossiblity' of the circle quadrature, but he could not produce any really satisfactory proof of it, and it took him a long time to discover the error in Gregory's reasoning[6] – thus confirming Huygens' opinion. Far more important than these fruitless investigations was the closer general acquaintance with the contents of Gregory's works – difficult and hard to penetrate, yet extraordinarily profound.

The *Vera quadratura* contains an interesting generalization of Archimedes' quadrature of the circle.[7] Gregory starts with the sector AOB of a central conic (fig. 15), to which belong the 'inscribed' triangle $AOB = f_0$ and the 'circumscribed' tangent-quadrilateral $ATBO = F_0$; OT meets the arc at C, N bisects the chord and we have therefore $OC^2 = ON.OT$. If now the inscribed quadrilateral $OACB$ determined by C be designated by f_1, the corresponding tangent-pentagon $OAUVB$ by F_1, then f_1 is the geometric mean of f_0 and F_0, while F_1 is the harmonic mean between f_1 and F_0, or

$$f_1^2 = f_0 F_0, \quad 2/F_1 = 1/f_1 + 1/F_0.$$

[4] *HO* 7: 247 = *HO* 20: 388.

[5] The excerpts from Huygens' *De circuli magnitudine inventa* (1654) (*C* 503, 505) and from Gregory's *De vera circuli et hyperbolæ quadratura* (1667) (*C* 504), were probably made already early in 1674; for we read in Leibniz–Wallis, 28 May (7 June) 1697 (*LMG* IV: 24) that Leibniz had but a brief chance of looking at the two pieces so that much contained in them had remained a puzzle to him.

[6] In his first draft of early March 1675 (*C* 922 B) of the letter to Oldenburg, 20 (30) Mar. 1674/5, Leibniz had remarked that he too had something to say on this problem. In the unpublished portion of his second draft mid-March 1675 (*C* 922 C), Leibniz agrees with Huygens' judgment about the *Vera quadratura* but in the actual letter he has omitted this paragraph obviously so as not to offend anyone in London. In his letter to Oldenburg 17 (27) Aug. 1676 (*LBG*: 198) he emphasizes his opinion that Gregory's 'impossibility-proof' is incomplete. Similarly, in the 'Præfatio opusculi de quadratura circuli arithmetica' (autumn 1676) (*LMG* V: 97) he remarks again that Gregory had probably made some mistake. He is still more outspoken in a statement contradicting Tschirnhaus, September 1684 (*LBG*: 457–61) which originated in a controversy between the two about the possibility of the algebraic quadrature of algebraic curves; this was never published. *See* ch. 13: note 27.

[7] For the following discussion *see* also Dehn–Hellinger in *GT*: 468–78; Scriba (1957): 13–27 and Whiteside (1961): 226–7.

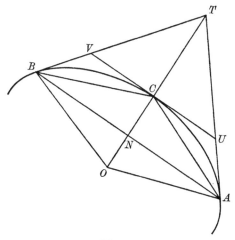

Fig. 15

When the procedure is repeated indefinitely we obtain the pair of sequences $f_0, f_1, f_2, \ldots$; $F_0, F_1, F_2, \ldots$, where for the ellipse

$$f_n < f_{n+1} < F_{n+1} < F_n$$

and for the hyperbola

$$F_n < F_{n+1} < f_{n+1} < f_n.$$

Proving that

$$|F_{n+1} - f_{n+1}| < \tfrac{1}{4}|F_n - f_n|$$

Gregory concludes that the double sequence converges to a limit Φ, the area of the sector. Gregory elsewhere makes use of a rational parametric representation from which the preceding inequality is an immediate deduction:

$$f_0 = u^2(u+v)\,t, \quad F_0 = v^2(u+v)\,t,$$

whence
$$f_1 = uv(u+v)\,t, \quad F_1 = 2uv^2 t$$

and
$$\frac{F_1 - f_1}{F_0 - f_0} = \frac{uv}{(u+v)^2}.$$

If now the sector Φ could be computed 'analytically' from the initial terms f_0, F_0 in the form $w(f_0, F_0)$ then it could similarly be generated from the terms f_1, F_1 and we ought to have

$$w(u^2[u+v],\ v^2[u+v]) \equiv w(uv[u+v],\ 2uv^2)$$

and this would lead to an 'analytical quadrature' only for rational functions w in the two variables; but, says Gregory, this is impossible.

The idea is truly ingenious, especially when one remembers the primitive tools at Gregory's disposal. That the conclusion in this form will not be successful, Gregory could not know. All his life he was convinced that his approach had been correct and became quite abusive even at well-meaning criticism. The dispute about the validity of his reasoning did not lead to any positive result. Yet Gregory was the first to elaborate the concept of convergence clearly and distinctly, and this precise achievement was rightly rated by Leibniz as one of his greatest merits.[8]

Apart from this Gregory supplied a highly important contribution to the systematic computation of a central-conic sector, namely approximations of circular functions on the one hand, and approximations to logarithms on the other. In this instance he was very probably influenced by Huygens' *De circuli magnitudine inventa*, whose content we may accordingly briefly summarize. As is well known, computation of arc-length or area of the circle by Archimedes' method requires very lengthy extractions of square roots; to secure Archimedes' own approximation $3\frac{10}{71} < \pi < 3\frac{10}{70}$ by his method it is necessary to go as far as the polygon of 96 sides. By an ingenious intuition, Nicolaus of Cusa[9] had, it is true, used the approximation $3 \sin \theta/(2 + \cos \theta) \approx \theta$, for which he even, by primitive considerations of proportionalities, produced an abortive 'proof'; but there was certainly no true insight anywhere into the complexities of the matter. Rather better (though still not fully proved) is the approach by Snell[10]

$$3 \sin \theta/(2 + \cos \theta) < \theta < \tfrac{1}{3}(\tan \theta + 2 \sin \theta);$$

that is, using our earlier nomenclature

$$3f_1^2/(f_0 + 2f_1) < \Phi < \tfrac{1}{3}(F_0 + 2f_1);$$

this Huygens transforms in his book[11] into the better form

$$\tfrac{1}{3}(4f_1 - f_0) < \Phi < \tfrac{1}{3}(f_0 + 2F_0).$$

His method of deduction follows Archimedes in his *Quadrature of the parabola*. In and about a circle-segment he systematically describes

---

[8] This becomes evident from the notes in *C* 1102, 1453 and 1455. Already in the first draft in early March 1675 (*C* 922 B) of his letter to Oldenburg, 20 (30) Mar. 1674/5, he comments on the coining of the concept of 'convergence' as a particularly happy one.

[9] *De mathematica perfectione* (1458), first theorem = *Opera* (1514) II, fol. ₂101v = German translation (1952): 162–3, 246.

[10] *Cyclometricus* (1621), props. 28 and 29.

[11] *De circuli magnitudine inventa* (1654), props. 4 and 5.

sequences of chords and tangents, at
each successive stage doubling the
number of sides, and in this way the
resulting areas are approximated.
Gregory had already pointed out[12]
that Huygens' method could immediately be interpreted as a comparison of the segments of parabola
and circle. Barrow goes further into
this:[13] let ABC be a segment less
than a semicircle (fig. 16). Then the
segment of the parabola which
touches the circle at C and passes
through A and B is less than the
circle-segment; on the other hand the
segment of the parabola which
touches the circle-tangents at A and B is larger than the circle-
segment. The second parabola bisects TN. By 'squaring' the
parabola-segments in the usual way, we can write

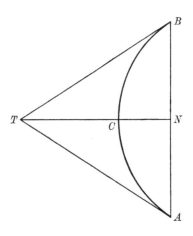

Fig. 16

$$\tfrac{4}{3}\triangle ABC < \text{segment } ABC < \tfrac{2}{3}\triangle ABT.$$

Gregory knew that it was always possible to express rationally in
terms of any f_n, F_n any preceding pair f_k, F_k; on this basis he derived
his approximations in the *Vera quadratura*. But there we have to deal
with inequalities for an arbitrary sector of a central conic, that is to
say with formulae of far vaster reach.

To indicate the procedure, say, for the sector of an ellipse, we
eliminate F_0, F_1 from Gregory's first three formulae, and obtain

$$(f_1 - f_0)/(f_2 - f_1) = 2f_1(f_1 + f_2)/f_2^2 < 4,$$

therefore

$$\tfrac{1}{3}(4f_1 - f_0) < \tfrac{1}{3}(4f_2 - f_1) < \ldots < \Phi.$$

The best approximation which he finds in this way, and for which he
confesses that he does not as yet possess a satisfactory geometrical
proof, comes out to be the 'triplicating inequality'

$$\tfrac{1}{5} \cdot \tfrac{1}{3}(4f_1 - f_0) + \tfrac{4}{5} \cdot \tfrac{1}{3}(f_1 + 2F_1) > \Phi.$$

This method cannot easily, in fact, be carried very far, for the
computational difficulties would become too great.

[12] *Exercitationes geometricæ* (1668): 2.   [13] *Lectiones geometricæ* (1670): 94.

3-2

A year later Gregory gave[14] a number of much narrower approximations. There they are expressed as approximations to a circular arc, but we may, in the present context, formulate them as propositions on areas, as for instance,

$$f_0 < \tfrac{1}{3}(4f_1 - f_0) < (64f_2 - 20f_1 + f_0)/45$$
$$< (4096f_3 - 1344f_2 + 84f_1 - f_0)/2835 < \dots < \Phi.$$

In this group of formulae Gregory proceeded as far as expressions of six terms and asserted that he was in possession of a general rule for forming them. These strange relationships thoroughly intrigued Leibniz; he was never able to demonstrate them and at one time intended to ask Oldenburg about them.[15] The modern reader can easily verify the formulae from the sine-series; since

$$f_1 = r^2 \sin \theta, \quad f_2 = 2r^2 \sin \theta/2, \quad f_3 = 4r^2 \sin \theta/4, \dots,$$
$$\Phi = r^2\theta;$$

but this expedient was not yet at Gregory's disposal at that time. We know, however, that he had the general formula of interpolation at equal intervals already in the spring of 1668[16] – perhaps the first suggestion had come to him from similar investigations by Briggs[17] (well before Mercator's *Logarithmotechnia*) – and he also knew the expansion[18] of $\sin n\theta/\sin \theta$ as a function of $x = 4 \sin^2 \theta/2$. From this there follows, on setting $1/2^{n-1} = t$,

$$f_n = f_1\left(1 + \frac{1-t^2}{3!}x + \frac{(1-t^2)(4-t^2)}{5!}x^2 + \frac{(1-t^2)(4-t^2)(9-t^2)}{7!}x^3 + \dots\right)$$

and so, when $n \to \infty$,

$$\Phi = f_1\left(1 + \tfrac{1}{3}\frac{x}{2} + \tfrac{1}{3}\cdot\tfrac{2}{5}\left(\frac{x}{2}\right)^2 + \tfrac{1}{3}\cdot\tfrac{2}{5}\cdot\tfrac{3}{7}\left(\frac{x}{2}\right)^3 + \dots\right).$$

[14] *Exercitationes geometricæ* (1668): 6–8.
[15] Leibniz–Oldenburg, unpublished second draft, mid-March 1675 (*C* 922 c) for the letter of 20 (30) Mar. 1674/5.
[16] Collins for Moray, mid-January 1668/9 (*HO* **6**: 374).
[17] *Arithmetica logarithmica* (1624); *Trigonometria Britannica* (1633); for an analysis of the essential content *see* Whiteside (1961): 233–5. It is interesting to see that Leibniz notes in mid-February 1672/3 (*C* 339) after a conversation with Pell on 2 (12) Feb. 1672/3 a related reference in the difference scheme of sines where the argument advances in arithmetic progression taken from the *Astronomia Britannica* by 'Briggs' (in fact by John Newton, 1657). It would be surprising if these three works had been unknown to Gregory.
[18] Gregory–Collins, 23 Nov. (3 Dec.) 1670 (*GT*: 119–20); *see* Hofmann (1950): 23–4.

Incidentally, Gregory introduced the versiera into his *Exercitationes*,[19] but Leibniz had not at the time noticed it.

Leibniz possessed a copy of the *Vera quadratura* of 1668, bound up with Gregory's *Geometriæ pars universalis*.[20] In this work Gregory systematically collected all he had learnt of significance in infinitesimal mathematics during his sojourn in Italy (1664–8). He declares in the preface that he is not much concerned with being original; the specialist reader will soon single out the author's own contribution from his borrowings of others.[21] In consequence only very few works are directly quoted, such exceptions are Wallis' *Arithmetica infinitorum* and the *Tractatus de cycloide*,[22] the *Opus geometricum* of Grégoire de Saint-Vincent,[23] Angeli's treatises on parabolas and hyperbolas[24] and Schooten's commentary to Descartes' *Geometria*,[25] also Renaldini's *Ars analytica*.[26] Gregory prides himself on having considerably simplified the centroid methods propounded by Valerio, La Faille, Guldin and Tacquet and having summarized the contents of Archimedes' treatises *De sphæra et cylindro* and *De conoidibus et sphæroidibus* into a few comprehensive propositions.[27] His teacher, Angeli, had been Cavalieri's pupil (and incidentally a friend of Sluse's), thereby establishing a connection with the circle round Torricelli, to whom a succession of Fermat's and Roberval's results had come by way of Carcavy, Mersenne and Ricci.[28] No doubt Gregory learnt through Angeli of a number of methods deriving from still unprinted papers and, in addition, from published works by other Italian scholars on topics on infinitesimals – for instance, the *Exercitationes geometricæ* of Michelangelo Ricci which Gregory singles out for praise in a letter to Collins:[29] Gregory's own tangent-method is a combination of that given by Ricci and certain ideas of Fermat. What makes the *Geometriæ pars universalis* a difficult book to read is its almost exclusively verbal discussion of extremely complicated

[19] Only one half of the curve is used and only 'its beginning' is drawn in his Fig. 7; the point of inflexion is not yet recognizable. *See* ch. 7: note 15.
[20] About this copy *see* p. 70.
[21] Fol. 2v.
[22] Prop. 62, p. 114.
[23] Prop. 46, p. 87.
[24] Prop. 66, p. 122.
[25] Prop. 48, p. 90. The reference is to a construction of tangents to conics. Algebraical topics and methods are introduced neither here nor elsewhere in the *Geometriæ pars universalis*, possibly for reasons of unity of style.
[26] Prop. 70, p. 132.
[27] Prop. 66, p. 123.
[28] For details *see* Hofmann (1963): 151–61.
[29] Gregory–Collins, 26 Mar. (5 Apr.) 1668 (*GT*: 50).

diagrams in which there are a number of minor mistakes.[30] Gregory had at that time still done little algebra and apparently was not fully aware of the scope of Descartes' *Geometria*, but even that is not remarkable for a disciple of the Italian mathematical school; in their libraries he would find works by men of that country and by Jesuits, but otherwise only few French ones. At that early time Gregory was informed only broadly of the main achievements of Huygens, and this knowledge might have come to him through his Florence colleagues or even through his friend Collins, whom he had met already in the summer of 1663 before he set out for Italy. Looked at from this point of view there remains very little in the *Geometriæ pars universalis* that is really Gregory's own independent work. This judgment is, however, in no way derogatory, but only repeats what Gregory himself indicates in his preface; in compiling this book he has learnt to think systematically and has digested all that had come his way consistently, systematically and rigorously. That by itself is no mean feat! Barrow later took over into his more famous *Lectiones geometricæ* much more of Gregory's work than has hitherto been suspected.[31]

Leibniz was immensely interested in the *Geometriæ pars universalis*, though he had not as yet acquired a personal copy of the book during his years in Paris. Of the two copies now in Hanover, one came after Huygens' death from his library,[32] the other deriving similarly from Knorre. Leibniz' marginal notes in them belong to a later period.[33] We may indeed sympathize with Leibniz when he did not work through the difficult proofs, but only individual propositions themselves and the general remarks occasionally inserted; he soon discovered that Gregory had been fully conversant with the characteristic triangle and its uses (although there is no direct mention of it anywhere), and that therefore most of the propositions which Leibniz had found by its aid, and which he believed to be new, had been known before. In regard to his own analytical method Leibniz learnt

[30] For mathematical details of this work *see* Prag, in *GT*: 487–509, and Scriba (1957): 28–52; also the references below bearing on relations to Leibniz and contemporary problems.

[31] He mentions by name in Lectio XI, §10 Gregory's transformation of an integral from prop. 11 (which in fact goes back to Roberval) – *see* above, ch. 5: note 44 – and in Lectio XII, appendix 3, probls. 9–10, Gregory's theory of 'involution and evolution' from his props. 12–18, pp. 29–41.

[32] His personal copy is listed as No. 81 in the sale catalogue after his death (*HO 22*) amongst the 'Libri mathematici in quarto'. There are no entries by Leibniz in this copy.

[33] Listed in Mahnke (1926): 29–30.

a great deal from the interesting integral-transforms of Gregory. We shall look again later at certain other details.

Only a few months after the *Geometriæ pars universalis* had appeared the second, improved edition of Sluse's *Mesolabum* was published with a long appendix, the *Miscellanea*, on current topics regarding infinitesimals. Sluse was an acquaintance of Huygens, highly esteemed by him both for his charm, sweet temper and modesty and because of his lucid style of writing.[34] He had studied for many years in Rome, was on the best of terms with Ricci and knew personally the whole circle of disciples of Cavalieri and Torricelli. On the other hand, through Oldenburg he was in touch with the Royal Society and at regular intervals received detailed reports from him about new researches and results in England.[35] His lively and extensive correspondence had to compensate him for the narrow provincialism of the small cathedral town of Liège where there was hardly anyone who cared much for scholarship and research.

Leibniz did not possess a copy of the *Mesolabum* while he was in Paris, but his numerous excerpts from the book testify to his thorough study of its contents. He worked carefully through the graphical treatment of cubic and biquadratic equations set out geometrically in the first two chapters and then analytically by considering the intersections of a circle with a conic and of two conics.[36] Sluse's painstakingly contrived investigations, aiming always at the simplest and most elegant methods, obviously made a deep impression on him.[37] Even two decades later he still remembered them and reports that he had passed on Sluse's method to Ozanam, who was delighted with it.[38] Through Sluse, Leibniz was introduced to the problem of finding the normal to a conic,[39] which was to occupy him over a long period; and

[34] The correspondence with Huygens covers the period from 1657 to 1668 (*HO 2-3*) after which date it is interrupted on account of wartime difficulties, to be resumed only through Oldenburg as an intermediary.
[35] The correspondence with Oldenburg begins, on the latter's initiative, in 1667 and continues till Oldenburg's death ten years later; it has fortunately been preserved in the archives of the RS and is currently being edited (*OC*, 1965–); only excerpts or portions of it having been known hitherto.
[36] *Mesolabum* (₂1668): 1–95. Leibniz' work on *C* 786 concerns the solution of a cubic equation by obtaining two quadratic equations in two unknowns (and cube root extraction): *Mesolabum*, 89–90; in *C* 790, 853, 857 he deals similarly with the biquadratic (*Mesolabum*: 90–93), and in *C* 1041 with angle-trisection (*Mesolabum*, 84–6).
[37] Thus he declares in *C* 861 (*LMG* VII: 252) that Sluse is superior in this respect to all predecessors.
[38] Leibniz–Foucher, January 1692 (*LPG* I: 404); also Leibniz–Wolff, late January 1707 (*LWG*: 70); and Leibniz–Varignon, 1710 (*LMG* IV: 169).
[39] *Mesolabum*: 130–7; to this refer a number of notes written in 1674, as *C* 855–6, 861 (*LMG* VII: 251), 864.

71

he knows and appreciates his determination of the point of inflexion on the conchoid[40] (this is the place where Sluse had originally intended to introduce his tangent-method[41] but, aware of Ricci's similar studies,[42] had desisted).[43] No less attractive were Sluse's investigations into the quadrature of spirals,[44] the determination of their centroids,[45] the finding of extreme values[46] and the solution of a few individual problems in number theory.[47]

When, later, Ricci was appointed to high office in the church and had to give up mathematics completely, Sluse, yielding to his wishes, now decided to bring his tangent-method before the public.[48] Lacking, however, the time needed to produce a whole book he chose to have his rules printed in the *Philosophical Transactions* in the form of a letter to Oldenburg. This letter was read by Oldenburg at one of the two meetings of the Royal Society at which Leibniz was present while in London, but at that time he was as yet unable to take in properly what was read. Leibniz personally passed on this issue of the *Philo-*

[40] Sluse deals on his pp. 117–22 first with the common conchoid (on a rectilinear base) then with the curve $(c-x)(x^2+y^2) = ax^2$, sometimes called 'Sluse's conchoid' (122–9).

[41] Sluse–Oldenburg, 10 Mar. 1670 (*OC* 6: 521) and 9 Mar. 1671 (*OC* 7: 477).

[42] Ricci's method of finding extreme values and determining tangents to parabolas $(y/b)^n = (x/a)^m$ or hyperbolas and ellipses $(y/b)^{m+n} = (x/a)^m \cdot (c \pm x)^n/(c \pm a)^n$ is analysed in Hofmann (1963): 170–6. The reference to 'similar studies' occurs in the brief preface to the *Exercitatio* of 1666.

[43] *See* the reference in Sluse–Oldenburg, 28 Feb. (10 Mar.) 1669/70 (*OC* 6: 521) which caused Collins to urge his friends – Wallis, Barrow, Gregory, Newton – to communicate their own tangent methods as soon as possible; *see* also Hofmann (1943): 26–7, 31, 39, 45, 52–3, 56–61, 70–4, 81–8, 90–4, 113–16.

[44] *Mesolabum*: 99–103, 112–14.

[45] *Ibid.* 103–12, 137–71.

[46] *Ibid.* 114–17. An analysis of this part of Sluse's work which goes decisively beyond Ricci, is given in Hofmann (1963): 177. Sluse anticipates the so-called Jakob Bernoulli inequality: Barrow in his *Lectiones geometricæ* (1670, Lectio VII, §§13–16) had taken it from Sluse, and Bernoulli in the first dissertation on series, prop. 4 (*BKC*: 380) from Barrow whose work had been his introduction to infinitesimal mathematics. A first reference occurs in *AE* (January 1691): 13 (*BKC*: 431). Sluse's method for extreme values is mentioned by Leibniz in *C* 500 and it is his starting point for the work of *C* 1139. The note *C* 1233 D contains a draft for a brief enumeration of all previous studies which he considered as especially important for his 'Quadratura circuli arithmetica'. It was his intention to include here the method of Ricci and Sluse, but eventually he omitted it from the final version, the 'Compendium' (autumn 1676) (*LMG* v: 99–113).

[47] *Mesolabum*: 171–9.

[48] Oldenburg in his letter 26 Mar. (5 Apr.) 1670 (*OC* 6: 597) had drawn Sluse's attention to the tangent treatment in Barrow's *Lectiones geometricæ* (1670, Lectio X, §14, pp. 80–1) and had suggested that he should himself communicate his own related researches. Barrow's work, however, came into Sluse's hands only later. For Leibniz' information about Sluse's method *see* ch. 3: note 13, and note 50 below.

sophical Transactions to Huygens, but did not apparently discuss this paper with him; otherwise he would have heard that Hudde and Huygens, too, had progressed to essentially the same computational rule. Sluse's precept refers only to 'geometrical' curves and presupposes that the equation is already free from radicals and fractions and written in the form of a polynomial $f(x, y) = 0$. What follows we will show on his own example $x^3 = by^2 - xy^2$. First, the equation must be prepared in such a way that one has on the left all terms in x, on the right all those with y; terms with both x and y have to be written both on the left and on the right, with their correct signs. Thus a symbolic expression (it cannot properly be called an equation any longer) is reached – in the present case $x^3 + xy^2 \sim by^2 - xy^2$. Now on the left multiply each term by the index of x in it and by t/x (where in fact t is the length of the subtangent $y\,\mathrm{d}x/\mathrm{d}y$ at the point (x, y)), on the right multiply each term by the index of y in it. By equating the two expressions thus obtained, the subtangent will be determined: in the present case from $(3x^2 + y^2)t = 2y^2(b - x)$. This rule is no more than an algorithm – basically one finds

$$f_x\,\mathrm{d}x = -f_y\,\mathrm{d}y$$

and then evaluates $t(= y\,\mathrm{d}x/\mathrm{d}y) = -yf_y/f_x$. What is awkward in this rule is the unorganic use of the symbolic expressions; but the way in which the sign of t is taken notice of, is praiseworthy. Sluse carefully exemplifies it on three equations of circles, properly recognizing even tangents parallel to the axes.

This rule agrees in its substance, if not in its form, with the famous tangent-rule which Newton gives in a celebrated letter to Collins;[49] it is more skilfully formulated by Newton who is interested only in the length of the subtangent t. The discovery of this tangent-rule played an essential rôle in Newton's accusations during the subsequent priority quarrel. Newton all his life maintained that it was by reading this letter that Leibniz was stimulated to develop the fundamental idea of his differential calculus and that he therefore owed essentially everything in this field to him. Against this Leibniz always insisted that he had already taken the rule earlier, and independently of Newton, from Sluse. This is the historical true state; for in his papers there exists an excerpt from Sluse's treatise, hitherto unnoticed,[50] though it is correctly dated under the year 1673 in the

[49] Newton–Collins, 10 (20) Dec. 1672 (*NC* 1: 247–8).
[50] *C* 616. The excerpt significantly confines itself to the text as reproduced in *PT* 7 No. 90 of 20 (30) Jan. 1672/3: 5043–7; the example and the calculations for it have

Catalogus criticus II (but without reference to the excerpted original text which eluded the otherwise well-informed compiler). Newton indeed presented the matter in such a way as if Sluse, too, had been dependent on him,[51] but this is also incorrect; on the contrary, Sluse had already developed his rule between 1655 and 1660, that is to say, at a time when Newton had still not seriously begun to occupy himself with mathematics.[52] That Leibniz was searching already in the summer of 1673 for a simple tangent-rule and searching through the relevant literature in so far as it was accessible to him – that is, for instance, Wallis' tract on tangents,[53] but not the related places in Barrow's *Lectiones geometricæ* – becomes clear from an important manuscript[54] in which Leibniz already attempts to tackle the inverse problem of tangents.[55]

It is, of course, not possible to state exactly to the day when precisely Leibniz began or ended the study of each of the works and treatises here quoted, but this much can be said with certainty, that he received the decisive impulse not from Barrow and Newton, but from Huygens, Pascal, Grégoire de Saint-Vincent, Mercator, Gregory and Sluse. Infinitesimal problems were being hotly pursued simultaneously in France, Italy and England; the improved concept of indivisibles was being used as a guiding principle by Fermat, Pascal and Huygens equally as well as by Torricelli, Ricci, Angeli and Sluse. Gregory got to know it in Italy, and perhaps Barrow had it from Gregory. The characteristic triangle – to take up a particular point – was known already to Fermat, Torricelli, Huygens, Hudde, Heuraet, Wren, Neil, Wallis and Gregory long before it was made

been omitted. After this fashion did Leibniz make excerpts at the start of his mathematical studies: it is safe to conclude that this piece was written already in the spring of 1673. A supplementary calculation comes in *C* 1209, where Sluse is not named.   [51] Note to Gregory–Collins, 5 (15) Sept. 1670 (*CE (1722)*: 95–6).
[52] *See* also Rosenfeld (1928).
[53] Wallis–Oldenburg, 15 (25) Feb. 1671/2 (*WO* II: 398–402); first printed in *PT* 7 No. 81 of 25 Mar. (4 Apr.) 1672: 4010–16.
[54] *C* 575; *see* Mahnke (1926): 45.
[55] Where Leibniz determined tangents to algebraic curves he employed Sluse's method even after he had invented his own symbolism for infinitesimals, as in *C* 1120 (*LBG*: 162). Only in a letter to Oldenburg, 18 (28) Nov. 1676: (*NC* 2: 199) (of which a portion only is known) does he remark that Sluse's rule is still lacking ultimate completeness; similarly in the letter to Oldenburg, 21 June (1 July) 1677 (*NC* 2: 213). There the tangent-rule using differential calculus is communicated for the first time though it had been previously developed in an example in a note of November 1676 (*LBG*: 230). In his letter to Tschirnhaus, early 1680 (*LBG*: 413), Leibniz expressly emphasizes that Sluse's rule is restricted to algebraic curves whose equation has been freed from irrationalities; later on Sluse's rule is hardly ever mentioned again.

public by Barrow. Each of these predecessors had used it, but nobody wanted to expose the jealously guarded secret by which he had found his results. Leibniz learnt of the characteristic triangle not from Barrow's book but, as has been briefly explained earlier on, from reading Pascal's *Lettres*. Nor, as Child[56] has asserted, is the transmutation identical with the area transform in Barrow's *Lectiones geometricæ*. It is true we find there the area of a segment expressed, in modern terms, by a change to polar coordinates, and in present-day nomenclature this would amount to a quadrature by means of

$\frac{1}{2}\displaystyle\int_0^\theta r^2 . d\theta$; but this, and the equivalent quadrature in the form

$\frac{1}{2}r^2\theta - \displaystyle\int_0^r r\theta . dr$, had been used already by Fermat, Roberval and Torricelli, and the transform had been cleverly and skilfully handled by Gregory in the *Geometriæ pars universalis*. To be sure, in Cartesian coordinates Leibniz' transformation amounts to the same as Barrow's quadrature in polar coordinates, and Child has concluded from this that Leibniz took his method from Barrow.[57] But in reality, though Leibniz' transformation is originally expressed in rectangular co-ordinates, it has nothing essentially to do with Barrow's idea.[58] True, during his first visit to London, Leibniz had bought a copy of Barrow's *Lectiones geometricæ* bound up with his *Lectiones opticæ*; he did not, however, then study it with any care but merely leafed through it.[59] At the time, it would seem, the *Lectiones opticæ* alone caught his interest,[60] and when he came to work seriously through the

---

[56] *Mathematical manuscripts of Leibniz* (1920): 173–5. Child had not looked at *LBG* (1899), and since he could not know the unpublished Leibniz manuscripts he reached fundamentally wrong conclusions which were fully refuted by Mahnke (1926).   [57] Lectio XI, probl. 24.

[58] This can be seen very clearly in prop. 1 of the 'Quadratura circuli arithmetica' (autumn 1676) – *see* Scholtz (1934): 16–17, 41–2. Child might have gathered the correct facts from the 'Compendium' where in *LMG* v: 99 both the text of prop. 1 and the diagram are to be found. *See* further text and fig. 9 on p. 55.

[59] He had a copy with the 1672 titlepage. During his first stay in London Leibniz had bought many books that interested him for an outlay of forty *thaler*, as he records in a letter to Habbeus, 5 May 1673 (*LSB*: I. 1 418). Among them were Mercator's *Logarithmotechnia* (1668), Gregory's *Exercitationes* (1668), Barrow's *Lectiones opticæ et geometricæ* (1669, 1670). He refers to this purchase also in Leibniz–Oldenburg, 16 (26) Apr. 1673 (*LBG*: 92).

[60] *See* Leibniz–Oldenburg, 16 (26) Apr. 1673 (*LBG*: 92) where he points out an error in Barrow's *Lectiones opticæ* (1669), corrected by Huygens, in Lectio XVIII, p. 125. We know of this through Huygens' letter to Oldenburg, 12 (22) Jan. 1669/70 (*HO 7*: 3). The details are given in a note of 1692 (*HO 13*: 775–7). Since nothing of all this had been printed we may conclude that Leibniz had been privately told by Huygens himself, probably mid-March 1673 when he delivered the letter from Oldenburg of 9 (19) Feb. 1672/3.

Lectiones geometricæ he had already made his own decisive discoveries in infinitesimal calculus.[61]

That Leibniz turned his attention only later to the *Lectiones geometricæ* will not surprise anyone who knows the book. For a beginner – and that is what Leibniz still was in the spring of 1673 – Barrow's work is still far more difficult to read than Gregory's *Geometriæ pars universalis*. It presupposes in particular deep familiarity with Euclid's *Elements*, and there Leibniz' knowledge at that time was most deficient. Moreover Leibniz could not divine just how valuable the contents of the *Lectiones geometricæ* actually were. Huygens, the mentor of his first independent mathematical studies, had never mentioned this particular book to him. The reason for this can easily be surmised. Huygens was severely ill[62] in the spring of 1670 and suffered an almost complete loss of memory;[63] for several

[61] Leibniz had been referred to Barrow's *Lectiones* already in Oldenburg's letter of 10 (20) Aug. 1670 (*LSB* II. 1: 61), but had not read them at that time. In *C* 1106, of 1 Nov. 1675 (*LBG*: 159), where he is experimenting with his symbolisms for infinitesimals, Barrow's name is grouped together with all those other authors who have worked on infinitesimal problems. In *C* 1120 of 11 Nov. 1675 (*LBG*: 161) Leibniz reduces the finding of a curve with a given subnormal $n(x)$ to an equation $dy/dx = n(x)/y$ and gives the result as $\frac{1}{2}y^2 = \int n(x).dx$, though in fact his symbolism is still imperfect so that instead of dx he conceived as an 'infinitesimal unit'. In the *AE* (June 1686): 292–300 (*LMG* v: 231) Leibniz points out that just this problem also occurs in Barrow's *Lectiones* (Lectio XI, prop. 1, p. 85). In his personal copy he inserted here between the statement of the theorem and its demonstration the words *novi dudum*: 'I have known it all along'. Then in the letter to Jakob Bernoulli, April 1703 (*LMG* III: 67) Leibniz writes that he had already before Barrow's *Lectiones* were ever published filled some 100 pages with deductions on ordinary and characteristic triangles: this is incorrect as it stands; what he means to say – and then it would be right – is obviously 'before he (Leibniz) had read Barrow's work'. In his cancelled postscript to this letter (*LMG* III: 73) Leibniz adds that he later found printed in Barrow's book a great deal of his own results of that time. It would appear then that Leibniz looked more closely at the *Lectiones geometricæ* at the earliest in the winter of 1675; see Mahnke (1926): 24. Mahnke's conjecture, Leibniz might have again been directed towards the *Lectiones geometricæ* by Tschirnhaus, gains substance when we remember that Tschirnhaus too had acquired Barrow's book in London, in fact the slightly augmented printing with the 1674 titlepage: this follows from Collins–Gregory, 19 (29) Oct. 1675 (*GT*: 342). Tschirnhaus himself got to work intensively on the *Lectiones*; a first reference occurs in his letter from Paris to P. van Gent, 6 Nov. 1675 (MS: Amsterdam, Wisk. G. 49f); more references come in his letters to Leibniz, 10 Apr. 1678 and March 1679 (*LBG*: 358, 388–90). In this last letter he mentions Lectio IV, §§16–17 and Appendix III, probl. 3. At this place Tschirnhaus makes it fairly clear that he cannot see anything essentially new in Leibniz' symbolism with its '*monstræ characteres*' in comparison with Barrow's text that to him (if not to us!) appears to be so much more readily intelligible.

[62] Huygens–Oldenburg, 12 (22) Jan. 1669/70 (*HO* 7: 2) at first mentions nothing but a heavy indisposition caused by the long, cold winter.

[63] Vernon–Oldenburg, 15 (25) Feb. 1669/70 (*HO* 7: 9–10).

months he had to live extremely carefully, and recovered only very slowly in spite of the devoted care of his brother Lodewijk who had gone to Paris for the sole purpose of helping the invalid;[64] obeying the doctors' advice Huygens returned in the autumn of 1670 to his parents' home in the Hague.[65] Justel brought him a copy of the *Lectiones geometricæ*[66] there in July 1670, and in the first letter he wrote to Huygens at the Hague,[67] Oldenburg had drawn his attention to Barrow's remark (which we mentioned earlier) concerning the approximation of a circle-segment by inscribed and circumscribed parabolas, but Huygens was still in so weak a state of health that he could not apply himself to these matters.[68] Before the year 1691 we find in his correspondence and in his notes no indication at all from which we might conclude that he had earlier occupied himself with the *Lectiones geometricæ*.[69] Now Tschirnhaus had indeed already in 1678 stressed the close links between Barrow's and Leibniz' own ideas[70] – a view repeated by Jakob Bernoulli in his article on 'Specimina calculi differentialis'[71] in the year 1691 – and ever since this connection has been adduced again and again in very different form and contexts. Against this Leibniz always contended that he had received his decisive stimulus not from Barrow but from reading Pascal, and the papers of Leibniz surviving from this time where the name of Barrow hardly appears in connection with the development of infinitesimal methods, confirm this statement. For the present-day

[64] Lodewijk Huygens had been in Paris since April 1670 (*HO* **22**: 657). The first signs of an improvement in health, noticeable by a renewed interest in scientific matters, are joyfully reported by Lodewijk Huygens to their brother Constantijn in May 1670 (*HO* **7**: 26–7).
[65] The two brothers reached the Hague on 9 Sept. 1670 (*HO* **7**: 37, note 9).
[66] Oldenburg–Huygens 20 (30) Sept. 1670 (*HO* **7**: 38). Surprisingly the printing of 1670 is not listed in the sale catalogue (*HO* **22**); instead the 1672 title is given among the 'Libri mathematici in quarto' as No. 59, together with the *Lectiones opticæ* of 1669.
[67] Oldenburg–Huygens, 8 (18) Nov. 1670 (*HO* **7**: 46).
[68] In his letter to Oldenburg, 5 (15) Oct. 1670 (*HO* **7**: 41) Huygens writes that he had been dangerously ill and still needed to husband his strength: so he could not yet give an opinion on Barrow's *Lectiones geometricæ*. On 21 (31) Oct. 1670 (*HO* **7**: 43) he reports further slow recovery and mentions the place with Barrow's conjecture about an approximate quadrature of a circle-segment; but he had still not looked at it closely (*see* above, note 13). Huygens did not go into this question more thoroughly later on either.
[69] He only began to occupy himself more closely with Barrow's *Lectiones geometricæ* in 1691. The immediate cause were the studies, then started together with Fatio, of problems on tangents and quadratures; this led to his adopting several characteristic theorems (*HO* **20**: 507, 509, 513).
[70] *See* note 61.
[71] *AE* (January 1691): 13 (*BKC*: 431–2). For details *see* Hofmann (1966): 241–3.

researcher, now removed from the affairs of those times by fully 300 years, it might well be very tempting to import into this tangled development the idea of a continuous growth and progress in insight and knowledge. If indeed one were to judge merely by the chronological sequence of publication of the individual works, then a well-nigh palpable relationship between Barrow the predecessor and Leibniz the successor appears to obtrude itself: but the facts deny it, and we must definitively reject the hypothesis of Leibniz' indebtedness to Barrow as erroneous.

7
FIRST COMMUNICATIONS
ABOUT THE NEW RESULTS

At last the calculating machine was completed. A long and thorny road had led from the rough model of 1672 to the proper working instrument of the summer of 1674. Again and again an earlier model had to be altered and improved; the craftsmen lost interest in the troublesome and thankless task and expected a quick reward for their work, and if Leibniz had not found in Olivier a true artist of his trade with his heart in the job, the machine might never have been finished.[1] Now Leibniz has the first and, for the present, only example of the hotly desired instrument in his rooms; occasionally he demonstrates it to a friend or acquaintance and is not averse to hearing good reports of his work spread in the circles of the Paris scientists. The calculating machine brings him into contact with Étienne Périer, Pascal's eldest nephew, who a short while before had brought out a second edition of the *Pensées*. Périer is on a brief stay in the capital and calls on the young German to inspect the machine.[2] His curiosity is understandable: the great Blaise Pascal himself had had a number of calculating machines built, very modest instruments indeed by whose aid one can only add and subtract,[3] while Leibniz' machine can also perform multiplication and division.[4] The visitor is most amiably received; the inventor proves to be a man of the world well informed in many fields of knowledge and possessed of enchanting manners; he acknowledges himself to be an ardent admirer of Pascal's writings, thanks Périer for his efforts in publishing the

[1] The improvements of this version compared with the rough wooden model first shown in London, together with the still remaining faults (which Leibniz never succeeded in eliminating) are described in detail by von Mackensen (1969): 60–8.
[2] É. Périer–Leibniz, early June 1674 (*C* 1351) suggests that the visit took place in the middle of June.
[3] Pascal's machine was begun in 1642; the dedicatory letter to the chancellor P. Séguier is dated 1645, the privilege 22 May 1649. For details of the construction and operation of the machine *see* von Mackensen (1969): 47–9; concerning the still extant eight models *see* Payen (1963).
[4] Toinard–Leibniz 20 June 1674 (*C* 683) where Toinard records Périer's great admiration for Pascal's machine.

Pensées, and regrets that so much of value in Pascal's posthumous work has remained unprinted. Périer himself intends to get his uncle's unpublished papers into print, and it is a pleasant surprise for him to find in this German, barely four years his junior, a collaborator who is moreover particularly well versed in the field of mathematics, his own knowledge of which is rather scanty. Though he is afraid that nothing ready for publication may be found amongst them, he is nonetheless willing to hand over Pascal's mathematical papers for Leibniz' perusal.[5] Yet he pleads for patience because Périer is fully conscious of his obligations as head of this famous family and will transmit the valuable manuscripts only through the most reliable intermediaries to Leibniz, and the latter must undertake full responsibility for the careful custody and scrupulous treatment of the precious papers.

What an unexpected success! In his first access of joy at the fine prospect that has opened before him, Leibniz writes to Oldenburg[6] telling him of the completion of his calculating machine and the widespread stir it has provoked, so redeeming his promise of long before. He will himself demonstrate the successful instrument to the Royal Society and hopes he can do so soon – though at the moment he does not know when he can come; his days are amply filled with political and literary work for patrons[7] and in conversation with close friends; and so but little time is left for scientific pursuits, but at least in the field of mathematics, so he reports, he has, more by happy intuition than by lengthy study, gathered a number of handsome pieces of knowledge. As a sample Leibniz gives his method for the rational quadrature of the cycloid-segment,[8] not without referring to Huygens' well-known similar quadrature of a cycloidal zone.[9] He has found certain important theorems, he writes, and among others, for instance, the exact representation of the area of a full circle and its general sector by an infinite series of rational numbers, and moreover far-reaching analytical methods, which he claims to be more im-

[5] Leibniz–Oldenburg, 5 (15) July 1674, hitherto unpublished postscript (*C* 687).
[6] *Ibid.* also *LBG*: 104–5.
[7] This is an allusion to the relations with Duke Johann Friedrich of Hanover who inclined to the French side, and to the conciliation attempts on behalf of the bishop Franz Egon Fürstenberg of Strasbourg who tried to obtain the release of his brother Wilhelm Egon, arrested on the Emperor's orders in the spring of 1674 at Cologne on a charge of acting as an enemy of the Empire and now imprisoned in Vienna, not to be set free until 1679. For details *see* Leibniz' letters on the matter traceable from the register in *LSB* I. 1; *compare* also Wiedeburg (1970): I. 1: 508–38.
[8] Leibniz–Oldenburg, 5 (15) July 1674 (*LBG*: 105).
[9] *Horologium oscillatorium* (1673): III, prop. 7, p. 69.

portant than individual propositions, however subtle their con-
trivance. The last is evidently an allusion to his transmutation
theorem.[10] There follow some polite remarks on Leibniz' acquaintances
in London and finally, in a postscript which has not so far been
printed a reference to his recent meeting with Périer and his promise
of a detailed account regarding the remains of Pascal in his possession.
Leibniz did not wish to send this letter through the post but entrusted it to a friend who shared his quarters in Paris, the Danish
nobleman Walter.[11] The latter had been in England twice before and
the letter he carried was to serve as his introduction to Oldenburg.

In his letter to Oldenburg, Leibniz mentions that he had shown his
theorem on the rational cycloid-segment to the 'most eminent
geometers' in Paris and that they had acknowledged its originality.[12]
He probably alludes by this remark to a compilation of his geometrical discoveries to date written out in a very clear and careful
hand which probably went to Huygens.[13] It contains the rational
quadrature of the cycloid-segment, the rational quadrature of the
circle by means of the Leibniz-series and a summary of his summation in the *Accessio ad arithmeticam infinitorum* of the reciprocal
figurate numbers in the form of what he calls the 'harmonic triangle'
placing it in deliberate correspondence with Pascal's arithmetical
triangle.

$\frac{1}{1}$	$\frac{1}{1}$	$\frac{1}{1}$	$\frac{1}{1}$	$\frac{1}{1}$	$\frac{1}{1}$	...
$\frac{1}{2}$	$\frac{1}{3}$	$\frac{1}{4}$	$\frac{1}{5}$	$\frac{1}{6}$	$\frac{1}{7}$	...
$\frac{1}{3}$	$\frac{1}{6}$	$\frac{1}{10}$	$\frac{1}{15}$	$\frac{1}{21}$	$\frac{1}{28}$	...
$\frac{1}{4}$	$\frac{1}{10}$	$\frac{1}{20}$	$\frac{1}{35}$	$\frac{1}{56}$	$\frac{1}{84}$	...
$\frac{1}{5}$	$\frac{1}{15}$	$\frac{1}{35}$	$\frac{1}{70}$	$\frac{1}{126}$	$\frac{1}{210}$	...
...	...	...	...	...	...	...
$\frac{1}{0}$	$\frac{2}{1}$	$\frac{3}{2}$	$\frac{4}{3}$	$\frac{5}{4}$	$\frac{6}{5}$	...

To Huygens this matter of the sums of these series presented
nothing new in principle and the theorem on the segment, if a surprising detail, was immediately evident; the arithmetical quadrature
of the circle, however, whose result only was enunciated, remained
unintelligible to him, and so he asked Leibniz for a more detailed
description of it with a proof.

Of this elaboration[14] we know both a Latin and a French draft, and

[10] *LBG*: 105.   [11] Leibniz–Oldenburg, 6 (16) Oct. 1674 (*LBG*: 106).
[12] Leibniz–Oldenburg, 5 (15) July 1674 (*LBG*: 105).
[13] Leibniz for Huygens, summer 1674 (*C* 691).
[14] Leibniz for Huygens, October 1674 (*C* 773, 797).

the fair copy, made by a scribe but afterwards corrected by Leibniz, which was presumably intended as the printer's copy for publication in the *Journal des Sçavans*. Huygens has looked through the article and added some pencil-notes, making improvements in the text and the proof. In a most appreciative accompanying letter[15] Huygens proposes for the quadratrix $z^2/a^2 = x/(2a-x)$ (namely the versiera, which Leibniz had simply called the 'anonymous curve'), the designation 'cyclocissoid' since the curve can be *ordinatim* composed from these two curves. This induced Leibniz to add in the main text the observation that we have here $2z = y+t$, where $y^2 = 2ax-x^2$ defines a circle and $t = x^2/y$ defines a cissoid. Huygens saw in the simple form of the Leibniz-series support of his hope for a ruler-and-compass construction for squaring the circle. He points out that the versiera had already been adduced in Gregory's quadrature of the cissoid in his *Exercitationes geometricæ*[16] and he therefore thinks it right to leave it to him as the first inventor to give the curve its permanent name. He reminds Leibniz that he himself in fact had been the first to succeed in squaring the cissoid[17] as could be verified from a pertinent place in Wallis' *Tractatus de cycloide*.[18]

This remark touches on an older affair which we may here briefly sketch.[19] Huygens had again been drawn to the cissoid, whose normal he had in 1653 determined by a mechanical application of Descartes' rule,[20] by an observation by Sluse that the volume of the solid generated by rotating the cissoid-area $2 \int_0^{2a} t \, .\mathrm{d}x$ (with t and y defined as above) about the asymptote depends on the quadrature of the circle.[21] In fact the volume in question is

$$K = 2 \int_0^{2a} 2\pi(2a-x) t \, .\mathrm{d}x = 4\pi \int_0^{2a} xy \, .\mathrm{d}x:$$

this integral can be interpreted as the moment of the semicircle $\int_0^{2a} y \, .\mathrm{d}x$ about the origin, and is therefore equal to $a(\tfrac{1}{2}\pi a^2)$, so that $K = 2\pi^2 a^3$. Sluse expresses this according to Guldin's rule as the

15 Huygens–Leibniz, 6 Nov. 1674 (*LBG*: 566–7).
16 *Exercitationes geometricæ* (1668): 23–4.
17 Huygens–Wallis, 27 Aug. (6 Sept.) 1658: *HO* 2: 212.
18 Wallis, *Tractatus duo* (1659): 81 (*WO* i: 545).
19 Hofmann (1941).
20 Huygens, note of 1 Sept. 1653 (*HO* 12: 76–8).
21 Sluse–Huygens, 4 Mar. 1658 (*HO* 2: 144).

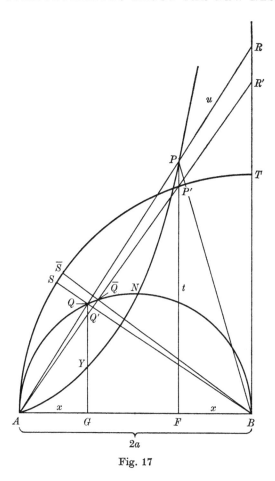

Fig. 17

volume of the torus generated by rotating the circle πa^2 about one of its tangents.[22] Thus it was made clear that the cissoid has a finite area and that this can be found once its centre of gravity has been determined. This was enough to stimulate Huygens to a marvellously simple observation to this effect:[23] let (fig. 17) $t = FP$, $y = GQ$ and next to the radius $AQPR$ draw the radius $AQ'P'R'$ slightly rotated from it; we then derive the ratios

$$\text{trapezium } P'R'RP/\Delta AQ'Q = [(2a-x)+2a]/x$$
$$= [4a^2 + 2a(2a-x)]/2ax.$$

22 Sluse–Huygens, 14 Mar. 1658 (HO 2: 151–2).
23 Huygens, note of 18 Mar. 1658 (HO 14: 309–12).

83

Let the radius AR' meet the semicircle in $\bar{Q}$ and with B as centre draw the circle through A meeting BQ in S and $B\bar{Q}$ in $\bar{S}$, and then we have the ratios

$$\text{trapezium } P'R'RP/\Delta AQ'Q \approx (\Delta BS\bar{S}+\Delta BQ\bar{Q})/\Delta AQ\bar{Q};$$

hence

$$\text{trapezium } P'R'RP \approx \Delta BS\bar{S}+\Delta BQ\bar{Q},$$

and so

cissoid-area $ABRPN$
$$= \text{circle-sector } BST + \text{circle-segment } BQ.$$

Expressed in modern terms the argument would, on setting

$$r = 2a \sin^2 \theta/\cos \theta$$

in polar coordinates, go as follows

$$f = 2a^2 \tan \theta - \tfrac{1}{2}\int_0^\theta r^2 . \mathrm{d}\theta$$
$$= 2a^2 \int_0^\theta (1+\sin^2 \theta). \mathrm{d}\theta = \tfrac{1}{2}(2a)^2 \, \theta + \tfrac{1}{2}a^2(2\theta - \sin 2\theta).$$

Huygens gives a variety of forms of a fully rigorous indirect proof which we need not consider here. He adds that the infinitely extended area between the cissoid and its asymptote equals $3\pi a^2$, and that therefore the centroid divides the x-axis in the ratio $5:1$ measured from the cusp. Correspondingly one can find the volume of the solid generated by rotating the cissoid-area $2 \int_0^{2a} t. \mathrm{d}x$ about the parallel to the asymptote through the cusp.[24]

In the autumn of 1658 Huygens had sent his main result, the quadrature of the whole cissoid $2 \int_0^{2a} t. \mathrm{d}x$, to Wallis with the remark that he did not believe this could also be done by the method of the *Arithmetica infinitorum*.[25] Wallis was away from Oxford for some time and only received Huygens' letter after considerable delay, but at once made great efforts to demonstrate the effectiveness of his method on this problem too. After only a few days he succeeded in this.[26] His reasoning is again based on an induction which for us is

[24] Huygens–Sluse, 5 Apr. and 28 May 1658 (*HO* 2: 163–4; 178–9).
[25] Huygens–Wallis, 27 Aug. (6 Sept.) 1658 (*HO* 2: 212).
[26] Wallis–Huygens, 22 Dec. 1658 (1 Jan. 1659) (*HO* 2: 299–301).

equivalent to the observation that integration by parts applied to

$$f(p, q) = \int_0^{2a} x^p (2a - x)^q \,.\, dx \text{ will give a recursion}$$

$$qf(p+1, q-1) = (p+1) f(p, q);$$

and hence it follows that for the required integral

$$\int_0^{2a} x^2 \,.\, dx / \sqrt{[x(2a - x)]} = \tfrac{1}{3} \int_0^{2a} \sqrt{[x(2a - x)]} \,.\, dx,$$

Wallis immediately inserted this result into the *Tractatus de cycloide et cissoide* which he was just then compiling.[27] He wrote down his proof in great haste and was himself aware of certain formal defects in his exposition; he promised a strictly geometrical demonstration[28] but never came back to implement his promise later on. Huygens in his reply[29] had to acknowledge that Wallis had achieved more than he expected by his method; still, the procedure did not appear to him as logically rigorous as the syllogisms of the ancients. He himself had now arrived at the equality of the area of the sector *ABPN* with three times the segment *BQ*. At a later date Huygens sent his own proof of April 1658 to Wallis who printed the piece in the second part of his *Mechanica*.[30]

Huygens received Wallis' treatise on the cycloid and cissoid[31] only in the spring of 1660, mentioning it in a letter to Carcavy[32] where he also notes in passing that he himself was the first to discover the quadrature of the cissoid. It seems that this remark was only considerably later transmitted to Fermat, who thereupon sent a proof of his own.[33] This deserves special notice because, contrary to his usual habit, Fermat merely gives the preparatory reasoning by infinitesimals which shows how subtly he was able to employ the characteristic triangle. In modern terms, his argument amounts to bringing certain symmetries into evidence by setting

$$\int_0^{2a} t \,.\, dx = \int_0^a t \,.\, dx + \int_a^{2a} t \,.\, dx = \int_0^a [x^2 + (2a - x)^2] \,.\, dx / y$$

$$= 2 \int_0^a [a^2 + (a - x)^2] \,.\, dx / y = 2 \int_0^{\frac{1}{2}\pi a} a \,.\, ds + 2 \int_0^a (a - x) \,.\, dy.$$

---

[27] Wallis, *Tractatus duo* (1659): 82–4 (*WO* I: 545–7).
[28] Wallis–Huygens 18 (28) Feb. 1658/9 (*HO* 2: 359).
[29] Huygens–Wallis, 21 (31) Jan. 1658/9 (*HO* 2: 330).
[30] Huygens, note of late autumn 1658 (*HO* 2: 170–3) sent to Wallis perhaps in 1660; printed in his *De Motu* III (1671): 754 in a scholium to prop. 29 of ch. v, *compare WO* I: 906.   [31] Huygens–Wallis, 21 (31) Mar. 1659/60 (*HO* 3: 58).
[32] Huygens–Carcavy, 27 Mar. 1660 (*HO* 3: 56).
[33] *FO* I: 285–8 = *HO* 4: 4–6, enclosure to Carcavy–Huygens, 1 Jan. 1662 (*HO* 4: 1–3).

Huygens rightly considered his own proof as clearer and more complete than Fermat's.[34] The method of utilizing symmetry is indeed most interesting; Huygens in fact uses it too. Fermat had conceived of the idea as early as 1638 and had used it in checking Roberval's quadrature of the cycloid;[35] Huygens had rediscovered it independently and had employed it in many places with complete mastery, but hitherto nothing on it had been published.

In his *Exercitationes geometricæ* Gregory, feeling strongly provoked by the scornful tone of Huygens' review of his *Vera quadratura*[36] and willing to take up the proffered challenge, had made a renewed attempt at the quadrature of the cissoid. All he knew was Wallis' demonstration, which he considered unsatisfactory; he had heard nothing of the line Huygens had taken in his solution – only that the problem originated with his great adversary was known to him. He now felt tempted to show how easily the whole problem could be tackled by the general methods he himself had developed in his *Geometriæ pars universalis*. He found it unnecessary to refer to

---

[34] Huygens–Lodewijk Huygens, 19 Apr. 1662 (*HO* **4**: 111–12).

[35] Roberval had found the quadrature of the common cycloid sometime between 1634 and 1636 but had kept it secret in order to present it as a striking novelty at the public disputation for the Ramus-professorship (*see* p. 161) at Easter 1637 (*MC* **7**: 57). As there was no other competitor he retained the chair for another three years (and subsequently indeed to the end of his life) and was now at liberty to divulge his new result to his correspondents from early 1638 on, as he recalls, for instance, in a letter to Torricelli, 1 Jan. 1646 (*TO* iii: 349). We do not unfortunately now possess the communication of January or February 1638 in which Mersenne told Fermat of Roberval's discovery; in his answer of February 1638 (*MC* **7**: 52) Fermat considers Roberval's statement to be erroneous and doubts if he can sustain it. We have Roberval's proof from his 'Traité des Indivisibles' contained in the *Ouvrages* (1693): 191–3; it is based on his use of the *compagne* of the cycloid, a versine-curve where the area between the curves becomes equal to half the generating circle so that it is easy to find the cycloid-area itself. Roberval in his letter to Fermat, 1 June 1638 (*MC* **7**: 249) re-asserted the correctness of his result while Fermat continued to doubt it; he voiced his reservations in a letter to Mersenne, c. 20 July 1638 (*MC* **7**: 377–80), but freely acknowledged in his next letter, 27 July 1638 (*MC* **7**: 397) that he had been wrong, adding (p. 398) a proof of his own based on expressing the ordinate of the cycloid by the ordinate $\sqrt{(a^2 - x^2)}$ of the circle with the circle-arc $\frac{1}{2}\pi a + s$ added on; the area between the cycloid and the circle is then determinable from

$$\int_0^a ([\tfrac{1}{2}\pi a + s(x)] + [\tfrac{1}{2}\pi a - s(x)]) \, . \, \mathrm{d}x.$$

Nobody at the time had noticed that a similar method of symmetrization, albeit disguised in purely geometrical form, is to be found in Kepler's *Astronomia nova* (1609), chs. 40 and 48.

[36] Huygens for Gallois: *JS* No. 5, 2 July 1668 (*HO* **6**: 228–30).

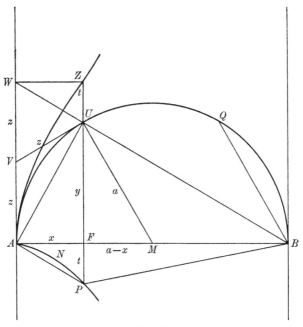

Fig. 18

Huygens by name, assuming that every active mathematician – and others had nothing to do with this debate – was of course informed regarding Wallis' treatise.

He deduces[37] the relationship sector $ABPN = 3$ segment $BQ = 3$ segment AU purely geometrically in the following way: as can be seen from the diagram (fig. 18), AP is parallel to UB and so trilineum $ABP =$ trilineum AUP. The trilineum AFP is now moved *ordinatim* (that is, affinely parallel to the ordinate FU) till it coincides with the trilineum AUZ, so that altogether trilineum $AUP =$ trilineum AFZ. In this way the curve AZ is generated with ordinate $FZ = y + t = 2z$. Now the triangles UWZ, PAF are congruent, hence $A\widehat{U}W$ is a right angle and $AV = VU = VW = z$ equals the tangent to the circle from V. Gregory now constructs a cylinder perpendicular to the circular base AUB and intersects it by a plane through AW inclined at 45° to the base. The portion of the cylinder surface on the base AU equals $\int_0^s x \, . \, ds$. Since further

[37] *Exercitationes geometricæ* (1668): 23–4.

$\mathrm{d}x/y = \mathrm{d}y/(a-x) = \mathrm{d}s/a$ and $x/y = z/a$, we deduce

$$\text{trilineum } AFZ = 2 \int_0^x z \,.\, \mathrm{d}x = 2 \int_0^s x \,.\, \mathrm{d}s = 2a(s-y) = 4 \text{ segment } AU,$$

and the rest is obvious. It is remarkable how close Gregory came in this sequence of deduction to Leibniz' own result, though by a methodologically different route. To him the geometrical structure was still the main point; he could not do without the geometrical interpretation of $\int_0^s x \,.\, \mathrm{d}s$ as part of a cylinder surface. His deductions proceed without any literal algebra purely verbally and hence appear clumsy and rigid; only by Leibniz' notational artifice will they become easy to grasp and pliable in expression. This very example shows most impressively the progress achieved in the transition from the geometrical method of the old school to Leibniz' analytical representation, and gives us some idea of the immense intellectual effort that was involved in the creation of the Calculus. Huygens, the peerless chief proponent of the Archimedean type of deduction, still thinks strongly with a visual imagination; he had not fully understood Leibniz' underlying transmutation theorem in its true significance and proposes to shorten Leibniz' deduction by combining it with Gregory's method for the determination of the trilineum AFZ – yet what is important here is not the formal reduction, but the novel, comprehensive methodological concept. For the first time we feel a resistance developing in Huygens' mind towards Leibniz' intentions – which was later to become an insurmountable prejudice against his analysis. At forty-five years of age he lacks the agility of mind to see the advance which Leibniz' line of thought is inaugurating. An interesting index of Huygens' inbred honesty is his change in attitude towards Gregory's quadrature of the cissoid, which he had originally – annoyed by the strong recriminations against him in the preface to the *Exercitationes geometricæ* – pronounced to be a rather feeble effort.[38]

Huygens' remarks showed once again that Leibniz was still far from being adequately familiar with contemporary mathematical literature. The paper on the quadrature of the circle which he had originally considered ready for publication but which in fact was methodologically still rather imperfect, was therefore kept back by him and once more revised. Somewhat later a new final paragraph[39]

[38] Huygens v. Gregory, late 1668 (not sent): *HO* 6: 321–3.
[39] Leibniz for Huygens, October 1674 (*C* 797).

was added in which Leibniz talks of the limits of the literal method of Viète and Descartes; by this indeed every 'geometrical' (that is algebraical) problem could be reduced to equations, but not general questions in the geometry of curves, whose arithmetization was performed only by Leibniz' own method. And we read here the first hint of his dissatisfaction with Viète's cumbersome method for the approximate solution of numerical equations.[40] In a short list of authors he names as teachers and predecessors Brouncker, Huygens, Mercator, Wren, Wallis, Cavalieri, Fermat, Grégoire de Saint-Vincent, Heuraet and Pascal; he singles out especially the merits of Brouncker and Mercator in squaring the hyperbola, and concerning the 'anonymous' curve (versiera) he refers to Gregory's quadrature of the cissoid (not accepting the name cyclocissoid which Huygens had proposed); finally he speaks extremely lucidly of the value of his new analytical notation.

Already in the summer of 1674 Leibniz had again met Ozanam, whose problems in number theory still remained unsolved. In the meantime Mengoli had tried his hand at the problem of six squares[41] and had been tempted to pronounce it insoluble,[42] only to be refuted in a pamphlet by Ozanam himself.[43] But Ozanam was ready only to communicate a solution, not at all to divulge the method he had used.[44] This we know, however, from his later publication of it in his *Dictionnaire mathématique*.[45] There he starts from the three numbers

$$u = 2abxy, \quad v = a^2x^2 + b^2y^2, \quad w = b^2x^2 + a^2y^2.$$

[40] *Numerosa resolutio* (1600) = *Opera* (1646): 162–228.
[41] *See* Hofmann (1958). The problem had been posed by Ozanam in a Paris pamphlet, spring 1674 (C 704) in a form slightly different from the text of early March 1673 (*see* ch. 4: note 4): to find three numbers u, v, w so that $u-v$, $u-w$, $v-w$; u^2-v^2, u^2-w^2, v^2-w^2 all are squares.
[42] *Theorema arithmeticum* (pamphlet) Bologna (Spring 1674) (C 705).
[43] *Theorema arithmeticum* (pamphlet) Paris, 18 Apr. 1674, recapitulating Mengoli's exposition (C 706). The numerical example here given is

$$u = 2 \cdot\cdot 399 \cdot 057, \quad v = 2 \cdot\cdot 288 \cdot 168, \quad w = 1 \cdot\cdot 873 \cdot 432;$$
hence
$$u^2 - v^2 = 720 \cdot 945^2, \quad u^2 - w^2 = 1 \cdot\cdot 498 \cdot 575^2, \quad v^2 - w^2 = 1 \cdot\cdot 313 \cdot 760^2;$$

$$u - v = 333^2, \quad u - w = 725^2, \quad v - w = 644^2.$$

Leibniz has worked through the calculations and checked them by 'casting out nines'. His own efforts at a solution (C 719, 721, 723) remained unsuccessful.
[44] The basic problem of finding three numbers u, v, w so that $u \pm v$, $u \pm w$, $v \pm w$ all are squares was treated by Ozanam algebraically only in a Paris pamphlet in Latin, 30 Jan. 1677, of which a manuscript copy has survived.
[45] The French version in his *Dictionnaire* (1690 = 1691): 90–1 is somewhat abbreviated (and marred by misprints) compared with the original pamphlet.

89

With this $v \pm u$ and $w \pm u$ are each already squares; the six-square problem will be solved, when $v \pm w$ can also be made squares. To achieve this he sets

$$x = z - ay/b \quad \text{and} \quad v + w = p^2/b^2, \quad v - w = q^2/b^2.$$

He then gets

$$p^2 = [(a^2 + b^2)\, y]^2 - 2ab(a^2 + b^2)\, yz + (a^2 + b^2)\, b^2 z^2,$$

$$q^2 = [(a^2 - b^2)\, y]^2 - 2ab(a^2 - b^2)\, yz + (a^2 - b^2)\, b^2 z^2.$$

With

$$(a^2 + b^2)^2\, q^2 - (a^2 - b^2)^2\, p^2 = 2ab^3(a^4 - b^4)\, z\, (bz/a - 2y)$$

he obtains an expression that can easily be split into the difference of two squares and finds further

$$p = (a^2 + b^2)(y - bz/2a) + ab^3 z/(a^2 + b^2),$$

and so on, and hence finally

$$x : y : z = a(3b^8 - 6a^4 b^4 - a^8) : b(b^8 + 6a^4 b^4 - 3a^8) : 4a(b^8 - a^8).$$

For $a = 1$, $b = 2$ Ozanam obtains

$$u = 2\,399\,057, \quad v = 2\,288\,168, \quad w = 1\,873\,432,$$

which is the example he had previously given. All Leibniz knew at the time was that the solution had been reached by algebraic reasoning, but he was unable to rediscover it.

Gregory – to anticipate – heard of the problem in the summer of 1675 through his friend Frazer.[46] He succeeded in a very short time in finding an extremely subtle solution,[47] as we know from the notes he left behind.[48] He proceeds by a methodologically superior way, by setting

$$v - w = a^2, \quad v + w = p^2,$$

$$u - w = b^2, \quad u + w = q^2,$$

$$u - v = c^2, \quad u + v = r^2,$$

and observing $a^2 = b^2 - c^2 = r^2 - q^2$ which permits him to take

$$b = (a^2 + x^2)/2x, \quad c = (a^2 - x^2)/2x,$$

$$q = (a^2 - y^2)/2y, \quad r = (a^2 + y^2)/2y.,$$

[46] Frazer–Gregory, June (?) 1675, *compare GT*: 311.
[47] Lost enclosure to Gregory–Frazer, 13 (23) July 1675 (*GT*: 311–12), recognizable from the reply, 10 (20) Aug. 1675 (*GT*: 323).
[48] *GT*: 430–3.

The identity $v^2 - w^2 = (u^2 - w^2) - (u^2 - v^2)$ gives by substitution

$$(ap)^2 = (bq)^2 - (cr)^2 = a^2(x^2 - y^2)(a^4 - x^2y^2)/4x^2y^2.$$

Gregory now sets $x^2 - y^2 = z^2$ and $a = t + x$ and obtains

$$(2pxy)^2 = x^2z^4 + 4tx^3z^2 + 6t^2x^2y^2 + 4t^3xz^2 + t^4z^2.$$

Then, following the general rules given in Billy's *Inventum novum* for the treatment of biquadratic expressions of just this type he equates this in turn to

$$(xz^2 + 2tx^2 + t^2z)^2 \quad \text{and to} \quad (xz^2 + 2tx^2 + t^2x[x^2 - 3y^2]/z^2)^2.$$

In the first case he ends up with

$$t = (x^2 + y^2 - 2xz)/2z,$$

whence $\qquad a = (x^2 + y^2)/2z, \quad p = (4x^2y^2 - z^4)/8xyz$

(this yields Ozanam's triplet for $x = 5$, $y = 3$, $z = 4$); in the second case Gregory obtains finally

$$t = 4x(x^4 - y^4)/(-3x^4 + 6x^2y^2 + y^4),$$

and so on. The high esteem in which Leibniz later held Gregory's achievement in number theory appears to be founded on what he had learnt (though probably only in general terms) of this solution of the six-square problem.

A second, much simpler problem which Ozanam had put forward in a single-leaf pamphlet[49] in February 1674 demanded three numbers whose sum shall be a square and the sum of whose squares shall be a fourth power. It arose as an extension of the question in Billy's *Inventum novum* I, 45: to find the numbers x, y so that $x + y = a^2$, $x^2 + y^2 = b^4$. This question comes from Fermat[50] who had submitted it to many of his correspondents without ever receiving a solution. He himself begins with a rightangled triangle $x^2 + y^2 = z^2$ and sets $x = u^2 - v^2$, $y = 2uv$ and hence

$$x + y = (u + v)^2 - 2v^2 = a^2; \quad x^2 + y^2 = (u^2 + v^2)^2 = b^4.$$

[49] Extant in *C* 703A. Leibniz has written in the solution in small numbers found by him. For the subject matter *see* Hofmann (1969a).

[50] Details on the earlier history of the problem and the gradual construction of the solution will be found in Hofmann (1969b), more on the structure of the system of solutions in Heller (1970). Leibniz would only have had before him what was said in Billy's *Inventum novum* (1670): I. 45, p. 13 and III. 32, pp. 31-2; on both occasions the solutions – Fermat's own – are given as 13-figured positive numbers, these being in fact the smallest possible. But we know nothing about any further, more thorough occupation with this problem by Leibniz.

Now the first equation is satisfied by

$$u+v = 2\xi^2+\eta^2, \quad a = 2\xi^2-\eta^2, \quad v = 2\xi\eta;$$

and this leads to $\quad u = (\xi-\eta)^2+\xi^2, \quad v = 2\xi\eta,$

or $\quad 4\xi^4-8\xi^3\eta+12\xi^2\eta^2-4\xi\eta^3+\eta^4 = b^2 = (2\xi^2-2\xi\eta+\eta^2)^2+(2\xi\eta)^2.$

This again can be satisfied by taking $2\xi^2-2\xi\eta+\eta^2 = \eta^2-\xi^2$, so that we obtain $3\xi^2 = 2\xi\eta$, and from this follows $\xi = 2$, $\eta = 3$ and thus $u = 5$, $v = 12$; $x = -119$, $y = 120$.

To proceed, let now $u = p+5q$, $v = 12q$, then this will give

$$p^2+34pq+q^2 = a^2$$

$$p^2+10pq+169q^2 = b^2,$$

and on eliminating q^2 we have finally

$$14q(12p+2868q/7)$$

$$= (13a)^2-b^2 = (13p+1434q/7)^2-(p-1434q/7)^2.$$

From this follows

$$p = 2\,048\,075, \quad q = 20\,566,$$

and so on.

Ozanam's problem is far simpler than Fermat's. In conversation with Ozanam, Leibniz remarked he did not believe the solution to be all that difficult; it seemed to him mainly a question of patience.[51] Ozanam at once took him at his word and he was forced to try and tackle the calculation. After several abortive attempts Leibniz eventually succeeded in the following manner[52] (which we reproduce in modern notation).

Let $\quad x+y+z = p^2, \quad x^2+y^2+z^2 = q^4$

$$x = a(a+b), \quad y = a(2p-b), \quad z = p^2-2ap-a^2;$$

then, on substituting,

$$p^4-4ap^3+6a^2p^2+4a^2(a-b)p+2a^2(a^2+ab+b^2) = q^4;$$

and if further $q = p-a$, one obtains

$$p = (a^2+2ab+2b^2)/4(b-2a),$$

$$q = (9a^2-2ab+2b^2)/4(b-2a) = (17a^2+[2b-a]^2)/8(b-2a),$$

[51] Leibniz–Oldenburg, 6 (16) Oct. 1674 (*LBG*: 107); and Leibniz–Magliabecchi. 8 Nov. 1691 (*LSB* I. 7: 420).
[52] *C* 712, 714–15, 724. For details *see* Hofmann (1969*a*): 108–13.

and so

$$x = a(a+b), \quad y = a^2(a+6b)/2(b-2a), \quad z = q^2 - 2(p-q)^2$$

for the solution. Leibniz notes that the values will be positive for $b > 2a$, that is for $p > 1 + \sqrt{2}$. From $a = 1$, $b = 3$ he obtains after clearing of fractions $x = 64$, $y = 152$, $z = 409$; $p = 25$, $q = 21$. This also follows from the readily understood formula

$$(p^2 - 2p - 1)^2 + (2p - 3)^2 + 4^2 = (p-1)^4$$

which yields $p = \frac{25}{4}$, and which is itself a special case of a solution in three homogeneous parameters u, v, w: on setting

$$x = p^2 - 2pu - v^2, \quad y = p(u+v) + \tfrac{1}{2}(v^2 - w^2),$$

$$z = p(u-v) + \tfrac{1}{2}(v^2 + w^2)$$

we have at once $x + y + z = p^2$. If we further demand that

$$x^2 + y^2 + z^2 = (p-u)^4,$$

we obtain p from u, v, w.

This access of knowledge in a new branch of mathematics leads Leibniz into the field of number theory which he had scarcely explored before, and brings him into renewed contact with the philosopher, physicist and mathematician, Mariotte, in whose experiments he was strongly interested throughout his life. A short summary of his latest results from the autumn of 1674 seems to have been destined for Mariotte; this might have been compiled somewhat later than the paper on the arithmetical quadrature of the circle addressed to Huygens.[53] First he deals with the problem by Ozanam which we have just mentioned, next he gives the sum of the inverse figurate numbers, and lastly presents a general survey of the two most recent big discoveries. One of these, he says, is of a geometrical nature and relates to the 'mechanical' (i.e. transcendental) problems that arise in conjunction with quadratures; hitherto it had been possible to find the quadrature of higher parabolas and hyperbolas, but not those of the circle and ellipse. For this difficulty Leibniz claims to have found

[53] Leibniz for Mariotte (?), October 1674 (*C* 796). Leibniz often mentions Mariotte's numerous physical investigations and papers, always favourably and approvingly. He had a high opinion of Mariotte as a sincere and reliable person; see his letter to Tschirnhaus, late June 1682 (*LBG*: 437). Further evidence comes from Mariotte's cooperation when Olivier was working on Leibniz' calculating machine in Paris during the latter's absence; see the references in the register for *LSB* I. 2 and I. 3. Mariotte obviously also was knowledgeable in the current methods of indeterminate analysis or he might not have chosen to edit Frénicle's posthumous *Traité des triangles en nombres* (1676); see further p. 201.

93

the decisive remedy through his arithmetical transformation of the quadrature into that of an equivalent rational figure. The other discovery, he continues, is of an arithmetical nature and concerns the treatment of problems in number theory when solutions in fractions are known. Leibniz professes to have a general method which is capable of characterizing solutions in integers and which enables him to determine solutions in smallest numbers or (where the range of validity is limited) largest possible numbers. Here Leibniz alludes to his contemporaneous researches on indeterminate linear equations in several unknowns. They had been provoked by a problem extensively treated by Bachet:[54] to find a number which on being divided by certain given divisors will leave prescribed remainders.[55] The piece we have cited here is typical of Leibniz' mode of argument, which is frequently based on extremely bold generalizations. In the present case, what he has actually achieved bears no relation to what he had imagined to be within his reach. But such excessive optimism is necessary to keep up one's courage if one is to continue with a difficult piece of research whose full significance cannot be foreseen at the outset.

Leibniz saw weeks and months go by as he waited – in vain – for an answer from Oldenburg.[56] So he decided to write to London again himself.[57] This letter too was brought over by a friend – perhaps the doctor Le Vasseur.[58] In it Leibniz reports a conversation he has had with Malebranche's disciple, Prestet,[59] who was then engaged in

[54] *Problèmes plaisans*, ₁1612, probl. 5; ₂1624, probls. 18, 21.
[55] Contemporaneous notes: *C* 742–4, 757–9.
[56] Oldenburg's answer is now lost, though we can tell the date from a note which he made in French at the end of Leibniz' letter of 5 (15) July 1674 (unpublished, *C* 687): 'received 12 (22) July 1674, brought over by Walter, answered on 15 (25) July'. Nothing, however, is known about the contents of the answer.
[57] Leibniz–Oldenburg, 6 (16) Oct. 1674 (*LBG*: 106–8).
[58] Le Vasseur's name has been written on the envelope by Oldenburg which leads us to assume that he had delivered the letter.
[59] *LBG*: 106. This refers to a letter by Prestet, September 1674 (*C* 1279). He had misunderstood Descartes' procedure in the *Géométrie* (1637): 383 for reducing a biquadratic without the second term to a cubic, and, further misled by an error in his own working, had believed Descartes to be wrong. Leibniz remarks that Descartes' rule is really the same as one given by Viète (*Æquationum recognitio* (1591) = *Opera* (1646): 144–5). Since Descartes merely states the result, we cannot tell whether he has taken his rule from Viète or has developed it independently from the well-known report Cardan gives in his *Ars magna* (1545): cap. xxxix, regula ii (on fol. 22v) of Ferrari's own procedure. Leibniz further refers to De Beaune's explanation in the 'Notæ breves' (*Geometria* (1659): 137–9) where in fact Viète's procedure is inverted to afford a method of proof. He does not mention Schooten's considerations in the 'Commentarii' (*ibid.* 315) but expressly refers to Hudde's excellent exposition in his 'Reductio æquationum' (*ibid.* 494–5).

94

collecting all contemporary algebraical knowledge in a new textbook; somehow he had become obsessed with the idea that Descartes' method for reducing a quartic equation to a cubic is not generally applicable, and could in nowise be diverted from his error. Next follows a report on Ozanam, who recently had shown Leibniz the manuscript of his 'Diophantus promotus':[60] this was without doubt a very deserving book on number theory, using literal algebra throughout. What had been missing in Bachet's edition was here largely supplied, and apart from this an interesting further chapter had been added. Leibniz explains that he feels a particular obligation to mention this impending publication to Oldenburg, because he believed a similar book was about to be printed in England. This is an allusion to Kersey's *Algebra*, the second volume of which, containing chapters on number theory, came in fact into the bookshops early in October.[61]

Now follows a very significant reference to Leibniz' new studies on problems of quadrature which were of the greatest importance for mechanics;[62] Brouncker and Mercator had accomplished the 'rational' quadrature of the hyperbola, but nobody had been able to effect the equivalent for the circle, since even Wallis' infinite product and Brouncker's infinite continued fraction were nothing but approximations. Leibniz himself, however, so he now repeats, had given a rational series for the quadrature of the circle, which stands in a remarkable correspondence to the quadrature of the hyperbola. This did not claim to represent π as the quotient of two finite numbers – that, according to Leibniz, was probably impossible – but here was a representation by means of an infinite series, and by the same method the arc could be found from its sine without first going back to the whole circle. Wallis had drawn attention to Brouncker's arithmetical quadra-

[60] According to Varignon–Leibniz, 25 May 1714 (*LMG* IV. 196), Ozanam would not give the 'Diophantus promotus' to the printers without a proper fee. The manuscript is now lost.
[61] Compare Collins–Gregory, 25 Sept. (5 Oct.) 1674 (*GT*: 285). The list of contents in Oldenburg for Leibniz, 6 (16) Apr. 1673 (*C* 409A) is presumably identical with the publishers' advertisement asking for subscriptions, *see* Newton–Collins, 25 May (4 June) 1672 (*NC* 1: 161–2). The work appeared in parts: Collins–Gregory, 23 Sept. (3 Oct.) 1672 (*GT*: 245). A preliminary notice appears in *PT* 7 No. 90 of 20 (30) Jan. 1672/3: 5152–4; a review of the first volume in *PT* 8 No. 95 of 23 June (3 July) 1673: 6073–4; both these texts are probably from Collins' pen, perhaps also the publishers' advertisement. Tschirnhaus in declaring (*C* 1055) *tout court* the work to be valueless, certainly was far too derogatory. Leibniz' judgment in 'De ortu, progressu et natura algebrae' (1685–6) (*LMG* VII: 215–16) where he praises the author's industry can still be repeated today without reserve.
[62] Leibniz–Oldenburg, 16 Oct. 1674 (*LBG*: 107).

ture of the hyperbola (to which Leibniz refers in the letter) long ago in the dedication of his polemical tract against Meibom's theory of proportions, as well as in a letter to Digby of roughly the same date.[63] When Sluse in the autumn of 1658 received from Huygens a copy of Wallis' *Commercium epistolicum*,[64] he immediately expressed the hope that Brouncker might soon decide to communicate his result.[65] He probably knew that Soverus had already in 1630 promised something on the quadrature of the hyperbola but had died before he had completed his work – and that nobody cared to look through his papers preserved at Padua[66] because they feared to be reproached with having appropriated some of Soverus' thoughts should they themselves ever later happen to publish anything on the subject.

But ten years were to go by before Brouncker decided to reveal his quadrature, and even then only because the problem of the quadrature of the hyperbola had come to the forefront of public interest through Gregory's *Vera quadratura* and so had become the object of lively discussion, and because moreover the same topic was also dealt with in Mercator's *Logarithmotechnia*, then in course of being printed, and to delay further would have cost Brouncker the prestige of priority. Brouncker constructs[67] the hyperbola $y = 1/(1+x)$ in the square $0 \leqslant x \leqslant 1$, $0 \leqslant y \leqslant 1$ and subdivides the area $\int_0^1 \mathrm{d}x/(1+x)$ by repeated bisection in the manner indicated in the diagram (fig. 19). He now asserts that the hyperbola-area (log 2) is greater than

$$\frac{1}{1.2} + \left(\frac{1}{3.4}\right) + \left(\frac{1}{5.6} + \frac{1}{7.8}\right) + \left(\frac{1}{9.10} + \frac{1}{11.12} + \frac{1}{13.14} + \frac{1}{15.16}\right) + \cdots$$

while he sees that the remaining area $(1 - \log 2)$ is greater than

$$\frac{1}{2.3} + \left(\frac{1}{4.5} + \frac{1}{6.7}\right) + \left(\frac{1}{8.9} + \frac{1}{10.11} + \frac{1}{12.13} + \frac{1}{14.15}\right) + \cdots$$

where the error on continuing the subdivision to $n = 2^k - 1$ points will be

$$\frac{1}{(n+1)(n+2)} + \frac{1}{(n+2)(n+3)} + \cdots + \frac{1}{(2n-1)2n}$$

$$= \frac{1}{n+1} - \frac{1}{2n} = \frac{n-1}{2n(n+1)}.$$

[63] *Compare* ch. 4: note 37. Wallis–Brouncker, 5 (15) Dec. 1656 (Wallis (1657): 3³).
[64] Sluse–Huygens, 11 Oct. 1658 (*HO 2*: 249).
[65] Sluse–Huygens, 19 Oct. 1658 (*HO 2*: 259).
[66] Gregory–Collins, 16 (26) Mar. 1667/8 (*GT*: 49–50).
[67] *PT* **3**, No. 34 of 13 (23) Apr. 1668: 645–9.

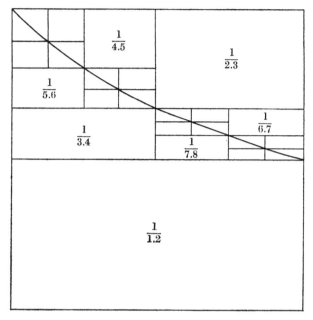

Fig. 19

When his procedure is generalised – and he indicates himself how this is to be done – one obtains some theoretically quite interesting but in practice poorly convergent (and so not really useful) expansions:

$$\log\,(1+x) - \frac{x}{1+x} = x^2 \left\{ \frac{1}{(2+x)(2+2x)} + \left[\frac{1}{(4+x)(4+2x)} \right. \right.$$

$$\left. \left. + \frac{1}{(4+3x)(4+4x)} \right] + \ldots \right\}$$

and

$$x - \log\,(1+x) = x^2 \left\{ \frac{1}{2(2+x)} + \left[\frac{1}{4(4+x)} + \frac{1}{(4+2x)(4+3x)} \right] + \ldots \right\}.$$

Brouncker as yet knows nothing of natural logarithms – this technical terminology was coined only by Mercator;[68] he is entirely concerned with finding common logarithms to base 10 from his work with hyperbola-areas. A considerable quantity of further interesting though unpublished work on the quadrature of the hyperbola had been done at the same period – the most important is Huygens'

[68] Supplement to the *Logarithmotechnia* in *PT* **3**, No. 38 of 17 (27) Aug. 1668: 759–64.

approach from an approximate value for the centroid[69] – but Leibniz could know nothing of all this; nor could he have heard that Mercator, too, had gone on to formulate series for the circle of which, unfortunately, our only information comes from meagre allusions in Collins' letters unsupported by any tangible evidence.[70] Of Newton's results in his *De Analysi* of 1669 he had indeed had a general indication[71] in Oldenburg's missive of 6 (16) April 1673, but he would not gather their precise character and significance from that.

In the *Commercium epistolicum* this report by Leibniz on his latest discoveries is printed[72] with the comment that,[73] in reality, Collins had already for some four years been passing Newton's series on to his friends,[74] and that Gregory for three years had made known his own results to his circle of acquaintances,[75] while Leibniz during his first visit to London in 1673 had communicated nothing in this field[76] and had sent the series for the circle only after he had received the result from Oldenburg. This account is correct in so far as Collins had already sent a general report on Newton's work on series to Sluse[77] in the autumn of 1669 and had passed on the series for the circle-zone to Gregory[78] in the spring of 1670; also, Gregory must have been in possession of his interpolation series[79] from 1668 onward although he did not send instances of them to Collins[80] before December 1670. But to conclude from this chronological sequence that Leibniz was dependent on the work of the English is going far too far. We know very accurately from his paper that he possessed his own quadrature of the circle already in the autumn of 1673. In the margin of his copy of the *Commercium epistolicum* Leibniz rightly noted[81] that the statement there was wrong, and that his

---

[69] Autumn 1661 (*HO* **14**: 451–7).

[70] *Compare* Collins–Gregory, 30 Dec. 1668 (9 Jan. 1669); 7 (17) Jan. 1668/9; 2 (12) Feb. 1668/9; 15 (25) Mar. 1668/9 (*GT*: 56, 60, 66, 71) and Hofmann (1943): 109–11.

[71] *See* ch. **4**: note 34.

[72] Excerpt from Leibniz–Oldenburg, 6 (16) Oct. 1674 (*CE* (*1712*): 37–8 = *LBG*: 107–8) taken from its first printed appearance in *WO* III, 618–19.

[73] *CE* (*1712*): 38.   [74] *See* ch. **4**: note 54.

[75] References to this occur in the *CE* (*1712*) in the following letters of Gregory to Collins: 19 (29) Dec. 1670 (*CE* (*1712*): 23 = *GT*: 148) (*see* ch. **4**: note 56 and ch. **10**: p. 135); 15 (25) Feb. 1670/1 (*CE* (*1712*): 25–6 = *GT*: 170–1) (*see* also ch. **10**: p. 140).

[76] In a marginal note to this Leibniz emphasizes that he was not in possession of the series for the circle area at the time.

[77] Collins–Oldenburg for Sluse, 12 (? 22) Sept. 1669 (*OC* **6**: 227–8), passed on in Oldenburg–Sluse, 14 (24) Sept. 1669 (*OC* **6**: 233–4) reproduced in *CE* (*1712*): 21.

[78] Collins–Gregory, 24 Mar. (3 Apr.) 1669/70 (*GT*: 89).

[79] *See* ch. **6**: note 16.   [80] *See* ch. **3**: note 47.

[81] Leibniz' note on *CE* (*1712*): 38.

series was different from Newton's. Concerning Leibniz' remark on finding the arc from its sine[82] the *Commercium epistolicum* refers to Leibniz' letter of 12 May 1676 in which he makes a point of asking for the process of finding Newton's series for the arcsine.[83] In his annotation on this[84] Leibniz states that all he wanted was Newton's method of proof. We shall look into this whole matter more closely later on.[85]

To avoid the further loss of a letter Oldenburg gave his answer, dated 8 (18) December 1674, to Walter on his departure for Paris;[86] he was also to take a letter to Huygens.[87] He there talks of the publication of the second volume of Kersey's *Algebra* and of Gregory's researches in number theory into simultaneous Diophantine equations[88] which Collins still hoped to have published in the *Philosophical Transactions* before Ozanam's edition of Diophant appeared;[89] the paper was, however, never printed and today its manuscript is lost. Regarding the calculating machine with which Leibniz had, in the eyes of the Royal Society, rather let him down, Oldenburg expresses himself cautiously:[90] it would be highly desirable if Leibniz, remembering the promise he had given at a public session of the Society, would on a convenient occasion deliver the 'little instrument' to them.

As for the theory of quadratures, writes Oldenburg,[91] both Newton and Gregory had progressed to a general method which was applicable to all geometrical and mechanical curves and in particular the circle, so that one could from the ordinate of a curve determine its arclength, its area and its centroid, the surface and volume of the solid generated by the rotation of its area, and their segments; moreover, it was feasible to calculate log sin, log tan, and log sec without prior computation of the natural functions, and vice versa. If the circle could really be exactly squared by arithmetic, then Leibniz was to be congratulated; but since on the contrary Gregory was then occupied with his proof for the impossibility of an exact quadrature of the circle, it might indeed be necessary to examine and consider this

[82] Leibniz–Oldenburg 6 (16) Oct. 1674 (*LBG*: 108 = *CE* (*1712*): 38).
[83] Leibniz–Oldenburg 2 (12) May 1676 (*LBG*: 167–8 = *CE* (*1712*): 45).
[84] Leibniz' note on *CE* (*1712*): 45.
[85] *See* p. 211 and ch. **15**: note 5.
[86] Oldenburg–Leibniz, 8 (18) Dec. 1674 (*LBG*: 108).
[87] Oldenburg–Huygens, 9 (19) Dec. 1674 (*HO 7*: 395–6).
[88] Enclosure (now lost) to Gregory–Collins, 15 (25) Sept. 1674, mentioned in Collins–Gregory, 23 Nov. (3 Dec.) 1674 (*GT*: 291).
[89] Collins–Gregory, 23 Nov. (3 Dec.) 1674 (*GT*: 291).
[90] Oldenburg–Leibniz, 8 (18) Dec. 1674 (*LBG*: 108).
[91] *Ibid.* 109, reproduced in *CE* (*1712*): 38–9.

matter most carefully. The letter ends with information on the latest work by Fellows of the Royal Society.

The reader who knows something of the circumstances immediately feels that Oldenburg is no longer as forthcoming and friendly as hitherto when writing to Leibniz. He is reserved and critical and it becomes distinctly noticeable that he is upset by Leibniz' long silence and has forgotten neither the overlong delay in the communication of the calculating machine nor the discussion about interpolation theory which had ended so unhappily for Leibniz. He naturally feels disinclined to become involved in further unpleasantness over the quadrature of the circle, particularly as there had been difficulties enough with Huygens just recently.[92] The reference to Newton's and Gregory's infinitesimal methods is couched in very general terms and just as vague and unforthcoming as the corresponding remark in Oldenburg's communication of 6 (16) April 1673. Leibniz has underlined its phrase *ignorare Te nolim* in the excerpt reproduced in the *Commercium epistolicum*,[93] and added *ergo antea ignorabat*. But he must have noticed at once that *ergo* was not the right word here and so he replaced it by the more appropriate *nempe*.

The reference (which Oldenburg owes to Collins) to Gregory's proof of the impossibility of squaring the circle[94] is not justified in the circumstances; Gregory's assertion of the analytical impossibility of the quadrature only meant to affirm that π could certainly not be the solution of an equation of finite degree with integral coefficients. We do know that Gregory was then preparing an improved edition of his *Vera quadratura*,[95] though the manuscript is not preserved with his other papers that have by now come to light. Interesting and perhaps hitherto not properly appreciated is the following circumstance: Oldenburg wants to restrain Leibniz and hence advances his own doubts about the possibility of the arithmetical squaring of the circle. He achieves the exactly opposite effect, because Leibniz was led to assume from this remark that his type of expansion was altogether unknown in England, and so in spite of its cool, even unfriendly, reserve he was given a fresh impulse to pursue his researches confirmed as he was in his opinion that he was advancing over completely untrodden ground: this turn of events had scarcely been intended by his English correspondents.

[92] This refers to the controversy on the first discovery of algebraically rectifiable curves; *see* ch. 8.   [93] Leibniz' note on *CE* (*1712*): 38.
[94] Oldenburg–Leibniz, 8 (18) Dec. 1674 (*LBG*: 108).
[95] Gregory–Collins, 23 Nov. (3 Dec.) 1670 and 15 (25) Feb. 1670/1 (*GT*: 118, 171).

8

THE QUARREL OVER RECTIFICATION

Here we must pause for a moment and concern ourselves with an issue that seems at first glance to be only loosely connected with Leibniz' affairs. It will however become evident that it is of the greatest significance for an understanding of his relationship with London. I refer to the dispute between Huygens and several Fellows of the Royal Society about the first discovery of the rectification of certain types of curves. It was caused by the historical account of the topic in the *Horologium oscillatorium* III, 7–9 and led to Huygens and Wallis breaking off their correspondence. Huygens was deeply annoyed; the ill-feeling generated was only slowly to be dispersed by Oldenburg's unceasing attempts at a reconciliation. The basis of the dispute – to us one mere problem in integration – was of such special importance to the mathematicians of the day because an axiom handed down through the centuries and thought unassailable by the Aristotelians was here knocked on the head.

We are in fact dealing here with a philosophico-mathematical question which is probably connected with the attempts already made repeatedly by the Greeks, though never successfully, to square the circle by geometrical construction. Aristotle maintains that there can be no rational proportion between the curved and the straight, and his commentator Averroes even denies the possibility of a rational proportion between two arcs.[1] It is very doubtful whether Archimedes too supported this Aristotelian opinion; at any rate it is to his treatise *On the sphere and cylinder* that we owe the fundamental axiom for convex plane figures – valid for arcs as well as for areas – on which all later rectifications are based: 'that which is included is the lesser'.

During the Dark Ages little more than the name of Archimedes survived; only when a Greek text came to light again in the age of Humanism were his thoughts resurrected in a major way. At the instigation of the scholarly pope Nicolaus V towards the middle of the

---

[1] *Compare* Hofmann (1942): 6.

fifteenth century the Greek codex was translated into Latin by a cleric, Jacob of Cremona, and in this form Nicolaus of Cusa came to know it. Though the Cardinal was interested in mathematics, he was primarily a philosopher and the multitude of the administrative problems he had to tackle prevented him from thoroughly studying this difficult work which was, to make matters worse, disfigured by many errors in translation. He read into Archimedes' text a link with his own philosophical principle of the *coincidentia oppositorum* and combined Archimedes' reductio ad absurdum with the Aristotelian opinion which had by then assumed almost the rank of a scientific dogma. He asserts that though a circular arc might be approximated by straight lines to any desired degree of accuracy, it could still never be exactly rectified[2] – an opinion which he himself and succeeding mathematicians extended to cover other convex arcs as well. Regiomontanus indeed raised objections against the doctrine of Aristotle and Averroes,[3] but to no avail. He had in 1462 revised Jacob of Cremona's Latin translation of Archimedes by collating it with a Greek manuscript, but he died before the planned edition was ready for printing. The revised Latin text was eventually edited, together with the Greek original, by Th. Gechauff (Venatorius) in 1544 and this became the basis for Viète's study of Archimedes (incidentally, he knew also Regiomontanus' *Trigonometry* in the edition of 1533).[4] Viète understood the importance of 'the axiom of Archimedes' concerning the included and the including, pointing out that it was at variance with assumptions of Euclid and his school.[5] The observations of Regiomontanus and of Viète failed to attract the attention of mathematicians in the first third of the seventeenth century. Thus Aristotle's dogma of the incomparability of curved and straight lines is still valid for Descartes, as when he says, for instance in the *Géométrie* that no geometrical curve can ever be

---

[2] He formulates his own concept most clearly in his essay 'De circuli quadratura' of 1450 (which is still unprinted but is now accessible in German translation (1952): 45–9). In Nicolaus of Cusa's *De mathematica perfectione* (1458) he repeats briefly that there cannot exist a ratio expressible in whole numbers between arc and chord of a circle (*Opera* (1514) II, fol. ₂101r = p. 161 of the translation). The passage was read and largely approved by his contemporaries and by most succeeding mathematicians.

[3] In an appendix of his *Trigonometry* (1462–4) Regiomontanus edited a number of papers by Nicolaus of Cusa and discussed them critically in 1464, all of which was included by Schöner in his edition of 1533. The Aristotelian dogma of the non-comparability of straight and curved lines is strongly denied there (p. 37).

[4] *Apollonius Gallus* (1600) = *Opera* (1646): 339.

[5] *Supplementum geometriæ* (1593) = *Opera* (1646): 240; *Variorum...liber* VIII (1593) = *Opera* (1646): 398.

geometrically rectified.[6] For mechanical curves he thinks it possible:
that much is evident from his example of the logarithmic spiral,
discussed by him in a 1638 letter, which he defines as the curve
making a constant angle with the radius vector at every point.[7] Not
much later Fermat, doubtless influenced by Viète, had in the course
of a tangent construction recognized the parabola-arc as rectifiable.[8]
Similarly Hobbes though not always well-informed on mathematical
questions had in 1640 even suggested that the Archimedean spiral
and the parabola might be related to each other under preservation of
arc-length.[9] Following Hobbes' suggestion[10] Roberval was able
sometime in 1642–3 to demonstrate this equality of arcs;[11] Mersenne
reports on all this in his *Cogitata* (1644): 129–31,[12] and when he stayed
at Rome during the winter 1643–4 he talked of this in conversation
with Ricci who passed it on to Torricelli.[13] The latter already at this
time knew the rectification of the logarithmic spiral,[14] which he had
accomplished, independently of Descartes, through the mechanical
generation of the curve as $r = ak^t$, $\theta = \alpha t$; he also knew that the
spiral approaches its pole through infinitely many revolutions, but
that the length of arc to the pole nevertheless has a finite value. He
had found this result by means of a construction involving a geometric progression, but this was published only posthumously[15] in
the year 1919, and more completely in 1955 from a recently found
better manuscript.

Torricelli could not at first justify the equality of the arc-lengths
of the spiral and the parabola, nor could he gather anything useful
from Mersenne's allusions;[16] it is noteworthy in any case that
Mersenne is obviously no longer convinced of the impossibility of a

[6] *Géométrie* (1637): 340 = *Geometria* (1659): 39.
[7] Descartes–Mersenne, 12 Nov. 1638 (*MC* **8**: 78).
[8] Fermat–Mersenne, 22 Oct. 1638 (*MC* **8**: 158).
[9] *NP* **3**: 207: note 442. The suggestion became a firm assertion in the *Examinatio* (1660): 122 though the attempted proof is fallacious.
[10] *Compare NP* **3**: 309–10: note 704.
[11] *See* Møller-Pedersen (1970): 26–43.
[12] *See* Krieger (1971): 101–3.
[13] In his letter to Ricci of 17 Jan. 1645 Torricelli asks Mersenne for the name of the French mathematician who had established the equality of arcs for the Archimedean spiral and the common parabola: *see* Tannery (1933). In reply, Mersenne's letter to Torricelli, 4 Feb. 1645 (*TO* III: 269) gives Roberval's name.
[14] The most important references are to be found in Torricelli's letters to Ricci, 17 Jan. 1645 (*see* note 13); to Carcavy, February 1645; to Ricci, 17 Mar. 1646 with a more elaborate account; to Cavalieri, 23 Mar. 1646; to Roberval, 7 July 1646; and to Cavalieri, 15 Aug. 1647 (*TO* III: 280, 360, 364, 391–2, 470).
[15] *De infinitis spiralibus*: *TO* I. 2 (1919): 349–73; augmented (1955): 17–76.
[16] Mersenne–Torricelli, 4 Feb. 1645 (*TO* III: 269).

rectification of the parabola and presses Torricelli to tell him if it really could not be done. A little later Mersenne again, urged by Fermat, points to the rectification of the parabola as a still unsolved problem of great significance which Fermat had not hitherto mastered.[17] At that time Torricelli had in fact already proved the theorem on equality of arcs which he had found first for the Archimedean spiral and the Apollonian parabola,[18] but later generalized[19] to the spirals $(r/a)^m = (\theta/2\pi)^n$ and the parabolas $(y/b)^n = (x/a)^{m+n}$ by means of the transform $x = r$, $y = \int_0^\theta r \, d\theta = mr\theta/(m+n)$ on setting $b = 2\pi am/(m+n)$. This 'rectifying' transformation,[20] $x = r$, $y = \int_0^\theta r \, d\theta$, occurs again, in more general form in Gregory's *Geometriæ pars universalis*,[21] and after this in Barrow's *Lectiones geometricæ*.[22] Gregory, of course, received many suggestions and comments from his teacher Angeli, himself a pupil of Cavalieri's and an admirer of Torricelli whose *Opera* (1644) he knew well.[23] Torricelli's result was not at all properly appreciated – Carcavy appears not to have passed on what he had been told – and so it was possible that Fermat, ignorant of any predecessor, could rediscover it believing it to be quite new.[24]

When Huygens was twenty-one years old he worked through Torricelli's *Opera geometrica* and came out strongly against the author's use of Cavalieri's indivisibles.[25] As a counter–example – one from which he had hoped originally to gain the rectification of the parabola – he gives the following.[26] Let ABD (fig. 20) be the parabola

[17] Mersenne–Torricelli, 26 Aug. 1646 (*TO* III: 411).
[18] Torricelli–Ricci, 7 Apr. 1646 (*TO* III: 368).
[19] Torricelli to Roberval, 7 Aug. 1646, to Carcavy, 8 July 1646; to Ricci, 24 Aug. 1647; to Cavalieri, 31 Aug. 1647 (*TO* III: 392, 407, 473–4, 476–7).
[20] *De infinitis spiralibus*: *TO* I. 2 (1919): 381–92.
[21] This is the fundamental idea of his general method of *involutio* and *evolutio* (props. 12–18). In prop. 64 it is applied to determine the equality of the arcs of higher spirals and parabolas.
[22] Lectio XII, append. 3, probls. 9–10; *compare* ch. 6: note 31.
[23] In the *Geometriæ pars universalis* (1668): 122, prop. 66, Gregory refers to four related papers (1659–67) and verbal communications by Angeli. In his paper of 1660 Angeli mentions a place in Torricelli's *Opera* (1644): ₂73–4 (*TO* I. 1: 154–5) as his starting-point. There we find the quadrature of the parabola achieved by means of a related Archimedean spiral.
[24] Fermat–Carcavy for Huygens, February 1660 (*HO* 3: 89–90), enclosure in Carcavy–Huygens, 25 June 1660 (*HO* 3: 85–6).
[25] The occasion is provided in the second part of Torricelli's *Dimensio parabolæ*: *TO* I. 1: 139–62.
[26] Huygens–Schooten, September 1650 (*HO* 1: 132–3).

$y/b = 1 - (x/a)^2$ and ABC a triangle
of equal perimeter with the parabola.
If the Cavalierian generation of an
area from the totality of all lines
(without breadth) contained in it
were correct, then the equality of
the areas of parabola and triangle
would follow on sub-dividing both
by lines which are similarly situated
round the mid-point M of the chord
towards the perimeter of the figure
$ADBC$. It is easy to prove that this
is not the case, says Huygens, but
far more difficult to find out the
reason for the mistake. Huygens was
already therefore at that time interested in the problem of the length
of the parabola; of any sort of inhibi-
tion by the Aristotelian prejudice
there is no longer any sign. The problem reappears anew in 1656; Hobbes

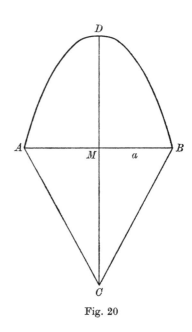

Fig. 20

had given a faulty rectification of the parabola in his 'Six Lessons',[27]
and Huygens at once refuted it;[28] as he did also an attempt by
Hobbes to correct his error.[29] A few months later Schooten learnt
through Mylon something of Roberval's proposition on the equality
of the arcs of a spiral and a parabola and passed it on to Huygens
who replied that he believed the proposition to be true having come
to the same conjecture himself, but that he had not yet found a
proof for it.[30]

This whole complex of problems had become a topic of wider
interest only with the 'Six Lessons'. Now almost all contemporary

[27] Hobbes, *De corpore* (1655): 157–61; altered in the translation, *Concerning Body*
(1656): 199–202; *Six Lessons* (1656): 50.

[28] Huygens–Wallis, 5 (15) Mar. 1655/6 (*HO* 1: 392), but we do not know the details
of this refutation.

[29] This renewed attempt is traceable from its refutation in Mylon–Huygens, 23 June
1656 (*HO* 1: 439–40). In the reply, Huygens–Mylon, 6 July 1656 (*HO* 1: 448) an
obscurity in Mylon's account is pointed out.

[30] Mylon–Schooten, June 1656 (lost). The contents are recognizable from Huygens–
Schooten, 6 Dec. 1656 (*HO* 1: 524) where Huygens states that he considers
Roberval's theorem to be probably true though he cannot acknowledge his
'mechanical' demonstration as conclusive and so would wish to see a purely
geometrical proof.

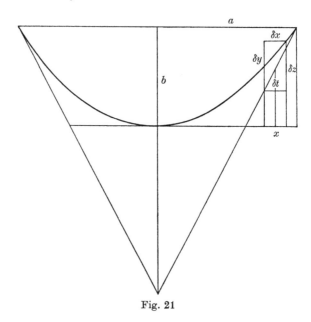

Fig. 21

mathematicians of repute occupied themselves with the rectification of the parabola. In this race Huygens was the first to reach the goal. He made his great discovery on 27 October 1657 and thereby finally outgrew the bounds of Cartesian thought, opening up the way to a systematic treatment of transcendental problems.

Huygens starts[31] from Archimedes' axiom, constructing a chain of tangents of the parabola bounded by equidistant lines parallel to the axis and translating them in the direction of the axis so that they become chords of the curve; this of course is possible only for the Apollonian parabola. Now the chain of chords is less, that of the suitably terminated tangents, more, than the arc of the parabola; while their difference can be made arbitrarily small by refining the division. Let $y/b = (x/a)^2$ be the equation of the parabola, then the tangent at (a, b) has the equation $z/b = 2x/a - 1$ (fig. 21). If now a distance $\frac{1}{2}\delta x$ is laid off in the axis on either side of x, then the difference of the ordinates to the parabola will be $\delta y \approx 2bx\delta x/a^2$ and that to the tangents $\delta z = 2b\delta x/a$; hence the intercepted section of chord or tangent at (x, y) on the parabola proves to be

$$\delta s \approx \sqrt{[1 + 4b^2x^2/a^4]}\,\delta x$$

[31] *HO* **14**: 234–53.

while along the tangent we have correspondingly

$$\delta t = \sqrt{[1 + 4b^2/a^2]}\,\delta x.$$

Hence by integration will follow

$$\frac{s}{t} = \frac{\int_0^a \sqrt{[1 + 4b^2 x^2/a^4]}\,.\,\mathrm{d}x}{\int_0^a \sqrt{[1 + 4b^2/a^2]}\,.\,\mathrm{d}x},$$

that is $= \dfrac{\text{area of a hyperbola}}{\text{area of a rectangle}}$.

Hence to find the length of arc of a parabola and the area of a segment of a hyperbola (or also the centroid of the hyperbola segment) are equivalent problems. The procedure is perfectly correct; the decisive element is the highly original idea of the 'translation' of the tangent chain; but it cannot be generalized. It is here shown for the first time that a geometrical curve can also be rectified, though not yet in geometrical manner but only with the area of a hyperbola as intermediary, that is to say mechanically (transcendentally).

There is something miraculous about men's thoughts. Once thought, they make themselves independent, as it were, and want to show their power in an effective way; willy-nilly they leave their initiator. Huygens, for example, had no intention of revealing his great discovery at once, but he did want to safeguard his priority by allusions in letters to his friends.[32] At the time Heuraet and Hudde, both students under Schooten, participated with ardent interest in all mathematical questions that were the subject of debate between Schooten and Huygens. Heuraet in particular laboured continually to get to the bottom of the theorem on the parabola-arc, of which he was ever given only the bare statement, but Huygens merely vouchsafed the remark that he had simultaneously found the surface area of the paraboloid of revolution,[33] so that Heuraet complained bitterly about the insoluble puzzle.[34] Heuraet had indeed progressed to a comprehensive method of quadrature[35] by the end of 1657, and

---

[32] *Compare* the references in Huygens' letters to Sluse, 2 Nov. and 20 Dec. 1657 (*HO* **2**: 80, 104); and to Schooten, 23 Nov. and 28 Dec. 1657 (*HO* **2**: 90, 112).
[33] Huygens–Schooten, 28 Dec. 1657 (*HO* **2**: 112).
[34] Heuraet–Huygens, 24 Feb. 1658 (*HO* **2**: 138–9).
[35] What this might mean in detail can only be conjectured without any certainty from Schooten-Huygens, 22 Dec. 1657 (*HO* **2**: 105).

a few months later he had penetrated to the kernel of the matter and, rightly guessing the true character of the problem, he developed a general method of rectification. Even before he set out on his grand tour together with Hudde the crucial notion had come to him, but to put it into final shape became possible only when the two travellers settled for a while at Saumur to continue their studies there. His method of rectification is recorded in a letter to Schooten[36] which was at once appended to the Cartesian *Geometria*. Unlike Huygens, Heuraet uses the triangle formed by normal, ordinate and axis and by means of $\delta x/\delta s \approx y/n = c/z$ reaches $cs = \int_0^x z \, . \, dx$. He does not give a full existence-proof, but merely a general consideration by infinitesimals. The examples to which he applies the result are the semicubic parabola, the Apollonian parabola and a few other curves. The case of the semicubic parabola was especially significant as being the first occurrence of a geometrically rectifiable algebraic curve. By it Descartes' statement and the last remnant of the old Aristotelian objection was finally refuted.

Now, to be sure, Wallis reported in his *Tractatus de cycloide*[37] that Neil had known the rectification of the semicubic parabola already in the summer of 1657. He proceeded, so Wallis tells us, from the parabola $(y/b)^2 = x/a$, and coordinated to it another curve whose ordinate z he made proportional to the area $\frac{2}{3}xy$, so that $z/b = xy/ab$. Taking $c = \frac{2}{3}a$ and setting $t = \sqrt{(c^2 + y^2)}$ he observed that

$$\int \sqrt{(\mathrm{d}x^2 + \mathrm{d}z^2)} : a : b = \int_0^a t \, . \, \mathrm{d}x : ac : \int_0^a y \, . \, \mathrm{d}x.$$

This whole investigation was probably started off by Hobbes' assertions in the 'Six Lessons' and, according to Wallis, bases itself on a few individual propositions in the *Arithmetica infinitorum*. Neil, who was quite unknown at the time, had a chance to lecture on his discovery at Gresham College in the summer of 1658. Brouncker contributed a new and better proof and wished to have Neil's theorem included in Wallis' *Commercium epistolicum*, but since nobody seemed to know anything of Neil, the paper was not published after all. This explanation which was only given after the English mathematicians had received Huygens' *Horologium*, published in

[36] Heuraet–Schooten, 13 Jan. 1659 (*Geometria* (1659): 517–20).
[37] *Tractatus duo* (1659): 90–2 (*HO* 7: 309).

1673,[38] is couched in somewhat unusual terms[39] and makes one suspect that everything is not above board, even though Brouncker[40] and Wren[41] were ready to bear witness in favour of Neil, who had died meanwhile. Most striking is the long interval between the discovery and its first communication at Gresham College. Whether the date 1657 is correctly given or ought to be replaced by 1658 is not very material for a discussion in the history of ideas; perhaps the correct date will one day be ascertained when Wallis' papers, containing much material of interest to the historian of mathematics, are edited.[42]

In the meantime Wren had, in connection with Pascal's great prize competition about the cycloid, succeeded in finding the arc-length of the cycloid,[43] namely $s = 2t$. For this result he gives an indirect proof to this effect. Let the cycloidal arc (fig. 22) be divided from the right by equidistant ordinates up to a certain $x = x_0$, and at each point of division draw the tangent as far as its intersection with the two

[38] In his *Horologium* (1673): 71–2, Huygens attributes the first discovery of an algebraically rectifiable, algebraic curve to Heuraet with a remark that Neil's exposition is inadequate.

[39] Wallis–Huygens, 30 May (9 June) 1673 (*HO* **7**: 306–8); Wallis–Oldenburg, 23 June (3 July) 1673, partly contained in Oldenburg–Huygens, 27 June (7 July) 1673 (*HO* **7**: 324–5); Wallis–Oldenburg, 4 (14) Oct. 1673 (*PT* **8**, No. 98 of 17 (27) Nov. 1673: 6146–9 = *HO* **7**: 340–3).

[40] Brouncker–Oldenburg, 18 (28) Oct. 1673 (*PT* **8**, No. 98 of 17 (27) Nov. 1673: 6149–50 = *HO* **7**: 344–5).

[41] Wren–Oldenburg, October 1673 (*PT* **8**, No. 98 of 17 (27) Nov. 1673: 6150 = *HO* **7**: 345).

[42] Leibniz had already in the *AE* (June 1686): 292–300 (*LMG* v: 232), tacitly referring to suggestions in the *Horologium* (1673): 72, pointed out that Heuraet's discovery may have been brought on by some remarks of Huygens, as that of Neil by Wallis. When an unsigned review from Leibniz' pen of *WO* ɪ (1695: containing also the *Tractatus duo*) had appeared in the *AE* (June 1696): 249–59 Wallis wrote a letter of protest to Leibniz, 1 (11) Dec. 1696 (*LMG* ɪv: 10) in which he also touched on the question of Neil's discovery for which, Wallis says, the starting point had been the 38th proposition in his *Arithmetica infinitorum* (1656); Huygens had later retracted his statement made in the *Horologium* in a letter to Wallis himself; but this letter, adduced by Wallis in 1696, has not so far been recovered nor is it referred to elsewhere in the older literature. Replying on 19 (29) Mar. 1696/7 (*LMG* vɪ: 11) Leibniz does not acknowledge his authorship of the review; he repeats however the remarks on Neil from the *AE* 1686 and further refers to a letter to the editors of the *AE* which was promptly printed in the issue for June 1697: *see LD* ɪɪɪ, where on pp. 345–6 a reference to the above cited letter by Huygens occurs.

[43] Wren's result is demonstrated in Wallis' *Tractatus duo* (1659): 64–70, where we are also informed that Wren had already communicated his discovery to his friends early in July 1658. Mylon's proof proceeds from similar considerations; dated 26 Jan. 1659 (*HO* **2**: 335–6), it came as an enclosure with the letter to Huygens, 31 Jan. 1659 (*HO* **2**: 334); obviously it had been accomplished before Wren's method became known.

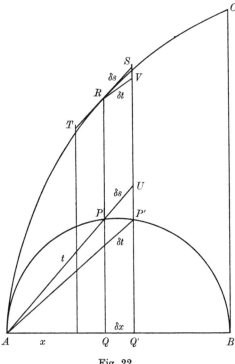

Fig. 22

neighbouring ordinates, then it follows from a well-known property of the tangent to the cycloid that $\delta s > \delta t$ where δt, as the increment next to δs, is measured along the tangent as one proceeds along the arc from left to right. The diameter of the semicircle is divided into equal parts, and δs_{n+1} is taken equal and parallel to δt_n. Hence the arc of the cycloid is less than the sum of the pieces δs but greater than the sum of the δt. These sums differ by the first tangent-piece δs and the last tangent-piece δt and this difference can be made arbitrarily small by increasing the number of dividing points. Thus the existence of an arc-length is secured. We have now $\delta s/\delta x = t/x = 2a/t$, hence

$$2ax = t^2, \quad t\,\delta s = 2a\,\delta x = (2t + \delta t)\delta t;$$

and so

$$\delta s \approx 2\delta t; \quad s - s_0 = 2(2a - t_0),$$

and so forth. This proof probably dates from the beginning of July 1658; it was sent to Pascal in the autumn of 1658 and is praised by him in December in his *Historia trochoidis*.[44] Huygens immediately

[44] Compare Pascal, *Œuvres* VIII (1914): 219.

110

reconstructed it from the result[45] $s = 2t$ and spoke very highly of Wren's discovery as being truly suggestive,[46] in particular since this was the first rational rectification of any curve.[47] That Descartes' and Torricelli's rectification of the logarithmic spiral had preceded it Huygens did not then know. Heuraet's general method was at that time too probably already developed in its essentials.[48]

And now at last Huygens decided to communicate his rectification of the common parabola. He did this in the following subtle form, which surely completely concealed his method from non-expert and expert alike.[49] Let $y/b = (x/a)^2$ be the given parabola and associate with it (fig. 23) the hyperbola $z^2 = a^2 + 4b^2x^2/c^2$, then the arc-length

$$s = (1/c) \int_0^c z \, . \, dx$$ will be equal to the height of a rectangle on the same base and of the same area as the hyperbolic strip.

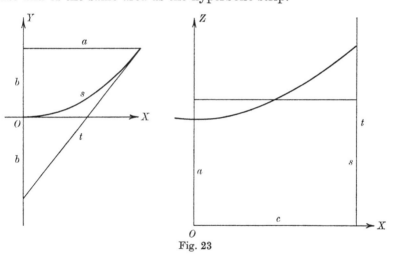

Fig. 23

45 The *Historia trochoidis* (1658) was sent with the letter Pascal–Huygens, 6 Jan. 1659 (*HO* 2: 310). Huygens' reconstruction occurs in a note written on 11 Jan. 1659 immediately on receiving the letter (*HO* 14: 363–7).
46 Huygens–Sluse, 14 Jan. 1659 (*HO* 2: 312–13).
47 Huygens–Carcavy, 16 Jan. 1659 (*HO* 2: 315); for a similar remark *compare Horologium* (1673): 72.
48 From Heuraet's dedication of 13 Jan. 1659 (*Geometria* (1659): 517) it is clear that the rectification had been found even before his departure for Saumur, while we conclude from Schooten–Huygens, 13 Feb. 1659 (*HO* 2: 353) that Heuraet and Hudde had left the Netherlands some eight months previously, and so Heuraet's discovery may be dated in the summer of 1658. We are in fact dealing with a method only – just as in Neil's case – not yet with its proof.
49 *Compare* the differing communications in Huygens' letters of 1659 to Carcavy, 16 Jan.; to Schooten, 7 Feb.; to Sluse, 10 (?) July; and to Grégoire de Saint-Vincent, 30 Oct. (*HO* 2: 316, 344, 418, 501). The first presentation in print is to be found in the *Horologium* (1673): 77.

111

Huygens had hesitated too long: in the eyes of his French correspondents he had forfeited his claim to priority of his discovery. As
Mylon wrote to him, Auzout had delivered[50] his opinion some four
weeks previously that if one could find the length of arc of the
parabola or Archimedean spiral, then one could also give the area
of the hyperbola, and vice versa. Moreover, Schooten reported a
similar, but altogether general result by Heuraet[51] (this refers to the
letter of 13 January 1659) which Huygens only saw fully in the
summer of 1659 in the new edition of the *Geometria*.[52] This was not by
accident: it was Heuraet's revenge for the game of hide-and-seek he
had played in earlier days concerning 'what Huygens had discovered
for the parabola'. The account of Wren's method in Wallis' *Tractatus
de cycloide* did not come into Huygens' hands[53] before the beginning
of 1660, and there he also found Neil's result quoted.

Amongst Pascal's prize-questions had been included Roberval's
statement that Archimedes' spiral and the common parabola have
equal arc-lengths. Mylon still reports at the end of January 1659 that
Auzout had been unable to contrive a proof of it.[54] Pascal's own
proof, published at about this same time in the *Lettres*,[55] did not reach

---

[50] Mylon–Huygens, 31 Jan. 1659 (*HO* **2**: 334). According to Mylon, Auzout also had
rediscovered Wren's method for the arc-length of the cycloid and had found the
rectification of other curves, but nothing factual is known of his methods.

[51] Schooten–Huygens, 13 Feb. 1659 (*HO* **2**: 353).

[52] When Huygens on 6 June 1659 (*HO* **2**: 412), had complained to Schooten that his
(1659) edition of the *Geometria* was not yet available at the bookshops, the latter
at once sent him his own copy which Huygens had seen during a visit to Leyden
(*HO* **2**: 415).

[53] Wallis letter to Huygens, 24 Nov. (4 Dec.) 1659 together with his *Tractatus duo*
arrived only on 20 Mar. 1660, according to Huygens–Wallis 21 (31) Mar. 1659/60
(*HO* **3**: 58).

[54] This appears to follow from the somewhat ambiguous text in Mylon–Huygens,
31 Jan. 1659 (*HO* **2**: 334).

[55] The proof is given in a letter of 1658 addressed to A.D.D.S. printed in the *Lettres*
(1659). Carcavy had mentioned this proof already in the letter to Huygens,
7 Feb. 1659 (*HO* **2**: 346), adding that Dettonville (= Pascal) had been moved to
this work by the mistaken idea of several authors about the length of arc of one
revolution of the Archimedean spiral. This appears to have originated early in the
seventeenth century and occurs, for instance, in Guldin's *Centrobaryca* I (1635):
39–57. It arose by the error of approximating the element of arc as $ds = r.d\theta$
instead of $ds = \sqrt{[dr^2 + (r.d\theta)^2]}$. Guldin in his book II (1640): 41–3 has, according
to J. H. Kinig, divided the arc of the spiral in 11 points so that the radii to these
points make angles of 30° with each other; it then follows that the corresponding
chord-polygon is longer than half the circumference of the circumscribed circle.
But Guldin did not succeed in finding the true length of arc of the spiral. In spite
of Pascal's publication the fallacious determination of the arc-length is still found
in many later authors, thus in J. Chr. Sturm (1689): 292. Further references to this
faulty rectification are given in *NP* **3**: 310–11, note 704, and Krieger (1971).

Huygens[56] until May 1659 when he at once raised some objections to it.[57] Pascal meanwhile had become so severely ill that he could no longer himself correct the insignificant mistake he had made.[58] Fermat, too, had discovered the fallacy and during the summer of 1659 gave to Carcavy a correction of it[59] which came to Huygens only after considerable delay.[60] Fermat mentioned on this occasion the general theorem about the equality of arc-length of the higher parabolas and spirals without being aware of his predecessor Torricelli; but this did not strike the expected echo from Huygens.[61] On the contrary, he criticizes the introduction of the higher spirals as a useless sort of game, and saw in the general theorem nothing but a very simple consequence of the special case of the Archimedean spiral. Why Huygens took this unexpected stance remains unexplained; the topic is barely touched on in his own mathematical notes. It must be remembered that Heuraet's proof of his general method was not completely satisfactory. The defect lay in the insufficient attention paid to the question of existence; it was only remedied by Fermat's ingenious method in his *Comparatio curvarum linearum* of 1660 which turns out to be a penetrating extension and generalization of Wren's procedure. Like Wren, Fermat uses a sawtooth scaffolding between equidistant ordinates (fig. 24). His main theorem for a convex monotonically increasing arc is contained in the inequalities

$$\text{tangent } UP < \text{chord } PQ < \text{arc } \overset{\frown}{PQ},$$

$$\text{arc } \overset{\frown}{PR} < \text{tangent } PV.$$

In his proof Fermat draws the tangent VW to the curve at V (it is still required to end within the limits of the monotonic arc), and thus obtains

$$\overset{\frown}{PR} + \overset{\frown}{RW} < PV + VW,$$

hence $\quad \overset{\frown}{PR} < PV - (\overset{\frown}{RW} - VW) < PV.$

[56] Huygens–Boulliau, 8 May 1659 (*HO* 2: 402). In *HO* 2: 396, note 2, the arrival is dated as 'early in May'.
[57] That a remark to this effect was made in a now lost letter to Pascal in mid-May follows from Huygens–Sluse, early June 1659 (*HO* 2: 418).
[58] Carcavy–Huygens, 14 Aug. 1659 (*HO* 2: 456); Bellair–Huygens, 22 Sept. 1659 (*HO* 2: 486).
[59] Carcavy for Huygens, 13 Sept. 1659 (*HO* 2: 534–40), forwarded to Huygens by Boulliau only on 26 Dec. 1659 as shown by his letter to Huygens, 2 Jan. 1660 (*HO* 3: 7). Fermat's remarks of late summer 1659 can be found in *HO* 2: 536–8.
[60] It seems to have arrived only towards the end of January 1660, *see* Huygens–Carcavy, 26 Feb. 1660 (*HO* 3: 26).
[61] Huygens–Carcavy, 26 Feb. 1660 (*HO* 3: 27).

113

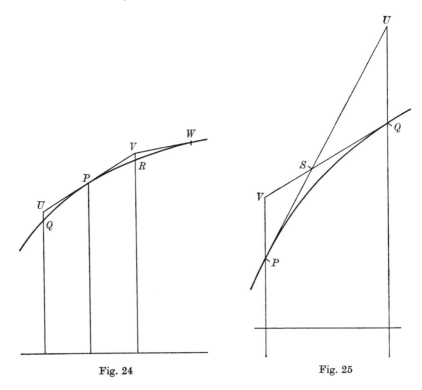

Fig. 24   Fig. 25

That a point W beyond the arc QPR itself has to be used is a formal blemish which only Gregory removed in his *Geometriæ pars universalis*. He too starts from Wallis' account of Wren's rectification – it is uncertain whether he knew Fermat's paper – but changes the main theorem into the form (fig. 25)

$$\text{tangent } VQ < \text{chord } PQ < \text{arc } \widehat{PQ},$$

$$\text{arc } \widehat{PQ} < PS + SQ < \text{tangent } PS + SU.$$

Only here does this method of the sawtooth pattern which will have to be combined with the infinitesimal theorem of the 'characteristic' normal–ordinate–axis triangle, find its most subtle and at that time unimprovable form. Both Fermat and Gregory were fully aware of the generality of their demonstrations; they could indeed give an existence proof for the rectification of any admissible arc. That it is possible in the case of the semicubic parabola to rectify one 'geometrical curve' by another 'geometrically', becomes now simply a

114

corollary to the main theorem. Nobody at the time suspected that the validity of Archimedes' axiom itself might, and ought to, be questioned.[62] The progress made beyond Huygens' method of translation – of which however no-one then knew any details – is significant. Fermat was certainly right when he emphasized that it is not easy to deduce the existence of arc-length by the method of inscribed and circumscribed polygons of chords and tangents; here only the sawtooth method achieved the desired effect.

When he came to write his *Horologium oscillatorium* Huygens gave in its third part a full theory of developing a curve and determining its evolute and thus he added a significantly new aspect to the existing method of rectification. In the seventh proposition he comes to deal with the rectification of the cycloid, which he now demonstrates no longer by the sawtooth pattern but by means of the involute of the cycloid through a vertex which he had deduced in the previous proposition. Wren is named with high praise and the relevant place in Wallis' treatise is quoted. Then in the eighth proposition Huygens finds the evolute of the common parabola, whence follows the rectification of the semicubic parabola as its converse in the ninth, with Heuraet now cited as its first inventor. Neil's and Wallis' account are also mentioned but Huygens believes that Neil had probably not understood the matter as thoroughly as Heuraet. Fermat is passed over briefly as his proof, though certainly independent, had appeared only in 1660 and hence too late to have any claim to priority. Concerning Heuraet's discovery Huygens refers to his own rectification of the parabola in the year 1657 and emphasizes that he had himself smoothed the way for Heuraet by his allusions to Schooten, since it should have been easy to find the basis of the rectification of the parabola from the determination of the surface of the paraboloid which he had at that time communicated.

Wallis received the personal copy of the *Horologium oscillatorium* sent him by Huygens during a short stay in London.[63] He felt his sense of national honour to be offended by Huygens' account and wrote at once to Huygens that the history of the discovery was incorrectly described: pride of place belonged absolutely to Neil, who had understood the geometrical rectification of an algebraical curve

[62] Bolzano (1817) was the first to investigate the possibility of 'proving' the axiom.
[63] Precisely on 29 May (8 June) 1673. Wallis' letter, dated the following day (*see* note 39), was therefore written in a state of considerable excitement and not at all as the outcome of careful thought.

no less thoroughly than Heuraet, but since at the time people put little store by single propositions it was not published. Huygens defended himself by the just remark[64] that he had known well enough of Neil's discovery; if only it had been fully published right-away, no difference of opinion over the points in question could ever have arisen. It may well be conceded that the twenty-year-old Neil was totally unknown in 1657, but his discovery was of such far reaching importance that there can be no excuse for not publishing it in Wallis' *Commercium epistolicum*. Even if one disregards the aspect that as a consequence of this discovery the Aristotelian dogma had been finally disposed of, yet by this very example a fundamental postulate of Descartes had been refuted and thence the English philosophers and scientists, who always had their reservations regarding Descartes (we shall have to deal with this in greater detail further on) were in fact presented with concrete evidence of his inadequacy.

Leibniz knew the account given in the *Horologium oscillatorium* and the letters printed in the *Philosophical Transactions*; he had also read Heuraet's paper in the *Geometria*, Pascal's *Lettres* and Wallis' *Tractatus de cycloide* and was therefore well informed about the successes of the various parties, though not accurately about the prehistory of the whole affair. Essentially he takes Huygens' point of view – we shall revert to this topic later – but he also tried to some extent to do justice to the claims of his English friends.

During this same time occurred the dispute between Huygens and Newton on the theory of colours; at first Huygens conceded that there was scope for discussion,[65] but later doubted even that.[66] In the introduction[67] of a long letter he wrote in defence of his theory Newton refers to the *Horologium* which Huygens had presented to him, offering to communicate his own general propositions on recti-fication since they were closely related to Huygens' theory of evolutes. But Huygens was in no mood to continue an English

[64] Huygens–Wallis, 30 June (10 July) 1673 (*HO* 7: 339–40).
[65] Huygens–Oldenburg, 30 Mar. (9 Apr.) 1672 (*HO* 7: 165).
[66] Some 23 letters exchanged in the course of this controversy between Huygens and Oldenburg, and Oldenburg and Newton from 21 June (1 July) 1672 until 4 (14) Aug. 1673 will be found in *HO* 7: 186–353 and *NC* 1: 207–94. For Leibniz' own strong interest in questions of optics *compare* for instance the 'Notitia opticæ promotæ' (1671) or the letter to Spinoza, 5 Oct. 1671 (*LSB* II. 1: 155), and above, ch. 6: note 60. We may assume that Huygens told him verbally the main issues of this discussion; subsequently he informed himself on the theory of colour through the papers published in the *PT*. For this *see* ch. 17: note 8.
[67] Newton–Oldenburg, 23 June (3 July) 1673) (*NC* 1: 290).

correspondence;[68] increasingly annoyed, he did not react at all to Oldenburg's reply[69] or to several further letters.[70] Not before May 1674 could he bring himself to write,[71] and then only to explain that he had indeed stopped writing to avoid further unpleasantness with the Fellows of the Royal Society, nor could he be easily conciliated even by Oldenburg's kindly persuasion.[72] The correspondence was only restarted in January 1675 when Huygens, who had invented the balance-spring for timepieces, sent an anagram regarding it to Oldenburg to secure his priority.[73] Unfortunately this was to give rise to an even more vehement dispute, this time with Hooke, eternally irritable, carping, envying, spoiling for a fight, as his adversary. And in this quarrel Leibniz too became involved.

[68] Huygens–Oldenburg, 30 June (10 July) 1673 (*HO* **7**: 338); Huygens–Wallis, 30 June (10 July) 1673 (*HO* **7**: 340).
[69] Oldenburg–Huygens, 4 (14) Aug. 1673 (*HO* **7**: 353–4).
[70] Oldenburg–Huygens, 3 (13) Nov. and 8 (18) Dec. 1673; also 2 (12) Mar. 1673/4 and 30 Mar. (9 Apr.) 1674 (*HO* **7**: 360, 364–5, 379, 380).
[71] Huygens–Oldenburg, 5 (15) May 1674 (*HO* **7**: 382).
[72] Oldenburg–Huygens, 25 May (4 June) 1674 (*HO* **7**: 385).
[73] Huygens–Oldenburg, 20 (30) Jan. 1674/5 (*HO* **7**: 400).

9

DISPUTES ABOUT CLOCKS

In December 1659 Huygens established the tautochronous motion of a body falling under gravity along a cycloid, and he hoped by means of this great theoretical discovery to make his pendulum clock into a precision-tool,[1] particularly when in August 1661 he also succeeded in regulating the movement of the pendulum by a small cursor.[2] But much was still needed before an accurately working pendulum clock was feasible, above all the anchor escapement – possibly an invention of Hooke's[3] – though it was employed in a turret clock manufactured by William Clement (for King's College, Cambridge) as early as 1671. Leibniz seems to have heard something of it during his first visit to London and to have passed his information on to Huygens[4] probably on 12 (22) January 1675 when Huygens came to talk to Leibniz in general terms about his most recent invention, that of the balance spring.[5] Here the isochronous property of elastic oscillations, whose theory was the subject of investigation by Huygens already in 1673–4, is used to regulate the movement of a clock. The clockmaker Thuret was immediately commissioned in the strictest secrecy to make a working model; at the next session of the Académie des Sciences Cassini, Picard and Mariotte were informed;[6] Oldenburg received a brief hint of the new discovery together with an anagram

[1] The second part of the *Horologium* (1673): 21–58 is concerned with the problem of tautochrone motion; a survey of Huygens' investigation is given in *HO* **16**: 344–9. The successful discovery on 1 Dec. 1659 is recorded in a lengthy note, *ibid.* 392–413, continued in *HO* **17**: 97–103.

[2] The invention of the 'cursor', recorded in *HO* **16**: 353–4 led to the discovery of the centre of oscillation (1661), fully dealt with in the fourth part of the *Horologium* (1673): 91–156. Huygens' first related notes of 1661 are reproduced in *HO* **16**: 415–31.

[3] *Compare* the account in *HO* **18**: 606.

[4] *HO* **18**: 605.

[5] About this invention, made on 20 Jan. 1675, and the subsequent controversy with the clockmaker Thuret we find information in Huygens' notes of 1675–6 (*HO* **7**: 408–16) written down so as to be prepared if legal action were to be taken.

[6] AS meeting, 23 Jan. 1675, *see HO* **7**: 410.

registering it,[7] and the model was shown to Colbert to secure a patent[8] which was readily granted.[9] Thuret turned out to be untrustworthy; he had demonstrated a clock with a balance-spring to Colbert as an invention of his own and claimed to have contributed to its discovery.[10] This led to a lengthy dispute during which Huygens wrote to his brother Constantijn in all seriousness, that he was prepared to leave France for ever unless this affair was settled in his favour.[11] Eventually, however, he was satisfied with a letter of apology from Thuret,[12] one all the more gratifying in that Thuret's timepieces were far more reliable than those of the other Paris clockmakers who quickly seized upon the new invention.[13] Another person to claim the principle of the hairspring as his own was the Abbé Hautefeuille, who quoted in his support a treatise he had some time previously submitted to the Académie;[14] indeed, preparations were made to take legal action,[15] but then the affair was suppressed.[16]

In London the new invention excited the liveliest interest from the beginning[17] because since Thuret had revealed his secret, Huygens soon sent there the text of his anagram[18] together with a copy of a brief description of his discovery that was to be printed in the *Journal des Sçavans*.[19] The letter and enclosure were read at the Royal Society[20] – the first meeting, as it happened, at which Newton was present. Hooke at once declared that he had already made this invention a long time before. Meanwhile Leibniz had seen the paper on clocks in the *Journal des Sçavans* and thought it proper to publish an invention of his own for regulating clocks which he had contrived some four years previously. He set out his communication in the form of a letter to La Roque, the editor of the *Journal*: its various extant drafts

---

[7] Huygens–Oldenburg, 20 (30) Jan. 1674/5 (*HO* 7: 400).
[8] *Compare* Huygens–Colbert, 5 Feb. 1675 (*HO* 7: 401).
[9] Colbert–Huygens, 15 Feb. 1675 (*HO* 7: 419–20).
[10] For details *see* note 5.
[11] Huygens–Constantijn Huygens (brother), 9 Aug. 1675 (*HO* 7: 484).
[12] Thuret–Huygens, 10 Sept. 1675 (*HO* 7: 498).
[13] Huygens–Oldenburg, 11 (21) Nov. 1675 (*HO* 7: 542).
[14] Hautefeuille for the AS, 7 July 1674 (*HO* 7: 458–60).
[15] *See* Hautefeuille's *Factum* (1675), also Huygens–Contesse, 5 May 1675 (*HO* 7: 457–8).
[16] *HO* 7: 453–5, note 23.
[17] RS meeting 28 Jan. (7 Feb.) 1674/5 (*BH* iii: 179) when Oldenburg read Huygens' letter 20 (30) Jan. 1674/5 (*HO* 7: 399–400) which contained the anagram.
[18] Huygens–Oldenburg, 10 (20) Feb. 1674/5 (*HO* 7: 422).
[19] Huygens for Gallois, probably mid-February 1675 (*HO* 7: 424–5); printed in *JS* No. 5 of 25 Feb. 1675: 68–70; also in *PT* **10**, No. 112 of 25 Mar. (4 Apr.) 1675: 272–3 (English translation).
[20] RS meeting 18 (28) Feb. 1674/5 (*BH* iii: 181–90).

allow us an interesting insight into the stages of its composition.[21] Leibniz refers purposely to the general appreciation accorded to Huygens for his new invention, and the terminology he uses throughout corresponds to that of Huygens. He is mainly concerned not, as Huygens was, with accounting for a new observational phenomenon (namely the constant time of oscillation of the spring) but with the purely mechanical result that the time of oscillation of a spring when released from the same state of tension until its return to the initial position is always the same. He wishes to regulate the motive power of a clock by a group of springs arranged in such a way that, once the movement is started, each successive spring is released by the preceding one and, on its return to its original position, itself sets off the next spring and is then immediately again held in its state of tension. In the first version it is stressed that no theoretical justification is required; this point of view is increasingly abandoned in the transitional versions right up to the one intended for the printer. In the first version Leibniz thinks of using six springs, in the final one, and likewise in a postscript directed to La Roque[22] but never printed, he confines himself to two. The paper ends with a discussion of possible disturbing factors – dependent on variation in temperature, air resistance, friction, irregular jerking effects due to the type of release, and so on. All in all, it presents a theoretically worthwhile idea in a technically immature form. Huygens probably arranged its publication,[23] but at all events he must surely have welcomed it. In an unfinished paper of the same time, perhaps intended for Huet,[24] Leibniz indicates how his system of springs might be combined with a pendulum. Originally Leibniz planned to send a copy of one of his draft versions to Oldenburg; but since printing went ahead rapidly, he was in fact able without delay to enclose the pertinent issue of the *Journal des Sçavans* itself in his letter.[25] A few weeks later Leibniz demonstrated a small model of his chronometer at the Académie des Sciences,[26] gaining friendly expressions of approval from Cassini and

[21] Leibniz for La Roque, mid-March 1675 (*C* 913 A–D).
[22] Leibniz for La Roque, late March 1675 (*C* 913 G).
[23] *JS* No. 7 of 25 Mar. 1675 (*LD* III: 135–7); Oldenburg's English translation in *PT* 10, No. 113 of 26 Apr. (6 May) 1675: 285–8.
[24] Leibniz–Huet (?), probably late March 1675 (MS: Hanover LBr 265, 2).
[25] In his letter to Oldenburg, 20 (30) Mar. 1674/5 (*LBG*: 110–12) Leibniz had asked him to arrange the printing of an English version of the French text he enclosed, but Oldenburg, judging this to be superfluous in the world of science, crossed this request out and so it does not appear in any printed versions of the letter. The letter was read at the meeting of the RS on 15 (25) Apr. 1675 (*BH* III: 216–17).
[26] AS meeting 24 Apr. 1675 (*C* 930).

Huygens but no more.[27] The construction of a working clock upon Leibniz' principle was, as far as we know, not attempted at the time. Hooke had a clock made to a now lost design[28] – we do not know its detailed construction – and when the clock was shown to the King in the summer of 1675 it worked well, though nobody was allowed to look inside. Subsequently Hooke accused first Oldenburg and then the Royal Society corporately of having betrayed his fundamental idea to Huygens. The president of the Society, Brouncker, was put in a difficult situation: he cannot have had much sympathy for the boasting, carping Hooke, nor was it certain whether he really had made a fundamental discovery or had simply had an impractical brainwave. To settle the affair, he arranged through Oldenburg's good offices for a watch to be made according to Huygens' design so that it might be compared with Hooke's.[29] How mistrustful people were of Hooke's assertions can again be judged from a contemporary letter by Collins, where, after reporting that Huygens had achieved something new in watches for which he had been granted a French patent, he adds that Hooke claimed to have made the same discovery previously.[30]

In spite of frequent requests by the English it was almost two months before the watch was made.[31] It was finally brought across in June by the Italian actor Biancolelli[32] who normally appeared on the Paris stage but frequently travelled to England on diplomatic service as a sideline. Huygens had sent full working instructions along, but as bad luck would have it a snag developed in the movement so that the watch stopped regularly at 12 o'clock.[33] In the interim Huygens

[27] Leibniz–Oldenburg, 10 (20) May 1675 (*LBG*: 123–4).
[28] Oldenburg told Huygens in his letter 28 June (8 July) 1675 (*HO* 7: 475) that Hooke also had constructed a pocket watch which had been presented to the King, though this report seems to have been premature; for Oldenburg writes on 15 (25) July 1675 (*HO* 7: 481) that the watch was still in the hands of the craftsman. Again, somewhat later, Papin tells Huygens, 10 Aug. 1675 (*HO* 7: 490–1) that, according to Oldenburg, the watch was still not ready and probably never would be. But on 13 (23) Sept. 1675 (*HO* 7: 499) Oldenburg reports that the watch is now finished and even shows seconds. He adds on 24 Sept. (4 Oct.) 1675 (*HO* 7: 506) that the King is pleased with the watch given to him; its case was however soldered up so that no-one could inspect the movement. A similar remark occurs in the letter Oldenburg–Leibniz 20 Sept. (10 Oct.) 1675 (*LBG*: 142). The watch has not survived.
[29] Oldenburg–Huygens, 27 Mar. (6 Apr.) 1675 (*HO* 7: 433).
[30] Collins–Gregory, 1 (11) May 1675 (*GT*: 299).
[31] Oldenburg–Huygens, 19 (29) Apr., 5 (15) May, 10 (20) May 1675 (*HO* 7: 455, 462, 464).
[32] Huygens–Oldenburg, 11 (21) June 1675 (*HO* 7: 471–2).
[33] Oldenburg–Huygens, 15 (25) July 1675 (*HO* 7: 481).

had had two magnificent watches executed (for the Dutch Stad-holder[34] William III and for King Louis XIV[35]) which went fault-lessly, and so the English mishap came as a particularly painful surprise to him.[36] The fault could not be remedied[37] in England and Oldenburg urged Huygens to send another watch to replace it.[38] Meanwhile Hooke's accusations against Huygens became more and more immoderate; Oldenburg had passed on only a few restrained allusions to them;[39] but verbal report may well have supplemented what was omitted from his letters. Huygens defended himself vigorously, stressing his utter ignorance of Hooke's alleged inven-tion,[40] and finally went over to the counter-attack,[41] comparing Hooke's actions to the activities of certain Paris clockmakers, each one of whom was laying a claim as co-inventor of its design by making minor alterations to the balance-watch.

The affair was heading to its dramatic climax. Hooke now publicly declared that it was all Oldenburg's fault: he had knowingly betrayed the secret of the balance spring to Huygens and was nothing better than a French spy and informer.[42] Though he was immediately persuaded to retract this monstrous accusation,[43] Hooke had created a very awkward situation for Oldenburg, who had always been politically somewhat suspect. He spoke out in no less offensive terms, we may add, in an appendix to his newly published *Description of helioscopes*;[44] on this occasion he also mentioned Leibniz' invention, brusquely disposing of it as of no practical value.[45] At Oldenburg's request Huygens immediately sent an objective statement of the circumstances to Brouncker, protesting emphatically against the

[34] Huygens–Constantijn Huygens (brother), 5 July 1675 (*HO* 7: 474).
[35] De Nyert–Huygens, 28 Aug. 1675 (*HO* 7: 493).
[36] Huygens–Oldenburg, 31 July (10 Aug.) 1675 (*HO* 7: 488–9).
[37] Papin–Huygens, 10 Aug. 1675 (*HO* 7: 490–1).
[38] Oldenburg–Huygens, 15 (25) July, 22 July (1 Aug.); 30 July (9 Aug.), 13 (23) Sept., 24 Sept. (4 Oct.), 27 Sept. (7 Oct.), 11 (21) Oct., 12 (22) Oct., 18 (28) Oct., 25 Oct. (4 Nov.), 1 (11) Nov. 1675 (*HO* 7: 481, 482, 486, 499, 506, 509, 514, 515, 527, 531, 532). The second watch, made by Thuret, was brought over by Thomas Ken with a letter from Huygens to Oldenburg, 11 (21) Nov. 1675 (*HO* 7: 542).
[39] Oldenburg–Huygens, 7 (17) June, 21 June (1 July) 1675 (*HO* 7: 469, 472).
[40] Huygens–Oldenburg, 1 (11) July 1675 (*HO* 7: 477–8).
[41] Huygens–Oldenburg, 2 (12) Oct. 1675 (*HO* 7: 510–11). Huygens had added a similar remark at the end of Oldenburg's letter of 13 (23) Sept. 1675 (*HO* 7: 499).
[42] Oldenburg–Huygens, 11 (21) Oct. 1675 (*HO* 7: 513).
[43] Oldenburg–Huygens, 15 (25) Oct. 1675 (*HO* 7: 516).
[44] *Helioscopes* [1675]: 26–31 (*HO* **7**: 517–26, particularly 520–2). (The year of publication is misprinted as 1676.)
[45] *Ibid.* (*HO* **7**: 519). Hooke's judgment is severe but entirely to the point; Leibniz' invention was indeed not fully thought out.

injustice of Hooke's allegations,[46] and Oldenburg himself in his review of the *Description of helioscopes* in the *Philosophical Transactions* wrote to similar effect;[47] he suggested also that a statement should be sent to the *Journal des Sçavans*,[48] but Huygens did not accede to this. The matter was dealt with at several meetings of the Royal Society[49] and Hooke was reprimanded, particularly after Constantijn Huygens (father) used his weighty authority as a highly respected diplomat to urge that the dispute be thoroughly cleared up in England.[50] But even now Hooke would not give in. In his *Lampas* of August 1676 he again inserted a criticism of Oldenburg's conduct,[51] one somewhat more conciliatory in its form but no less wounding in its substance than before. Thereupon the council of the Royal Society discussed the whole matter thoroughly[52] and inserted in the *Philosophical Transactions* an official declaration giving their full backing for their permanent secretary.[53]

It is very necessary to be aware of this highly charged atmosphere in order to understand the mood in which Leibniz' letters were received and assessed during the years 1675–6. The old mistrust of Oldenburg found new fuel to recharge it through Hooke's completely unjustifiable behaviour, and anti-German bigotry threatened at any moment to break out anew. Collins and his circle were certainly not willing to stand up for Hooke particularly after he also attempted to pick holes in Newton's theory of colours,[54] but they had become generally nervous and irritable. Since Leibniz had published his article on clocks, they seemed to have found evidence of the close connection between Leibniz and Huygens – though in reality it did not exist. To them it appeared as if Leibniz had somehow been pushed forward to sound them out – and after all one was concerned about the priority of a scientific achievement of the highest order, the theory of infinite series discovered by Newton. Collins felt a personal responsibility in so far as that Newtonian secret had been

[46] Huygens–Brouncker, 21 (31) Oct. 1675 (*HO* **7**: 529–30).
[47] *PT* **10**, No. 118 of 25 Oct. (4 Nov.) 1675: 440–2 (*HO* **7**: 536–8).
[48] Oldenburg–Huygens, 8 (18) Nov. 1675 (*HO* **7**: 535).
[49] RS Council meetings, 12 (22) Oct., 2 (12) Nov., 20 (30) Nov. 1676 (*BH* III: 319–24).
[50] Constantijn Huygens (father)–Oldenburg, 19 (29) Mar. 1675 (*HO* **7**: 431).
[51] Reproduced in *HO* **7**: 538–40.
[52] RS Council meetings, *see* note 49.
[53] *PT* **11**, No. 129 of 20 (30) Nov. 1676: 749–50 (*BH* III: 324); the 'declaration' reproduced *HO* **7**: 541–2.
[54] For Newton's original paper on colours and its reception *see NC* **1**: 92–102, 110–14, 171–88, 195–7; 362–86, 412–13, 416–17.

revealed to Gregory through his own clumsiness.[55] Newton was rather annoyed about it and initially intended to publish a comprehensive account of his new theory.[56] However it was impossible to find a publisher and he also grew increasingly disappointed, irritated and embittered by the never-ending niggling criticism of his theory of colours[57] – and so he resolved to publish nothing of his new discoveries in optics or in series[58] though it meant the sacrifice of his hopes and the renunciation of public appreciation of his discoveries. Should Newton now voluntarily tell a continental rival, a man a few years his junior, his latest thoughts and findings? Nobody will expect that. And when it seemed proved to Newton that Leibniz had profited by certain allusions all too freely transmitted to him, to re-invent his own most fundamental ideas, the young German must appear to him a plagiarist, though the unsuspecting Leibniz aimed only to acquire new knowledge, however brazen his manner might appear. There was the further unfortunate circumstance that by the word Analysis, Newton meant his theory of series, Leibniz his Calculus – and so a virtually inextricable tangle of misunderstanding was created. And the one person who was to be the mediator, however hard he might strive, was in no way expert, and indeed incompetent and ill-informed. Collins could just never envisage what ought or ought not to be done in the contest now about to begin between the two great

[55] Since Newton had in fact communicated his series for the circle-zone in a (no longer extant) letter to Dary, probably in the spring of 1670, Collins believed himself to be at liberty also to pass it on to Gregory, 24 Mar. (3 Apr.) 1669/70 (*GT*: 89). In the *CE* (*1712*): 22 this letter itself is not quoted, but merely an excerpt from Gregory's reply, 20 (30) Apr. 1670 (*GT*: 92) which in an unbiased reader might create the impression as if Gregory was not yet sufficiently informed about basic infinitesimal methods and had only learned these from the accounts of Newton's procedures. Turnbull in *GT*: 440–1 draws attention to the notes made by Gregory on Collins' letter which make it clear how Gregory came to send his communication to Collins, 19 (29) Dec. 1670 (*GT*: 148 = *CE* (*1712*): 23) in which the series for the circle-zone is correctly continued. We may add that it follows from the letter Collins–Moray, mid-January 1668/9 (*HO* **6**: 374–5, *see* ch. **6**: note 16), that Gregory already at that time was in possession of a general theory of interpolation. Why he did not at once recognize the correctness of the series for the circle-zone will be considered below, p. 134.

[56] *See NP* **2**: 168 and 288–90 for the various attempts to publish an account of the theory of series either as an appendix to the Latin translation of Kinckhuysen's *Algebra* or as a supplement to Newton's optical lectures which he originally meant to have printed at this time.

[57] *Compare* the allusion in Newton–Oldenburg for Leibniz, 24 Oct. (3 Nov.) 1676: (*LBG*: 224).

[58] This refers to a conversation Newton had with Collins during his stay in London in the spring of 1675. *See* also Collins–Gregory, 1 (11) May and 29 June (9 July) 1675 (*GT*: 299, 310).

rivals. Oldenburg, was undoubtedly a subtle diplomat, and an intelligent and warm-hearted man – if only he could have been advised by a more expert scholar, all might have been worked out for the best. Yet it was all to end in quarrel and confusion, because fate placed between the two geniuses a dwarf who was unable to manipulate the secret strings of this action.

10

LEIBNIZ RECEIVES FIRST DETAILS OF GREGORY'S AND NEWTON'S WORK

The year 1675 began for Leibniz with a weighty decision. He had demonstrated his calculating machine before the Académie des Sciences[1] but had seen only too clearly that from that quarter he might gain appreciation as a scientist but not an appointment by which he might earn his living. The instrument had cost him a fortune, and mathematics still attracted him, but practical necessity demanded that he should look for some diplomatic office or paid employment. His intention of entering the service of the Emperor had failed;[2] all that remained was an offer from Johann Friedrich, Duke of Hanover. Unwillingly and with some hesitation he wrote accepting a post he would have preferred to have foregone.[3] What was he to do in the small ducal capital where everything was dependent on the Duke's favour – and how easily that could be lost! Still he tried to safeguard his liberty, talked of his completed calculating machine, of which Colbert wanted one example made for the Paris observatory, and another for his financial bureau. Would the Duke, too, want to have one? And would it not be proper for Leibniz to start searching for rare books and manuscripts, antiques and inventions for his new master? He confessed, however, that he had for the moment given up all other studies to devote himself exclusively to his extended mathematical researches.

On what topics Leibniz' investigations at this time centred can to a certain extent be seen from his letter on 30 March 1675 to Olden-

---

[1] AS meeting, 9 Jan. 1675 (*C* 886).
[2] Leibniz–Lincker, February 1674 (*LSB* I. 1: 393).
[3] Leibniz–Duke Johann Friedrich, 21 Jan. 1675 (*LSB* I. 1: 492–3). In a letter to von Linsingen, late May 1675 (*LSB* I. 1: 498–9) Leibniz tried to settle more definite terms for his acceptance of the new office. On this occasion he referred, without giving any names, to the offer conveyed to him by Habbeus in a (no longer extant) letter of 25 Mar. 1673 to enter the personal service of Guldenlöw, the Danish prime minister of the time. The reply to Habbeus, 5 May 1673 (*LSB* I. 1: 416) explains why he did not accept this post: he feared to become dependent on the wishes and whims of a grandee of whose character and idiosyncracies he was ignorant, adding that he would find it difficult to conform with a rigid ceremonial of behaviour.

burg extant in three versions. We begin with the one[4] which he enclosed with a copy of his clock article from the *Journal des Sçavans*. Its first part refers to this article and the supplement to it intended for La Roque;[5] next follows a brief remark on the possibility of the arithmetical quadrature of the circle, which – so Leibniz maintains – is in no way contrary to Gregory's assertion of analytical impossibility; and then we have a general reference to Gregory's and Newton's methods, which Leibniz presumes to be methods of approximation. (This passage is also reproduced in the *Commercium epistolicum*.[6]) Leibniz asks explicitly whether the two Englishmen are in possession of a rectification of the hyperbola and the ellipse[7] – which Gregory manifestly had not, from proposition 66 of his *Geometriæ pars universalis*, achieved by the spring of 1668, and whether they had solved the problem 'absolutely' (Leibniz means by a closed algebraic expression), which appeared scarcely credible, or rather by one depending on the quadrature of the circle and the hyperbola. Leibniz offers to barter his quadrature of the circle for Collins' method for summing a finite number of terms of the series $\Sigma 1/k$, $\Sigma 1/k^2$, $\Sigma 1/k^3$: this he has been told, is already available in print, but not accessible to him.[8] His arithmetical quadrature of the circle is, he says, a far-reaching method which opens the way to many new propositions; hence he will in return expect a generous account of Gregory's and Newton's methods. He promises that he will not fail to mention both authors with due esteem and appreciation in the preface to his paper.[9] This last passage, one which shows with full clarity how far he must by now have advanced the new version of his arithmetical quadrature of the circle, was regrettably omitted in the *Commercium epistolicum* with the result that an extremely important assertion by Leibniz, if a somewhat naïve one, remained unpublished for almost two centuries.

In the first draft[10] no reference at all is made to the article on his clock-mechanism, but it contains an interesting classification of quadratures into approximative, mechanical, analytical and geometrical types. Approximate methods are divided into numerical

[4] *LBG*: 110–12.
[5] *LBG*: 111; *compare* ch. 9: notes 22 and 25.
[6] *CE (1712)*: 39. The full text has become available in print only through *LBG* (1899).
[7] *Compare* note 25 below regarding the attempt to deduce the rectification of the equilateral hyperbola from its quadrature.
[8] *See* p. 33.
[9] *LBG*: 112.   [10] *C* 922 B.

ones – such as the classical Archimedean[11] technique of circle quadrature, Ludolph van Ceulen's[12] computational procedure, and Wallis' approach by calculating finite portions of his infinite product for $4/\pi$ – and linear ones, such as the constructions of Snell[13] and Huygens[14] for the circle. In contrast, a mechanical quadrature would be exact but it requires a suitably constructed special curve: of this type are, for instance, constructions using string to develop a curve or the placing of a ruler in drawing tangents. (Leibniz instances the rectification of a circular arc by constructing the tangent to an Archimedean spiral $r = a\theta$ from the polar subtangent $r\theta$ – as in proposition 20 of Archimedes' De spiralibus – or by employing a cycloid,[15] presumed already drawn.) In arithmetical quadrature, Leibniz continues, an infinite sequence of numbers is employed rather than merely a single number; the first example of this was Brouncker's quadrature of the hyperbola,[16] later generalized by Mercator;[17] Leibniz himself is the first to have found the arithmetical quadrature of the circle. This was so marvellous, he insists, that one of the greatest geometers (he doubtless means Huygens) had declared that not even the discovery of an analytical quadrature could be more wonderful, since, if it were at all feasible, it would surely require complicated irrationals, whereas the Leibnizian arithmetical quadrature of the circle was simplicity itself. Analytical quadratures, lastly, were in their essence wholly different in type from arithmetical ones; they have to do with rational or irrational or even – the first time this technical term is employed – with transcendental expressions, and specifically with the simplest classification possible of a given problem. For instance, Gregory had asserted that π could in no circumstances be the root of a finite equation (with rational coefficients) – though Huygens' objections had in no way been refuted – and Leibniz promises to contribute something of his own

[11] Circuli dimensio.
[12] Fundamenta (1615) where π is calculated to 32 places.
[13] Cyclometricus (1621), props. 9, 18, 19.
[14] De circuli magnitudine inventa (1654), props. 7–9.
[15] If the cycloid is assumed drawn (fig. 12) then the arc-length of the generating circle, measured from O, is equal to the part PS of the ordinate between the circle and the cycloid.
[16] Compare pp. 96–7.
[17] Here Leibniz is wrong: Mercator's series in the Logarithmotechnia (1668) is not a generalization of Brouncker's procedure, though they agree in the presentation of log 2 by the series

$$\frac{1}{1.2} + \frac{1}{3.4} + \frac{1}{5.6} + \dots = \left(1 - \frac{1}{2}\right) + \left(\frac{1}{3} - \frac{1}{4}\right) + \left(\frac{1}{5} - \frac{1}{6}\right) + \dots$$

to the elucidation of this controversy. The subtlest achievement is that of geometrical quadrature, whereby the rectification of an arc or quadrature of an area is performed by some geometrical construction, not by ruler and compasses only but by aid of any curve at all. The means of construction admitted by Descartes (Leibniz alludes to his restriction to algebraical constructions) are too narrow; there exist after all analytical expressions that are not algebraic, and algebraic constructions that cannot be defined analytically. There Leibniz breaks off his sketch without completing it; however interesting in itself, its geometrical, mechanical and analytical concepts are not yet definitely enough distinguished and elaborated.

The second draft is introduced by a lengthy reference[18] to its planned enclosure, the original version of Leibniz' clock article, where special emphasis is given to there being no need of geometrical proof. From the remainder of its text we note in particular the reference, already quoted above, to Gregory's circle-approximations in the *Exercitationes geometricæ*,[19] and furthermore an allusion to Leibniz' London trip for the purpose of demonstrating his calculating machine from which it may be inferred that it was imminent.

Of late Leibniz' horizon had widened considerably, and in the geometry of curves especially he had made great strides forward. He is now at work studying general cycloids,[20] Cartesian ovals,[21] pedal curves of the circle and the parabola;[22] he seeks to construct the normals from an arbitrary point to the parabola and the general central conic,[23] and successfully applies himself to the study of inverse tangent problems,[24] striving in particular for a reasonable classification of quadrature problems. He adopts a unified approach to conics[25] by way of the defining equation $y^2 = 2ax + ax^2/q$ (where the origin is at a vertex), determining their arc-length in the form

$$\int_0^x \sqrt{[1 + a/q + a^2/(2at + at^2/q)]} \, . \, dt$$

[18] *LMG* i: 58–60.   [19] *Compare* pp. 67–8.
[20] *C* 775, 827, 902.   [21] *C* 832–3.
[22] *C* 826, 828–30.
[23] *C* 562, 847, 849, 851, 855–6, 861, 864–5.
[24] *C* 791, 840, 844–6, 849.
[25] *C* 773. Misled by his badly arranged computation Leibniz erroneously substitutes, in the case of the equilateral hyperbola where $q = a$, the value -1 for a/q with the result that he is henceforth convinced that the rectification of the equilateral hyperbola is reducible to its quadrature. This serious error is carried over into several manuscripts of similar character and gives rise to the query mentioned earlier (in note 7); it will occupy us again on several occasions.

(though he is as yet unable to distinguish the type of this integral). Again, still in 1674 he has plans for constructing an 'analytical' counterpart to his calculating machine,[26] an instrument for determining the solutions of equations, and he actually succeeds in doing so.[27] In the autumn of 1674 he is deep in the study of geometrical loci and graphical methods for solving equations.[28] An interesting instance is his approximate step-by-step determination of square roots[29] by taking $\sqrt{(a^2 \pm bc)} = a \pm z$ so that $z^2 \pm 2az = \pm bc$ and $z \approx bc/2a$. The mathematicians of the Renaissance take only this first step.[30] Leibniz who had probably first seen the procedure in Clavius[31] transforms it into a consistent method.[32] He works the example

$$\sqrt{(1+1)} = \frac{3}{2} - \frac{1}{4.3} - \frac{1}{8.3.17} - \frac{1}{16.3.17.577}\cdots,$$

where

$$3 = 2.1^2 + 1, \quad 17 = 2.3^2 - 1, \quad 577 = 2.17^2 - 1, \quad \ldots,$$

[26] C 772.

[27] C 815–16; for details see Hofmann (1970): 101–4. The instrument is mentioned in letters to Oldenburg, 2 (12) June, 2 (12) July, 18 (28) Dec. 1675 (LBG: 126, 131, 146); also in Leibniz–Huygens, mid-September 1675 (HO 7: 503). From Huygens' rather cautious reply, 30 Sept. 1675 (HO 7: 506) we may conclude that he has seen a sketch of the 'Compas d'equation', as had Tschirnhaus according to Leibniz' letter, 8 Jan. 1694 (LBG: 484). We cannot now tell what sort of improvement is referred to in the letter to Tschirnhaus, 31 Mar. 1694 (LBG: 492). A similar instrument is described in the *Encyclopédie*, Supplément II (1776): 834–5 (with an illustration in the *Planche* suppl. (1777)) under 'Algèbre'. This instrument works for quadratic equations only, but is justly claimed as a prototype capable of adaptation to higher equations. The inventor is not named: he might indeed have been inspired by the hint in LBG: 146 since this passage was available in print, in WO III: 620–2 where the letter in question is reproduced, and in CE (1712): 45 where an excerpt is given – but we do not in fact know his sources.

[28] C 787, 852–5, 857–60, 864–5.

[29] The method has been traced back, for special cases, to the Babylonians; it is given in Hero's *Metrica* I: 8; the cossists of the sixteenth century took it over from Islamic sources who themselves may have been in the chain of tradition by way of India and Greece, or may well have re-invented the method independently.

[30] For instance, Luca Pacioli (1494): 46.

[31] *Epitome arithmeticæ practicæ* (1583): chs. 27, 29.

[32] Leibniz could hardly have known that the procedure is given by Cataldi (1613). The partial sums of this sequence converge far more rapidly than those obtained from the binomial expansion. The comparable expression by continued fractions was indicated for the particular case of $\sqrt{13}$ by Bombelli (1572): 37. Perhaps this, or some communication during his visit to Italy (from early 1677) was the source for Tschirnhaus' account in his letters 17 Apr. 1677 and 27 Jan. 1678 (LBG: 334, 339) to Leibniz on approximations to $\sqrt{2}$ by means of continued fractions. Where the nth approximation is a_n/b_n and $a_1 = b_1 = 1$, we have the recurrence relations $a_{n+1} = a_n + 2b_n$, $b_{n+1} = a_n + b_n$ for the numerical sequence $\frac{1}{1}, \frac{3}{2}, \frac{7}{5}, \frac{17}{12}, \frac{41}{29}, \frac{99}{70}\ldots$. Leibniz' method yields the first, second, fourth, eighth... values of this progression.

but slips up in the details of his computation. The spring of 1675 saw the start of Leibniz' research into the most practical method of solution of the general biquadratic equation.[33] Probably in connection with this – and with the intention of travelling to London in the near future – he again looked at Oldenburg's letter of 16 April 1673, making careful excerpts from it.[34] We gain some information regarding the level of his algebraical knowledge at this period from a note of his, apparently sent to Huygens[35] in April 1675 and further corrected after its return, which deals with attempts at the solution of higher equations on the pattern of Cardan's formula for the reduced cubic.

Oldenburg's eagerly awaited reply to Leibniz' letter of 30 March 1675 is dated 22 April 1675; since he did not know enough about the mathematical questions touched on by Leibniz he had asked Collins for an appropriate briefing on them,[36] then subsequently translated Collins' English text into Latin, absorbing its content fully into his letter.[37] What appeared to English scientists to be the most significant passage was taken over into the *Commercium epistolicum*[38] and supplemented by annotations; in the margins of his own copy Leibniz subsequently made a number of comments to this excerpt, to which we shall later return. Oldenburg begins by telling Leibniz of the doubts raised in the Royal Society about his chronometer;[39] here he probably repeats Hooke's view in a form toned down so as not to offend Leibniz. Then comes his translation of Collins' report, one much more detailed than that of 16 April 1673.

First we hear of a new result by Gregory for the circle,[40] namely the series

$$\pi r = 4r^2 \Big/ \left(2d - \frac{1}{3}e - \frac{1}{90}\frac{e^2}{d} - \frac{1}{756}\frac{e^3}{d^2} - \cdots\right)$$

where $d = r\sqrt{2}$ is the side of the inscribed square and $e = r(\sqrt{2} - 1)$ the difference between the side of the square and the radius. This had already been alluded to during the debate between Huygens and Gregory about the *Vera quadratura*.[41] It proves not impossible to

---

[33] *C* 910–12. [34] *C* 935.
[35] *C* 937. The corrections probably date from May 1675.
[36] Collins–Oldenburg for Leibniz, 10 (20) Apr. 1675 (*C* 926).
[37] Oldenburg–Leibniz, 12 (22) Apr. 1675 (*LBG*: 113–20).
[38] *CE* (*1712*): 39–41, erroneously dated 15 (25) Apr. 1675.
[39] *LBG*: 113; *compare* ch. 9: note 45.
[40] Gregory–Collins, 23 Nov. (3 Dec.) 1670 (*GT*: 121).
[41] Gregory–Oldenburg, 15 (25) Dec. 1668 (*HO 6*: 310); printed in *PT 3*, No. 44 of 15 (25) Feb. 1668/9.

reconstruct the formula[42] by applying a direct development of his previously quoted expression[43]

$$\sin n\theta / \sin \theta = \binom{n}{1} - \binom{n+1}{3}x + \binom{n+2}{5}x^2 - \dots$$

where $\qquad x = 2(1 - \cos \theta) = 4 \sin^2 \tfrac{1}{2}\theta.$

We recall that he had earlier established the general difference formula

$$y_n = y_0 + \binom{n}{1}\Delta y_0 + \binom{n}{2}\Delta^2 y_0 + \binom{n}{3}\Delta^3 y_0 + \dots$$

and the quadrature series

$$\int_0^t y \, . \, dn = t\left[y_0 + \frac{t}{2!}\Delta y_0 + \frac{t(t-\frac{3}{2})}{3!}\Delta^2 y_0 + \frac{t(t-2)^2}{4!}\Delta^3 y_0 + \dots\right]$$

following from it: in the present instance this yields

$$\frac{1-\cos t\theta}{\theta \sin \theta} = \left[\int_0^t \frac{\sin n\theta}{\sin \theta} \, . \, dn = \right]\frac{t^2}{2}\left(1 - \frac{t^2 - 2}{1 \, . \, 2 \, . \, 3!}x + \frac{2t^4 - 15t^2 + 24}{2 \, . \, 3 \, . \, 5!}x^2\right.$$
$$\left. - \frac{3t^6 - 56t^4 + 294t^2 - 432}{3 \, . \, 4 \, . \, 7!}x^3 + \dots\right)$$

and hence, for $t = 2$,

$$\frac{1 - \cos 2\theta}{\theta \sin \theta} = \frac{2 \sin \theta}{\theta} = 2 - \tfrac{1}{3}x - \tfrac{1}{90}x^2 - \tfrac{1}{756}x^3 - \dots$$

and from this, when $\theta = \pi/4$, Gregory's own formula eventually follows. We observe that Gregory had at this time (spring 1668) mastered the theory of interpolation (for given equal intervals of argument); furthermore, he knew the expansions, for $t = 1$,

$$\tan \tfrac{1}{2}\theta = \tfrac{1}{2}\theta\left(1 + \frac{x}{12} + \frac{11x^2}{720} + \frac{191x^3}{60480} + \dots\right)$$

and $\qquad \sin \theta = \theta\left(1 - \frac{x}{6} - \frac{x^2}{180} - \frac{x^3}{1512} - \dots\right)$

and evidently used them to deduce the approximations that now caused so much trouble to Leibniz. From the isolated result that was sent to him Leibniz could not draw anything at all – especially since through a slip in Oldenburg's copy he was not even given the whole correct series: the missing term $e^2/90d$ was later inserted when part

[42] Gibson (1922–3): 9–10; *compare GT*: 124–5 for an amended interpretation.
[43] Gregory–Collins, 23 Nov. (3 Dec.) 1670 (*GT*: 120).

of the letter was reproduced in the *Commercium epistolicum* in a Latin version collated with the English original.[44] These series, Collins writes, had been communicated to him only after the appearance of Mercator's *Logarithmotechnia*[45] (in July 1668) and he had at once sent them on to Barrow.[46] From Cambridge then had come the response[47] that Newton had been in possession of Mercator's logarithmic series some time before the publication of the *Logarithmotechnia* and that he had applied it as a method for squaring all types of geometrical and mechanical curves. (This, of course, is an allusion to Newton's *De Analysi*, which Collins had, upon receiving it[48] in the summer of 1669, copied so that he later could produce it when asked for by Fellows of the Royal Society.) For example,[49] let $x = \sin z$; Newton's series are then

$$z = x + \tfrac{1}{6}x^3 + \tfrac{3}{40}x^5 + \tfrac{5}{112}x^7 + \ldots$$

and $\qquad x = z - \tfrac{1}{6}z^3 + \tfrac{1}{120}z^5 - \tfrac{1}{5040}z^7 + \ldots,$

where the continuation of both series is evident: from the first it was not hard to obtain Ludolph van Ceulen's decimal approximation to π by setting $x = \tfrac{1}{2}$ and so $z = \pi/6$. Leibniz was later to ask for details about the law for the coefficients of these series – and its derivation – but we shall take this point up below.[50]

Next, Collins cites for the area

$$2 \int_0^a \sqrt{(r^2 \mp x^2)} \, . \, dx$$

of a circle or hyperbola[51] the expansion

$$2ar \mp \frac{a^3}{3r} - \frac{a^5}{20r^3} \mp \frac{a^7}{56r^5} - \ldots,$$

[44] *CE (1712)*: 40.
[45] Compare the similar text in Collins–Gregory, 25 Nov. (5 Dec.) 1669 (*GT*: 74 = *CE (1712)*: 22).
[46] Mercator's *Logarithmotechnia* appeared in mid-July 1668; Collins sent it to Wallis on 14 (24) July 1668 (*CR* II: 490), and probably at the same time to Barrow.
[47] Barrow–Collins, 20 (30) July 1669 (*NC* **1**: 13–14 = *CE (1712)*: 1). This letter is therefore almost a whole year later.
[48] Barrow sent Newton's *De Analysi* to Collins with his letter of 31 July (10 Aug.) 1669 (*NC* **1**: 14 = *CE (1712)*: 2).
[49] *Compare* the similar wording in Collins–Gregory, 24 Dec. 1670 (3 Jan. 1671) (*GT*: 155); and the excerpt in *CE (1712)*: 24.
[50] Leibniz–Oldenburg, 2 (12) May 1676 (*LBG*: 167); *see* p. 211.
[51] Collins does not quote from the text in Newton's *De Analysi* (*NP* **2**: 216, 232), but from the version in Gregory–Collins, 19 (29) Dec. 1670 (*GT*: 149–50).

for the circle-segment $2 \int_0^a \sqrt{(2rx - x^2)} \cdot dx$ he gives[52] the series

$$\frac{4ab}{3} - \frac{2a^3}{5b} - \frac{a^5}{24b^3} - \cdots,$$

where $b^2 = 2ar$, and for the corresponding arc $4r \sin^{-1} (a/b)$ he quotes the series

$$2b + \frac{a^2}{3b} + \frac{3a^4}{20b^3} + \frac{5a^6}{56b^5} + \cdots,$$

adding that these last two series are due to Gregory, who had attained the method of series a few years later. Collins' account is, however, very superficial and omits to mention that Gregory originally started with series derived by finite-differences the theory of which he had already developed in 1668 – a time, that is, when no-one had heard of Newton's theory of series.[53] When Collins in his correspondence with Gregory had alluded to similar results obtained by Mercator,[54] he had in fact induced Gregory in several long letters at least to map out his general method and exhibit a few individual results gained thereby.[55] Amongst those produced by Gregory in response was, in particular, the general binomial expansion set out in the form of a logarithmic series[56] (this result also is referred to the year 1668); in modern terms, if

$$\log a = u, \quad \log x = u + y \quad \text{and} \quad \log (a + b) = u + v,$$

then

$$x \left[= a(1 + b/a)^{y/v} \right] = a + b \binom{y/v}{1} + \frac{b^2}{a} \binom{y/v}{2} + \frac{b^3}{a^2} \binom{y/v}{3} + \cdots.$$

As an example Gregory works out the first of 364 geometric means between 100 and 106 to determine the incremented value of £100 lent out for one day at an annual compound interest rate of 6%, using the standard formula £100$(1 + \frac{6}{100})^{1/365}$. We shall see later the importance of the extremely early date of this discovery: in 1670 Gregory was on the brink of discovering the whole method of power

[52] Gregory–Collins, 17 (27) May 1671 (*GT*: 190–1).
[53] *See* ch. 9: note 55.
[54] Collins–Gregory, 30 Dec. 1668 (9 Jan. 1669) (*GT*: 56); also 7 (17) Jan., 2 (12) Feb., 15 (25) Mar. 1668/9 (*GT*: 60, 66, 71).
[55] Gregory–Collins, 23 Nov. (3 Dec.) 1670 with enclosures; also 19 (29) Dec. 1670; 15 (25) Feb. 1670/1; 17 (27) May 1671 (*GT*: 118–22, 127–35, 148–50, 168–72, 187–91).
[56] Collins–Moray, mid-January 1668/9 (*HO* 6: 374–5); *see* ch. 9: note 55. The result is mentioned again in Gregory–Collins, 5 (15) Sept. 1670 (*GT*: 103). For the following, *compare* Gregory for Collins, 23 Nov. (3 Dec.) 1670 (*GT*: 131–2).

series when Collins sent him Newton's series of the circle-zone,[57] and the sole reason why he did not at once recognize its derivation was his belief that he ought to be able to compound it from his own interpolation-series – he succeeded only when he recognized that he had to deal with the quadrature of the binomial expansion of $\sqrt{(1-x^2)}$.

In his account for Leibniz Collins next sets down the famous series[58] for $\tan x$ and $\tan^{-1} x$ which Gregory, as we may now surmise from his manuscript notes, had determined by repeated differentiation. Namely, let r be the radius, s the arc and t the corresponding tangent of a circle, then $t/r = \tan(s/r)$ whence the arc

$$s = r \tan^{-1}(t/r) = \int_0^t r^2 . \mathrm{d}x/(r^2+x^2)$$

may be found by long division and a subsequent integration term by term, in an exactly analogous way to Mercator's, and the inverse series expansion of $t = r \tan(s/r)$ evaluated by repeated differentiation, computing

$$rt' = r\frac{\mathrm{d}t}{\mathrm{d}s} = r\left(1+\frac{t^2}{r^2}\right), \quad rt'' = r\frac{\mathrm{d}^2t}{\mathrm{d}s^2} = 2t\left(1+\frac{t^2}{r^2}\right),$$

$$r^2 t''' = 2r\left(1+\frac{t^2}{r^2}\right)\left(1+\frac{3t^2}{r^2}\right), \quad \dots$$

and then gathering their values for $t = 0$ in the 'Taylor' expansion

$$t = \frac{s}{r}(rt_0') + \frac{s^2}{2r^2}(rt_0'') + \frac{s^3}{6r^3}(r^2 t_0''') + \dots.$$

In the determination of $r^5 t^{\mathrm{vi}}$ Gregory made a typical arithmetical mistake: in its expansion the coefficient of t^2/r^2 arising as $960 + 272$, is taken as 987 $(= 960 + 27)$; consequently Gregory obtains for $r^6 t^{\mathrm{vii}}$ the value $272r + 3233t^2/r \dots$ instead of $272r + 3968t^2/r \dots$ and carries the mistake through into the sequel, obtaining in the final expression the erroneous term $3233t^9/181\,440r^8$. Subsequently, in the excerpt from Oldenburg's letter reproduced in the *Commercium epistolicum* this has been silently corrected to read $62t^9/2835r^8$. A remark at this point in the *Commercium epistolicum*[59] avers that Leibniz had produced his own discovery of the inverse-tangent series

[57] See ch. **7**: note 75.
[58] *LBG*: 115 = *CE* (*1712*): 41. The series are taken from Gregory–Collins, 15 (25) Feb. 1670/1 (*GT*: 170 = *CE* (*1712*): 25). For their deduction *see GT*: 350–3.
[59] *CE* (*1712*): 41.

to his French friends only after his receipt of this present letter from Oldenburg. In a private annotation on this passage Leibniz calls the imputation impudent and malicious, recalling that he had already drawn attention to his own discovery in the letters to Oldenburg of 15 July and 16 October 1674. Leibniz was right to stress that his wholly different method of deducing the arctan-series, in his arithmetical quadrature of the circle sent to Huygens in October 1674, was not indebted to Gregory's.

There are only brief references in Collins' account of Gregory's further results[60] regarding log tan and log sec, but more detail is given regarding Newton's quadrature and rectification of the quadratrix,[61] his determination of the surface of solids of revolution and of their second segments, and also his solution of equations by infinite series.[62] After considerable trouble Gregory had accomplished the same end-results but as yet he had not communicated[63] them because he wanted Newton, as the first discoverer to be the first to publish his results. The excerpt from Oldenburg's letter published in the *Commercium epistolicum* ends at this point.

The letter itself continues with the observation that as far as regards elementary algebra, Pell[64] had done something to fix limits to the roots of equations and was able by using logarithms to approximate every root; he could also approximately factorize irreducible equations of the third and fourth degree, without first employing Descartes' method or finding limits to the roots. According to Dulaurens it should be perfectly possible, within certain limits, to eliminate several intermediate terms of a general higher equation: if, for instance, an equation of sixth degree had four real roots, then two intermediate terms could be eliminated, but if it had two real roots, then four terms could be removed. Collins expresses the hope that Prestet might achieve further results in this area; Pell had promised similar results by means of a far-extended sine table.[65] Malebranche's

---

[60] *LBG*: 115, taken from Gregory–Collins, 15 (25) Feb. 1670/1 (*GT*: 170).

[61] Taken from the *De Analysi* (*NP* 2: 238–41 = *CE* (*1712*): 18); passed on in Collins–Gregory, 24 Dec. 1670 (3 Jan. 1671) (*GT*: 155).

[62] This brief account is similar to what Collins writes to Borelli, December 1671 (*NP* 3: 22 = *CE* (*1712*): 27). The reference is to the text in Newton–Oldenburg for Leibniz and Tschirnhaus, 13 (23) June 1676 (*NC* 2: 25–9).

[63] Gregory presumably expressed his wish not to anticipate Newton in a letter to Collins, 23 May (2 June) 1672, now lost: see *GT*: 236. The remark also occurs in Collins–Strode, 26 July (5 Aug.) 1672 (*CE* (*1712*): 29) and in Collins–Newton, 30 July (9 Aug.) 1672 (*NC* 1: 223 = *CE* (*1712*): 29).

[64] *LBG*: 116. Collins reports similarly in his letter to Gregory, 25 Mar. (4 Apr.) 1671 (*GT*: 180).   [65] *LBG*: 116.

recent assertion[66] that, if the irreducible case of the cubic could be solved, great progress in the theory of equations would follow provokes Collins briefly to survey his own method,[67] which consists in representing the cubic form $y = a_0 x^3 + a_1 x^2 + a_2 x$ graphically, noting the extreme values of y and determining the number of real roots in x of $y = 0$ from the limits of variation of y. A noteworthy supplement comes from Wallis:[68] to each arbitrarily chosen x_0 belongs a y_0 which can also be obtained from two other values x_1, x_2 which are roots of a quadratic and hence determinable, and in this way it will always be possible to approach as near to any prescribed value of y as desired. On the other hand, it is possible to reduce the equation $x^3 = qx + r$ by dividing its terms by appropriate terms of the geometric progression $1, \sqrt{q}, q, q\sqrt{q}$ – that is, by the substitution $x = y\sqrt{q}$ – to the standard form $y^3 = y + r/q\sqrt{q}$. and then to solve this by using tables of square and cube numbers like Guldin's[69] or those being prepared by John Smith.[70] Collins is, however, wrong when he assumes[71] that there will always be only one real root when $|r/q\sqrt{q}| > 1$, but three real roots when $|r/q\sqrt{q}| < 1$. A reference to his own method for solving cubic and biquadratic equations numerically by the use of spherical trigonometry on projecting a conic into a circle on a sphere concludes what Collins has to say on elementary algebra – a favourite subject of his, though he never achieved very much in it.

Collins next relates that Newton's and Gregory's studies on intersecting conics had started from the method of projection on to a sphere.[72] Their researches probably embraced the whole content of the lost eighth book of Apollonius' *Conics*[73] and also what Fermat and Roberval had discovered about higher loci; Kinckhuysen, too,

[66] Malebranche–Vaughan, spring 1675 (*LBG*: 116–17).
[67] *LBG*: 117.
[68] Wallis–Collins, spring 1668/9 (*CR* II: 601–4).
[69] *LBG*: 118. The tables come at the end of Guldin's *Centrobaryca* I (1635). *Compare* similar references in Collins–Gregory, 23 Nov. (3 Dec.) 1674; 1 (11) May, 29 June (9 July) 1675 (*GT*: 291, 298, 300, 310).
[70] Newton gave important theoretical help for the actual calculation of these (unpublished) tables in his letters to John Smith, 8 (18) May, 24 July (3 Aug.), 27 Aug. (6 Sept.) 1675 (*NC* 1: 342–4, 348–9, 350–1).
[71] *See* Collins–Gregory, 23 Nov. (3 Dec.) 1674 (*GT*: 291–2).
[72] *LBG*: 118. We have no evidence to substantiate this statement by Collins.
[73] Collins probably means Pappus' *Mathematical Collection* VIII where Pappus reports on several now lost Greek works on geometrical loci. This was in fact the basis for Fermat's reconstruction (*De locis planis*, *FO* I: 3–51) and for his further elaboration (*Ad locos...isagoge*, *FO* I: 91–110). Collins knew both these titles from the obituaries mentioned in ch. 4: note 41. Instead of to Roberval the reference should be to Fermat's *Isagoge*.

contributed something to these topics and there was hope of further advance in Prestet's new *Algebra*, which had raised considerable expectation in England. Both Newton[74] and Gregory[75] had furthermore long before developed more general methods for solving equations by higher parabolas. In communicating certain details from Newton's account which go a little further than in his earlier report on the same topic, Collins made a mistake recognizable from a dimensional discrepancy.

Now follows a detailed description of Newton's generation of a general conic by means of two rigid angles rotating about their vertices in such a way that one pair of their legs meet on a straight line:[76] the intersection of the other pair will then describe the conic. The construction of a conic through five given points, two of which are taken to be the vertices of the angles, is given in full; analogously, says Collins, the construction can be effected given four points and a tangent, and also three points and two tangents, and hence also a parabola may be determined through four given points, and similar problems resolved.

Collins next proceeds to another of his favourite topics, Davenant's problem[77] of finding four consecutive terms of a geometric progression, given the sums of their squares Q and of their cubes C. This, he says, is a question for experts, and Girard's rule for finding the sums of powers of the roots of an equation should play its part in it.[78] Concerning the series $\Sigma 1/k$, $\Sigma 1/k^2$ and $\Sigma 1/k^3$ he had circulated a short treatise thereon amongst his friends, but it had been lost:[79] with regard to the first he reports that he has endeavoured to bring out the symmetries in the general harmonic progression, as might be seen in the instance

$$\frac{1}{b-3c}+\frac{1}{b-2c}+\frac{1}{b-c}+\frac{1}{b}+\frac{1}{b+c}+\frac{1}{b+2c}+\frac{1}{b+3c} = \frac{7}{b}+\frac{28c^2}{b^3}+\frac{196c^4}{b^5}+\cdots.$$

Finally,[80] Collins repeats his remarks from his earlier report concerning Pascal and Desargues and on De Beaune's treatise on the

[74] A reference to Newton's lengthy manuscript of 1670 (*NP* 2: 450–517), mentioned previously in ch. 4: note 44 and note 46. Newton sent it to Collins with his letter, 20 (30) Aug. 1672 (*NC* 1: 231).

[75] Part of this manuscript is included in the (still unprinted) 'Historiola' (see p. 214) which Collins compiled in the early summer of 1675 from letters and papers received from Gregory.

[76] *LBG*: 119–20; *see* also ch. 4: note 45. [77] *See* p. 255.

[78] *Invention nouvelle* (1629): fols. F2r–2v. [79] *LBG*: 120–1; *see* above note 8.

[80] *LBG*: 121; *see* the text of the letter Oldenburg–Leibniz, 6 (16) Apr. 1673 (*LBG*: 87–8) to which reference is made in ch. 4: note 13 and notes 16–17.

solid angle, and asks for Leibniz' assistance in procuring a copy of a book by Gottigniez on elementary geometry which he had so far been unable to trace. This long, poorly arranged letter has many points of similarity with his earlier comprehensive report of 16 April 1673. For the first time Leibniz now learns the fine details about Gregory's and Newton's current researches, but is told only their results and not their methods. At first the atmosphere is one of friendly competition: essentially, Collins presents a survey not markedly different from the earlier one, but illuminates a couple of points by describing an example or two in more detail. For the first time Leibniz is aware that Collins has come to take him seriously and is consequently prepared to furnish information about Newton's and Gregory's unpublished studies. He naturally assumes that he is now hearing of the questions currently foremost in their mind, whereas in fact he is only being told of problems already solved and settled. This is surely not the consequence of any ill will on Collins' part towards him but simply manifests a certain reserve in the light of his earlier experience. On the whole Collins has passed on more in the way of general viewpoints than might reasonably be expected, though fewer methods and proofs than Leibniz wished. However that may be – Leibniz at that moment was about to embark on a deep study of fundamental problems in algebra and this letter of Collins gave him a considerable stimulus to his research. Straying from the field of the Cartesian *Geometria* and Schooten's commentaries, and ready for his own part to study the literature of the subject seriously, he began to look at the sources from which the leading scholars of his day had drawn their first knowledge.

With a good number of mathematically interested persons in Paris he enjoyed lively exchanges of ideas; thus, with Roberval he discussed certain oversights in the *Geometria* which the latter claimed to have spotted (and which already in 1656 had led to a lively interchange of letters[81] on the topic with Huygens); and with Ozanam he conversed about the construction of geometrical loci.[82] In the latter

[81] Huygens had met Roberval during his stay in Paris sometime between July and November 1655. Roberval, an ardent adversary of Descartes, may have drawn Huygens' attention to some inconsistencies in the *Géométrie*, so that Huygens in March 1656 (*HO* 1: 396) asked him for details, and a correspondence, involving also Schooten in particular referring to *Géométrie* (1637): 326 = *Geometria* (1649): 30, eventually amended in *Geometria* (1659): 26–7. Roberval told Leibniz of this error by Descartes, and Leibniz made a note of this conversation and its outcome in *C* 908.
[82] Leibniz–Foucher, January 1692 (*LPG* I: 404), also Leibniz–Varignon, 1710 (*LMG* IV: 169).

139

case[83] he was presented with the problem of constructing a curve (a hyperbola) such that the joins AP, BP of its variable point P to fixed points A and B make equal angles with the given lines AC, BC. Simultaneously Leibniz made excerpts from Wallis' *Mechanics*,[84] investigated the laws of impact and friction from a geometrical viewpoint,[85] and learnt from Foucher how to determine the change in weight of rarefied air.[86] These physical topics at first interested Leibniz more than the mathematical ones; hence he thanked Oldenburg only briefly for sending him Collins' series[87] and remarked that he had as yet been unable to compare them with his own result of several years before; as soon as he had done this he would write more on this matter. This was a most impertinent remark for Leibniz to make; it would have been better if he had bluntly confessed – what he here tried to conceal – that he could not in particular rightly understand Gregory's first series. The passage is singled out for comment in the *Commercium epistolicum*[88] with the rebuke that Leibniz never communicated this promised comparison of his own series with those sent from London, while when he was sent the sine and arcsine series a second time by Mohr[89] he had asked for information about Newton's method, and later he had appropriated for himself Gregory's arctan series[90] without a single reference to the letter of 22 April 1675 containing all the relevant details. As further evidence of his plagiarism Leibniz' introduction to his 'Quadratura arithmetica' in the *Acta Eruditorum*[91] is adduced: in a rather laboured interpretation of the text the assertion is made that Leibniz had not found the arithmetical quadrature before 1675 (that is, only after receiving the letter of 22 April 1675) and had made it known to his friends only subsequently; to the exposition intended for Oldenburg he had given a final polish only in 1676 and had then abandoned his idea of an immediate publication because, though the tract grew apace in his hands, his official duties in Hanover left him too little time to present it properly in print. His own Analysis, it was further alleged, had only been elaborated by Leibniz in sequel to these studies and in complete dependence on Newton.

This is how matters appeared to the British in 1712, depending

[83] *C* 940.
[84] *C* 941, 944.
[85] *C* 945–8, 965.
[86] *C* 949.
[87] Leibniz–Oldenburg, 10 (20) May 1675 (*LBG*: 122); excerpt *CE* (*1712*): 42.
[88] *CE* (*1712*): 42.
[89] Leibniz–Oldenburg, 2 (12) May 1676 (*LBG*: 167); excerpt *CE* (*1712*): 45.
[90] Leibniz–Oldenburg, 17 (27) Aug. 1676 (*LBG*: 195–6 = *CE* (*1712*): 60).
[91] *AE* (April 1691): 178 (*LMG* v: 128); excerpt *CE* (*1712*): 42.

almost exclusively for their documentary basis on the abridged versions of Leibniz' letters to Oldenburg printed in the third volume of Wallis' *Opera* and then on Leibniz' articles published in the *Acta Eruditorum*, but for the documents sent to Leibniz trusting in the essential accuracy of Collins' fuller English drafts. They were by then already fully convinced that Leibniz was guilty of conscious plagiarism – of Gregory in respect to the arctan series, and of Newton in respect to his general analysis – so that trouble was taken, not to check and establish the facts of a case which, in their opinion, was proven beyond the slightest doubt, but only to ascertain the relevant evidence. Yet it was not as simple as that; Leibniz' statements about the date of his discovery and communication of his arithmetical quadrature of the circle and his analytical developments from it were in reality correct. At the time, indeed, this could not be established, since the relevant letters were inaccessible among the papers of the various recipients, and though Leibniz himself possessed accurate copies of most of the papers exchanged and still had a fairly precise memory of their main details in his old age, the mass of material then in his possession had expanded beyond his control and he was accordingly unable to unearth the crucial pieces of evidence himself. Furthermore, a fundamental misunderstanding complicated the issue: Leibniz for the most part had asked not for results but for methods, intending by this not just outline-procedures but full demonstrations; yet what he had received had been single items only, and even these were considerably distorted in detail by confusing copyist's errors. This would not have been suspected at the time in London; since the English relied – fallaciously – entirely on the drafts preserved by Collins, believing that the Latin versions which Oldenburg had passed on to Leibniz were, certain details apart, accurate renderings of the essential content of the originals. What was sent by Oldenburg to Leibniz was frequently a shortened version of Collins' English draft, abridged at a number of crucial places, nor was it always accurately rendered into Latin. Conversely, the English did not consult Leibniz' original letters other than in the versions – full of misreadings – printed in Wallis' *Opera*, but tended rather to base their judgments on Oldenburg's copies in the Letter Book of the Royal Society. Now these copies not only are incomplete but in them all poorly legible sections are excised along with all purely personal phrases which Oldenburg had no wish to see preserved; omitted in particular are all passages where Fellows of the Society might feel in any way under attack. He made these cuts with the best of motives, actuated solely by the wish to remove irrelevancies or unfavourable expressions of opinion; but in these suppressed portions are several significant references from which the English could, had they so wished, have noticed that the selection of texts published in the *Commercium epistolicum* is in fact inadequate.

All this was disregarded at the time: Newton, himself the compiler of the *Commercium epistolicum*, was concerned only with establishing priority in individual topics and their fundamental procedures, as deducible from the results transmitted, and the question of method and related proofs is brushed aside as inessential. This (to an impartial observer) surprising attitude becomes partly explicable if we remember that it was then common practice to keep general methods of solution back when communicating results and that higher analysis was then passing through a severe crisis with regard to its rigorous foundation because of the forceful intrusions of the ill-defined concepts and techniques of indivisibles, so that no adequate methodological presentation of its advances would in any case have been at all feasible. Leibniz' stated wish to gain knowledge of the methods used by the English mathematicians and the relevant proofs was, to be sure, an unfair request to make. Newton did not want to oblige, but he would indeed have been unable to do so even if he had wished it. Nonetheless, Leibniz is correct to stress, in his marginal notes to the *Commercium epistolicum*[92] and elsewhere, that as far as methods were concerned he got nothing from the English; just like Newton and Gregory he had to hack his way to the crucial insights all by himself. As a matter of fact – we shall come back to this later on – each of the three rivals achieved his own method; none of them borrowed or took over, from either of the other two, more than certain incidental details. From this viewpoint the whole effect of the priority quarrel was to overvalue individual results without achieving any really comprehensive assessment of their significance. It was tragic that Newton and Leibniz each believed his own method to be the only possible one and would not understand that there are no rights of property in the realm of ideas; the time had grown ripe for higher analysis to develop, and so it sprang to life in three different minds, on each occasion a little differently formulated and reached by a slightly different method of approach.

[92] Annotation on *CE* (*1712*): 42–3.

11

STUDIES IN ALGEBRA

Leibniz had already interested himself in algebraical topics in the autumn of 1674. At that time problems in number theory to which Leibniz had been guided through his acquaintance with Ozanam, and also graphical methods for the treatment of 'solid' problems were to the fore. Towards the end of 1674 Leibniz had invented his *circinus æquationum*, the algebraic counterpart to his calculating machine. And when he worked through Descartes' *Geometria* his interest was roused by such items as its determination of factors of a given equation,[1] and discussion of the fundamental theorem of algebra,[2] but above all by its treatment of the irreducible case of the cubic and its reduction of the biquadratic equation to a cubic resolvent.[3] These questions of the day had already some time previously been extensively discussed by Viète, Stevin, Girard and also Descartes and his circle, but had not yet been adequately investigated – particularly the problem as to whether Cardan's formula was generally valid or applicable only to cubics with a unique real root. Leibniz believed the transition through the imaginary to be avoidable in the irreducible case,[4] so agreeing with Collins who repeatedly demanded that Cardan's somewhat restricted formula be replaced by a more general one.[5] Later, he convinced himself of the general validity of Cardan's formula, first by working through examples of equations of the form $x^3 - px - q = 0$ with an obvious rational root such as $x^3 - 13x - 12 = 0$ (already discussed by Girard)[6] or the instance $x^3 - 48x - 72 = 0$ where following a tentative method used by Bombelli (which he meant to consolidate) he could actually find the cube roots of the binomials in the expression

$$x = \sqrt[3]{\left[\frac{q}{2} + \sqrt{\left(\frac{q^2}{4} - \frac{p^3}{27}\right)}\right]} + \sqrt[3]{\left[\frac{q}{2} - \sqrt{\left(\frac{q^2}{4} - \frac{p^3}{27}\right)}\right]},$$

and then went on to verify that they yield a solution of the cubic

[1] *C* 821.　　　　　　　　　　　　[2] *C* 834.
[3] *C* 789–90, 848, 866, 923–4.　　[4] *C* 790, 1014.
[5] Oldenburg–Leibniz, 6 (16) Apr. 1673 (*LBG*: 87).
[6] *Invention nouvelle* (1629): fols. D2r–2v; *see* also note 8.

even when these roots are not real.[7] Leibniz also followed the problem in the literature and emphasized the somewhat neglected point that the treatment of cubics and biquadratics and of Cardan's binomials had been already standard in the work of the Italian school of the sixteenth century.[8] In particular, he stressed in a letter to Oldenburg that Descartes had had a well-informed predecessor in Ferrari.[9] The relevant details are not without their own interest. Ferrari began with the standard reduced form

$$x^4 + px^2 + qx + r = 0$$

and sought to determine an auxiliary quantity t so that the identity

$$(x^2 + t)^2 \equiv (2t - p)x^2 - qx + (t^2 - r)$$

may be satisfied; this implies that the righthand side is a perfect square, or

$$4(t^2 - r)(2t - p) = q^2.$$

(In Ferrari's day literal algebra had not been invented; hence when Cardan in his *Ars magna* deals with his gifted pupil's procedure, we find only numerical examples treated, but the rules he gives are

[7] *C* 1031 (*LBG*: 551, 558–61); also *C* 1032. Huygens probably received this draft in the summer of 1675 for his criticism while only a vague indication was given to Oldenburg, 2 (12) July 1675 (*LBG*: 132). More is to be found in Leibniz–Huygens, c. September 1675 (*HO* 7: 500–1). The paper *C* 1032 was handed over to Oldenburg and Collins during Leibniz' second visit to London 8 (18)–19 (29) Oct. 1676. A reference to the treatment of equations where cubic irrationalities can be removed occurs in the letter to Malebranche, 14 Aug. 1679 (*LSB* II. 1: 483–4).

[8] Leibniz, assisted by Ozanam, had in the late summer of 1674 (*C* 745–6), worked through the chapters on cubic and quartic equations in Descartes' *Geometria* (1659): 92–7 as well as in Schooten's commentary (*ibid.* 292–9, 302–9, 315, 324–44) and the relevant appendix (345–68). Here he found the reference to Girard's *Invention* (1629) leading him on to Bombelli's *Algebra* (1572). In the summer 1675 he drew Huygens' attention to this work (*HO* 7: 500–1), hitherto unknown to him. Yet Leibniz was entirely ignorant of the great achievements of the German 'cossists' whose best exponent Stifel is only marginally mentioned in Schooten's commentary and in the correspondence Frénicle–Mersenne. Hence Leibniz mistakenly saw in the works of the Italians the really decisive development before Viète and Descartes, as he expresses in *C* 1031 (*LBG*: 551), in a letter to Oldenburg, 2 (12) July 1675 (*LBG*: 131) and again in a letter to Molanus for Eckhard, early April 1677 (*LSB* II. 1: 309), as also later in his 'De ortu, progressu et natura algebræ' (1685–6) (*LMG* VII: 210–11), in letters to Bodenhausen, 14 (24) Mar. 1694/5 (*LMG* VII: 380) and to Magliabecchi 20 (30) Oct. 1699 (*LMG* VII: 315–16).

[9] References in the *Geometria* (1659): 93, 95 had induced Leibniz to look at Cardan's *Ars magna* (1545) where he found in cap. xxxix, regula II Ferrari's procedure. For the following *see* also ch. 7: note 59. Leibniz explicitly mentions Ferrari in *C* 747 B, 938, 1031 (*LBG*: 551, 562–3), also in his letter to Oldenburg 2 (12) July 1675 (*LBG*: 131).

unequivocal and completely general.) Descartes himself in his *Géométrie*[10] gives only a working rule without proof, asserting abruptly that we must solve the auxiliary equation

$$y^6 + 2py^4 + (p^2 - 4r)y^2 - q^2 = 0$$

in order to factorize the original quartic as the pair

$$x^2 \pm xy + \tfrac{1}{2}y^2 + \tfrac{1}{2}p \mp q/2y = 0:$$

this procedure is immediately deducible from Ferrari's on setting $2t - p = y^2$. The circle of Descartes' mathematical friends dwelt extensively on this method; De Beaune gave in his 'Notæ breves'[11] a not very satisfactory verification of the technique; Schooten's explanation,[12] finding Descartes' two quadratic factors of the biquadratic by the method of indeterminate coefficients appeared only in the *Geometria* of 1659. Many years before the appearance of the *Géométrie* Harriot had worked on transformations of biquadratics, printed posthumously in his *Ars analytica* of 1631,[13] and as Wallis (for once) acknowledges in his *Algebra*,[14] Descartes certainly contrived his own procedure quite independently of Harriot who does not solve the general quartic; both surely received an initial suggestion from reading Viète who himself drew on Ferrari.

Leibniz familiarized himself with the existing methods for solving equations, seeking to unify their approaches and then to extend them to higher equations. Thus we find him using the transform $x = u + v$ to reduce the standard form of the cubic[15] and also corresponding expressions in three or four terms in the case of quartics and quintics;[16] and again, employing the substitution $x = u + v\sqrt[3]{t} + w\sqrt[3]{t^2}$ to resolve the general cubic and an analogous expression in the extension to higher equations.[17] On the whole these are still immature probings –

[10] *Géométrie* (1637): 383–4 = *Geometria* (1659): 79–80.
[11] *Geometria* (1659): 137–9.
[12] *Geometria* (1659): 315.
[13] Harriot, *Praxis* (1631): 104–16.
[14] Wallis, *Algebra* (1685): 208–9; *see* also *WO* II: 212, 228.
[15] *C* 980, 1015 and *passim* elsewhere.
[16] *C* 1024, 1104₂ and *passim* elsewhere.
[17] *C* 1028 B, 1042, 1044 and *passim* elsewhere. *C* 1042 is entitled 'De bisectione laterum' and is continued in *C* 1138 and 1344. Leibniz presumably discussed the first part of this manuscript with Huygens in the summer of 1675; a piece has been cut off and is damaged but the full text can be restored from the English copies since Leibniz had given the complete manuscript to Oldenburg and Collins during his second visit to London, 18–29 Oct. 1676 and two copies are to be found amongst Newton's papers. The cut-off piece (*C* 1138) contains a reference to Tschirnhaus as the person who checked and confirmed Leibniz' calculations on angle-section:

he either reaches an illusory solution through error in calculations or has to break off its completion because of the increasing complexity of the computation.[18] Leibniz was firmly in agreement with the contemporary opinion that it should be possible to solve the general equation of nth degree by transforms involving radicals or iterated radicals.[19] Hand in hand with these considerations went investigations seeking to identify those higher equations which can be solved by particular types of substitution, such, for instance, as

$$x = \sqrt[n]{(u + \sqrt{v})} + \sqrt[n]{(u - \sqrt{v})}.$$

The family of equations so distinguished we shall refer to in the sequel as 'generalized Cardan equations'. To this family, according to Leibniz, belong quintics of the form $x^5 + px^3 + p^2x/5 = r$ and he gives examples of other higher equations up to the ninth degree. Here – depending on the sign of p and the numerical value of r – we have to do either with the problem of finding submultiples of an angle or that of inserting several geometrical means between two quantities.[20] Leibniz began with the substitution $x = u + v$, where $uv = p$ and then step-by-step determined the power sums $x_n = u^n + v^n$ by the recursion $xx_n = x_{n+1} + px_{n-1}$ as functions of x. Without knowing it Leibniz thereby found again a result which Fermat[21] had discovered in elaborating Viète's famous *Responsum ad problema Adriani Romani*. To this field of research belongs a particular result of

these calculations anticipate as it were De Moivre's formula – in modern transcription it is stated that where $x = \cos \theta$, $y = \cos \theta/n$, then

$$2y = \sqrt[n]{[x + \sqrt{(x^2 - 1)}]} + \sqrt[n]{[x - \sqrt{(x^2 - 1)}]}.$$

See Schneider (1968): 224–9. When it appeared in De Moivre's paper in the *PT* **20** No. 240, of May 1698 (published 1699): 190–3, Leibniz – quite modestly – put in his rightful claim of authorship in the *AE* (May 1700): 199–208 (*LMG* v: 346). He had in fact in the letter to Oldenburg, 17 (27) Aug. 1676 (*LBG*: 198) made a well-nigh incomprehensible allusion to this, and so he repeats his claim in a marginal note on *CE* (*1712*): 63. Huygens was probably in the summer of 1675 also shown *C* 1044 before it too was given to Oldenburg; in the two English copies which have survived among Newton's papers it has the title 'De circino æquationum'. Apart from an enumeration of higher equations solvable by a generalized Cardan formula it contains an approach by means of

$$x = \alpha_1 t^{(n-1)/n} + \alpha_2 t^{(n-2)/n} + \ldots + \alpha_{n-1} t^{1/n} + \alpha_n$$

applied to cubics and biquadratics. For Huygens' reaction to these papers *see* below p. 159.

[18] Leibniz–Oldenburg, 10 (20) May 1675 (*LBG*: 122).
[19] Leibniz–Oldenburg, 18 (28) Dec. 1675 (*LBG*: 147). Here too Leibniz mentions the difficulty of the calculations involved and indicates that others have progressed further than he has – probably referring to Tschirnhaus.
[20] What follows is merely a précis of Leibniz' paper *C* 1042.
[21] Fermat–Huygens, December 1661 (*FO* I: 189–94).

Leibniz which appeared truly surprising to his contemporaries, the equality[22] $\sqrt{[1+\sqrt{(-3)}]} + \sqrt{[1-\sqrt{(-3)}]} = \sqrt{6}$. Leibniz gained this in generalized form by setting

$$uv = p, \quad u^2 + v^2 = q, \quad \text{or} \quad u+v = \sqrt{(2p+q)},$$

where u and v are evidently the roots of the quartic $t^4 - qt^2 + p^2 = 0$; so that $t^2 = \frac{1}{2}q \pm \sqrt{(q^2/4 - p^2)}$, and at once

$$\sqrt{\left[\frac{1}{2}q + \sqrt{\left(\frac{q^2}{4} - p^2\right)}\right]} + \sqrt{\left[\frac{1}{2}q - \sqrt{\left(\frac{q^2}{4} - p^2\right)}\right]} = \sqrt{\left(2p+q\right)}.$$

The previous equation is the particular case $p = q = 2$; nobody noticed at the time that the identity is implicit in Euclid's book x, 47–54 (if $4p^2 < q^2$). The result was of the utmost importance for Leibniz; he already conjectured that for any algebraic function $f(x)$ the sum $f(x+iy) + f(x-iy)$ will always be real,[23] and from this insight derived his inner certainty that in the irreducible case of the cubic a real result will ensue even when it is impossible separately to extract the two cube roots of the Cardan binomials in real numbers – in other words, that Cardan's formula is generally valid and denotes a real root in every cubic equation with real coefficients.[24]

For most of the details of Leibniz' algebraical studies his letters of the period are once again our main source of information. In his answer to Oldenburg's[25] letter of 22 April 1675 – which we have already cited – Leibniz draws attention to his powerful, if computationally cumbersome, algebraic innovations. Regarding the achievements of the English mathematicians indicated by Oldenburg he was chiefly interested in the possible removal of intermediate terms in an equation and in the logarithmical solution of affected

---

[22] Contained in C 1031 (*LBG*: 563) and 1032 as well as in Leibniz–Huygens, mid-September 1675 (*HO* 7: 501). Huygens in his reply, 30 Sept. 1675 (*HO* 7: 505–6), acknowledges the identity as 'amazing', an opinion Leibniz quotes *verbatim* in a letter to Varignon, 2 Feb. 1702 (*LMG* iv: 93). The identity is repeated in Leibniz–Oldenburg, 17 (27) Aug. 1676 (*LBG*: 198–9); a copy of this letter was sent to Wallis, 14 (24) Sept. 1676 (*CR* ii: 598), who also expresses his astonishment when replying to Collins, 16 (26) Sept. 1676 (*CR* ii: 599) though he proposed to set as an alternative $\sqrt{[1+\sqrt{(-3)}]} + \sqrt{[1-\sqrt{(-3)}]}$ equal to $-\sqrt{(-6)}$.

[23] This is already expressed in the title for C 1031 (*LBG*: 550) where Leibniz labels the combination of complex terms into something real as a sixth species, elaborating this idea in more detail on pp. 553–4; the numerical examples cited above, p. 143, occur in this context, pp. 558–9.

[24] Leibniz–Oldenburg, 2 (12) July 1675 (*LBG*: 131) and Leibniz–Huygens, mid-September 1675 (*HO* 7: 501).

[25] Leibniz–Oldenburg, 10 (20) May 1675 (*LBG*: 122).

equations. Prestet has, he regrets, nothing to contribute here.[26] He had, to be sure, carefully studied Hudde's commentary on Descartes' *Geometria* and probably quite skilfully condensed it, but in neither geometry nor number theory nor, again, mechanics was he adequately competent, and was stubborn in his insistence that Descartes' solution of the biquadratic equation was insufficiently general; Leibniz has asked him in vain to say something more on this point, or about equations of the fifth and sixth degree. In contrast to Prestet, Ozanam was less well versed in elementary algebra but, in Leibniz' view, a good geometer and really expert in number theory; he was currently sucessfully attacking Diophantine problems, by means of literal algebra, while Billy and Frénicle usually stuck to pure arithmetic. There were, indeed, in this field some rather difficult problems – like the decomposition of a number into two squares[27] where to introduce algebra seemed useless; it would be interesting to hear English mathematicians' opinion upon this question. To our surprise there is no reference whatsoever at this place in Leibniz' letter to Fermat, who had succeeded in defining by his method of *descente infinie* the condition for the possibility of such a decomposition – though he had made public only his rule for it,[28] not the supporting proof (which he declared to be extremely difficult). Huygens had received from Fermat a description not in fine detail, but clearly outlined, of the technique of proof he had used,[29] but he was not particularly interested in questions of number theory and not properly appreciative of Fermat's research,[30] concurring more or less[31] in Descartes' rather disdainful dismissal of Fermat's discoveries as mere *gasconades*:[32] accordingly Huygens was unlikely to have specially drawn Leibniz' attention to Fermat. It is, indeed, a scarcely credible fact that during his stay in Paris Leibniz troubled himself neither about Fermat's published treatises and letters nor with his posthumous manuscripts (then in Carcavy's care and so easily accessible to Leibniz), even though he had several times been urgently requested to glance at Fermat's remains by his English

[26] *See* below, p. 155.
[27] Details are found in *C* 764, 961, 1061.
[28] Fermat's 'Observationes' to Diophantus III. 22 and v. 12 were accessible to Leibniz in the 1670 edition where they are printed on pp. 127–8 and 224–5.
[29] Fermat–Carcavy for Huygens, early August 1659 (*HO* 2: 458–62).
[30] *Compare* for instance Huygens' remarks in his letters to Mylon, 1 June 1656 (*HO* 1: 428) and to Wallis 27 Aug. (6 Sept.) 1658 (*HO* 2: 212).
[31] Huygens–Schooten, 4 Oct. 1658 (*HO* 2: 235).
[32] Schooten–Huygens, 19 Sept. 1658 (*HO* 2: 221–2).

correspondents.[33] Perhaps he held Ozanam's researches in number theory to be more important, directed as they were to elaborating an algorithmic treatment; Fermat had published only results, never methods, and then, puzzling though they might be, his results were regarded by his contemporaries as not derived by methodical reasoning but rather by systematic experimenting; like many others, Leibniz doubtless did not quite know what to make of them – and the English mathematicians had had nothing to say in answer to his remark about the possibility of decomposing a number into two squares.

Oldenburg, briefed by Collins, continued to contribute remarks on algebraical investigations pursued by English mathematicians throughout his correspondence with Leibniz; on the whole it had little result because all the power of the greatest minds was expended in the vain effort to solve higher equations by radicals. We may therefore confine ourselves to a summary survey of the topics covered in the correspondence during the year 1675.

We begin with a remark contributed by Oldenburg[34] that the gauger Michael Dary, a friend of Collins and, like him, lacking any proper scientific education, had found a quadratic construction of the binomial $\sqrt[3]{(u + \sqrt{v})}$ which was said to be valid even when the cube root could not be extracted; he had tried out his method on numerous examples and was thereby able to reduce all solid problems to plane ones. If this turned out to be right, it would be a very great advance. Collins, at any rate, had given it his full approbation. Leibniz, however, thought Dary's alleged discovery extremely improbable:[35] he could not yet, he wrote, give a proof of its impossibility, but he knew of a corresponding question in Diophantine equations (perhaps he means the problem of rendering an expression of the fourth or higher degree an exact square) which he had, however, not yet further pursued because of the computational difficulties involved. He asked to be given Dary's rule together with worked examples of it, in the

[33] Oldenburg–Leibniz, 6 (16) Apr. 1673 (LBG: 88); Oldenburg for Leibniz, same date: C 409 A; Oldenburg–Leibniz, 12 (22) Apr. and 24 June (4 July) 1675 (LBG: 118, 130). After Mersenne's death (1648) Fermat made a habit of depositing scientific papers with Carcavy whom he authorized in a letter, 9 Aug. 1654 (FO II: 299), to get them printed. We know of several attempts by Carcavy to achieve this; of his efforts to win Huygens' support the letters 22 June 1656 (HO 1: 432) and 14 Aug. 1659 (HO 2: 457) bear witness: but these and all later projects came to nothing. Though Leibniz at the time was in close contact with Carcavy there is no reference to all this amongst his notes.
[34] Oldenburg–Leibniz, early June 1675 (LBG: 125).
[35] Leibniz–Oldenburg, 2 (12) June 1675 (LBG: 125-6).

doubling of a cube, for instance, or the construction of a regular heptagon. In the meantime Collins had spotted the fallacy in Dary's reasoning and attempted to cast a veil over the whole awkward affair.[36] Of course, he writes, the solution of the cubic, except in reducible cases, was necessarily a solid problem in general not reducible to a plane one, or to a pure cubic; this could, for instance, be seen in the case of $x^3 - 21x^2 + 120x = y$ whose graphical representation as a Cartesian curve had turning points corresponding to $x = 4$ and $x = 10$. Leibniz at once took exception to this sort of proof.[37] He was fully convinced of the truth of the matter, but could only be satisfied with a rigorous demonstration in the style of Euclid's proof of incommensurability; what rôle was here played by the turning points, he failed, he says, to understand.

With regard to Dary's method Collins further reports[38] that having multiplied the cubic by x, Dary views it as a biquadratic, and then, adding a suitable constant term, he splits this into two quadratic factors, so avoiding the auxiliary cubic necessitated in Descartes' method. Collins confesses that he has not yet worked an example out himself, but that it is probably an approximative method. (In hindsight we can now easily see that the procedure is useless for practical calculation.) Leibniz could not, from the indications given to him,[39] quite visualize the procedure and so was promised by Collins a more elaborate explanation soon,[40] but no further mention of the matter is made. Leibniz presumably talked to Tschirnhaus about the method; subsequently, he made a note on Oldenburg's letter of 20 May 1675: '*nihil erat*'.

Far more interesting is another of Dary's rules, even though restricted by the assumption that the imaginary cube roots in Cardan's formulae can be extracted[41] (which in fact makes the cubic equation reducible). Let $x^3 = px + q$ and

$$\sqrt[3]{[\tfrac{1}{2}q + \sqrt{(q^2/4 - p^3/27)}]} = u + \sqrt{(-v)}$$

(so that $p = 3(u^2 + v)$ and $q = 2u(u^2 - 3v)$) and then

$$x_1 = 2u, \quad x_2, x_3 = -u \pm \sqrt{(3v)}.$$

[36] Oldenburg–Leibniz, 24 June (4 July) 1675 (*LBG*: 128).
[37] Leibniz–Oldenburg, 12 July 1675 (*LBG*: 131).
[38] Oldenburg–Leibniz, 24 June (4 July) 1675 (*LBG*: 128).
[39] Leibniz–Oldenburg, 2 (12) July 1675 (*LBG*: 131).
[40] Oldenburg–Leibniz, 30 Sept. (10 Oct.) 1675 (*LBG*: 141).
[41] Oldenburg–Leibniz, 24 June (4 July) 1675 (*LBG*: 128–9). Hitherto three possible solutions had been considered only in the context of angle-trisection.

The general rule is stated verbally only, and the calculation is illustrated on the example $x^3 = 21x + 20$ where $u = \frac{5}{2}$, $v = \frac{3}{4}$. We recognize here an early, if imperfect, attempt to generate from Cardan's formula not just one root of the cubic but all three.

Leibniz, in his reply to Oldenburg, expresses his opinion on the general validity of Cardan's formula and incidentally reminds him that it had first been derived by Ferro.[42] To his assertion that he himself could eliminate the imaginary by a direct root-extraction Oldenburg answered that this was well known in England,[43] where Wallis, too, was used to employing Cardan's formula to treat the real and imaginary cases formally exactly alike.[44] This, however, was an exaggeration; as a matter of fact, Wallis had been greatly surprised by Leibniz' equation $\sqrt{[1 + \sqrt{(-3)}]} + \sqrt{[1 - \sqrt{(-3)}]} = \sqrt{6}$ and could not immediately comprehend its truth.[45] The related problem of extracting the cube roots in the form $\sqrt[3]{(a + \sqrt{b})} = x + \sqrt{y}$ had been already discussed in Stifel's edition of Rudolff's *Coss* and extended to imaginary quantities in Bombelli's *Algebra* later;[46] it is mentioned in Descartes' letters[47] and expounded by him in Waessenaer's *Den on-wissen wis-konstenaer*, subsequently elaborated by Schooten in an appendix to his treatise on the organic construction of conics and then taken over by him into his edition of Descartes' *Geometria*. It had never, however, been resolved satisfactorily. Stifel and Bombelli came to be largely forgotten in the seventeenth century; Harriot knew – and cites – the latter and Leibniz himself was to refer to him again. Stifel also knew the deductions $x^2 - y = \sqrt[3]{(a^2 - b)}$ and $(2x)^3 = 3(2x)\sqrt[3]{(a^2 - b)} + 2a$, but confined himself to cases where $a^2 - b$ is a perfect cube; like Bombelli's his examples are manifestly contrived by constructing $a = x^3 + 3xy$, $b = (3x^2 + y)^2 y$ in reverse. Most later mathematicians did not advance much beyond this sort of systematic guesswork. So, for instance, Georg Mohr,[48] who had written a letter to Collins on this topic which Oldenburg copied and forwarded to Leibniz, had contributed some remarks on extractable cases of $\sqrt[3]{(\sqrt{a} + \sqrt{b})}$ of the kind previously considered by Stifel but

[42] *See* note 8 above.
[43] Oldenburg–Leibniz, 30 Sept. (10 Oct.) 1675 (*LBG*: 140–1).
[44] Wallis–Collins, 29 Mar. (8 Apr.) 1673 (*CR* II: 557–8); excerpt (*C* 1067) sent as enclosure with Oldenburg–Leibniz, 30 Sept. (10 Oct.) 1675.
[45] *See* note 22.
[46] *See* note 8.
[47] Descartes–Waessenaer, 1 Feb. 1640 (*DO* III: 23–4 and 29–30). Schooten, too, refers to Waessenaer's work, *Geometria* (1659): 369; *see* note 8.
[48] Mohr–Collins, 16 (26) Sept. 1675 (*C* 1066A).

his account is excessively complicated in expression, explained by numerical examples which are then generalized not algebraically but merely verbally. Leibniz replied to Wallis' and Mohr's letters in kind and friendly fashion though without any enthusiasm for their content:[49] this sort of observational approach made no effective advance over previous performances, and in particular Mohr quite unnecessarily restricted himself to the field of real numbers.

The other main goal for the algebraists of the time was the removal of intermediate terms in higher equations so as thereby to reduce them to simpler forms and ultimately to pure equations and so solve them by radicals. It was thought possible by particular transforms to change certain roots while preserving others unaltered and thus to approach the desired goal. Dulaurens, a mathematical amateur, had asserted the feasibility of this in the preface of his *Specimina mathematica* and had been confirmed in his belief by Frénicle[50] who, though a good arithmetician, was but a poor algebraist. Although Wallis had pointed out certain fundamental errors in the *Specimina*[51] and had replied to a not very skilful riposte by Dulaurens[52] with devastatingly renewed criticism,[53] Collins still considered Dulaurens' conjecture to be correct, the more so since Pell had also spoken out along these lines.[54] James Gregory had at first thought the elimination of the intermediate terms to be quite impossible,[55] but later on began to reconsider his opinion[56] and eventually wrote that he could in fact always eliminate the intermediate terms but usually only at the cost of raising the degree of the equation;[57] thus in order to remove the intermediate terms of a quintic it would be necessary to solve an equation of twentieth degree: and furthermore there were cases – when, for instance, the third term in a quartic needed to be removed – where the elimination could not always be achieved in real numbers.

[49] Leibniz–Oldenburg, 18 (28) Dec. 1675 (*LBG*: 148).
[50] For Dulaurens' assertion *see* p. 42, for Frénicle's approval *see* Collins–Gregory, 24 Dec. 1670 (3 Jan. 1671) and 21 Sept. (1 Oct.) 1675 (*GT*: 159, 332).
[51] Wallis for Oldenburg, 2 (12) July 1668 (*OC* **4**: 489–92). *Compare* also Wallis' remarks in *PT* **3**, No. 34 of 13 (23) Apr. 1668: 654–5 and No. 38 of 17 (27) Aug. 1668: 748–80.
[52] The *Responsio* probably came out in June 1668.
[53] Wallis in *PT* **3**, No. 38 of 17 (27) Aug. 1668: 744–7; No. 39 of 21 Sept. (1 Oct.) 1668: 755–9 and No. 41 of 16 (26) Nov. 1668: 825–38.
[54] Collins–Gregory, 1 (11) Nov. 1670 (*GT*: 111); Nov. 1671 (*GT*: 196).
[55] Gregory–Collins, 15 (25) Feb. 1670/1 (*GT*: 169).
[56] Gregory–Collins, 17 (27) May 1671 (*GT*: 187–8).
[57] Gregory–Collins, 17 (27) Jan. 1671/2 (*GT*: 211).

The notes on all this which were sent to Leibniz of necessity remained largely unintelligible to him;[58] in addition, in the text sent to him, Leibniz interpreted the adjective *arbitraria* which was meant to signify a 'general' equation as denoting an 'arbitrary' one, chosen at will, and in consequence could not see where the problem lay,[59] though Oldenburg quickly enlightened him.[60] What Gregory was trying to do has only become fully clear since the publication of some of his original papers, long considered lost.[61] We may now see that he believed he possessed a method of reducing an equation of nth degree, by means of a (solvable) equation of degree $n-1$, to a pure equation and hence of solving it by radicals. To cite the most interesting of his worked examples, namely the treatment of the quintic, starting from the standard form $x^5 + px^3 + qx^2 + rx + s = 0$, Gregory sets $x = u + v$ and then multiplies the lefthand side of the ensuing equation in u and v by an arbitrary form of fifteenth degree $v^{15} + a_1 v^{14} + \ldots + a_{15}$: the condition that in the result only terms in v^{20}, v^{15}, v^{10}, v^5 and a constant term are to appear while all the other terms are to vanish, determines sixteen equations for the sixteen unknowns a_k and u, and when these are solved we obtain v from an equation of degree four in v^5. Gregory believed that u and v are always expressible by radicals, reaching this conclusion in generalization of the forms of particular reductions in the case of cubics and quartics where it is in fact valid. He could not foresee that in the case of the general quintic we will obtain after eliminating the a_k, an irreducible equation of sixth degree in u. He was thwarted by the insuperable complexity of calculation which can be overcome only by the use of modern algebraic notation. Originally, Collins had planned to send an altogether different communication to Leibniz[62] which would have told him of Newton's logarithmic solution of simple algebraic equations in three terms.[63] There also he discussed the question of how to find the compound interest rate N of an annuity of present value A which is exhausted by n yearly payments of an amount B.

Leibniz had previously heard that Pell knew how to employ logarithmic and trigonometric tables to good effect in the solution of

[58] Oldenburg–Leibniz, 24 June (4 July) 1675 (*LBG*: 130).
[59] Leibniz–Oldenburg, 2 (12) July 1675 (*LBG*: 131).
[60] Oldenburg–Leibniz, 30 Nov. (10 Oct.) 1675 (*LBG*: 141).
[61] *See GT*: 384–7.
[62] Collins–Oldenburg for Leibniz, 15 (25) June 1675 (MS: London, B.M., Birch 4398: 139–40).
[63] Newton–Collins, 6 (16) Feb. 1669/70 (*NC* 1: 23–5); *compare* Hofmann (1943): 24–5, notes 69, 70; also *NP* 3: 566–7.

higher equations, but at the time all these hints remained vague and allusive in form.[64] Pell assiduously guarded his methods and was not easily persuaded to make a tangible statement regarding them. His papers are mostly unpublished, but fortunately preserved.[65] A close study of the mass of his preserved papers is much to be desired, for the historian of mathematics may expect to find in these manuscripts important references to the activities of Pell's circle of acquaintances. We now know that Pell was already at work on the theory of equations during his stay in the Low Countries (1643–52) and that he had, for instance, given his complete solution of an equation of sixth degree by means of sine-tables to Stampioen,[66] a surveyor teaching arithmetic at the Hague, the one who had given Huygens and his brothers their first lessons in computation. In the spring of 1671 Pell allowed himself to be drawn into conversation with Collins about his methods. He then confirmed Dulaurens' assertion that every equation of odd degree can be solved by an angular section or insertion of geometric means, and added, that this type of equation must always have at least one real root; he had himself made very good progress in dealing with higher equations and had composed a work on the subject, his bulky 'Canon mathematicus', in which he first determines limits to the real roots of an equation, and then approximates the roots by some kind of *regula falsi* facilitated by the use of logarithms: in a complete equation of eighth degree he could, for instance, with certain restrictions eliminate six intermediate terms – indeed with others four, with others again only two. Thomas Harriot (who had died in 1621), so Pell declared, already had carried out similar researches and had been successful in removing two intermediate terms in a complete equation. But when Collins proposed to Pell that he should publish his researches, he had refused this so vigorously that any further attempt to persuade him seemed hopeless.[67] Only in 1675 did Collins get any more information out of Pell:[68] he then reported that he could, by suitably varying the constant term, divide the lefthand side of an equation into a factor of degree $(n-2)$ and a quadratic one and could then approach the exact roots as near as he liked by assuming various constant terms –

[64] Oldenburg–Leibniz, 12 (22) Apr. 1675 (*LBG*: 116).
[65] Pell's manuscripts are now in the British Museum. Some of the letters exchanged with Charles Cavendish were published by Halliwell (1841), more by Harvey (1952), while his correspondence with Mersenne will be found in *MC* 8 ff.
[66] Collins for Gregory, 15 (25) Dec. 1670 (*GT*: 142).
[67] Collins–Gregory, 25 Mar. (4 Apr.) 1671 (*GT*: 180).
[68] Collins–Gregory, 1 (11) May 1675 (*GT*: 298).

this for cubics and quartics as well as for higher equations. A few months later Collins could supplement his account of this for Gregory[69] by saying that Pell was now able to find the roots of the irreducible cubic by means of sine-tables, those of cubics with a unique real root by using sec-tables and logarithms: in the former irreducible case he was able to construct the other two roots once he had the first one, and his methods were applicable to higher equations as well. At the time Leibniz learnt of all this – apart from what he had been told earlier[70] – only from Collins' passing remark that Pell was able to define the characteristics of a cubic from a (derivative) Apollonian parabola. To this Oldenburg later added that Pell had long ago promised something on the solution of equations by sines and logarithms and that they could only hope that he would eventually keep his word.[71] Leibniz replied[72] that Pell's method of factorization interested him very much; he could not suppose that it would be superseded by Prestet's *Algebra*, which, he reports, had at last appeared, for it contained only indeterminate arithmetic and elementary algebra and nothing at all on the newer methods – not a single hard problem was there tackled, no geometrical applications introduced, and only the equation of fourth degree was extensively treated, and then simply following Descartes' account.

The determination of limits to the roots of an equation is an expedient preliminary to their practical computation. When Collins drew his attention to this group of problems,[73] Leibniz once more returned to making himself thoroughly familiar with the relevant literature.[74]

Collins also sent Leibniz[75] an account of the contents of Wallis' treatise on *Angular sections*, which had been in his hands for several

[69] Collins–Gregory, 29 June (9 July) 1675 (*GT*: 310).
[70] *See* also p. 26, p. 136, and further p. 223.
[71] Oldenburg–Leibniz, 24 June (4 July) 1675 (*LBG*: 128–9).
[72] Already in the (unpublished) draft of his letter to Gallois of 2 Nov. 1675 (*C* 1107) Leibniz points out the rather elementary character of Prestet's *Elemens* (1675) and, as in Leibniz–Oldenburg, 18 (28) Dec. 1675 (*LBG*: 144), regrets the lack of interest in geometrical questions. The book was probably published early in October, since Collins can on 19 (29) Oct. 1675 (*GT*: 338) refer to accounts of it by Justel and others though the book itself has not yet arrived in London. Leibniz made excerpts from it, *C* 1170A, and wrote some criticism on it, *C* 1170B, 1278A. In the *PT* **11**, No. 126 of 20 (30) June 1676: 638–42 appeared a review, perhaps from Collins' pen, merely stating the contents. The elementary character of the work has perhaps been overstressed on comparing it with Kersey's *Algebra* (1673–4), doubtless a more interesting book.
[73] Oldenburg–Leibniz, 24 June (4 July) 1675 (*LBG*: 128).
[74] *C* 1004.
[75] Oldenburg–Leibniz, 24 June (4 July) 1675 (*LBG*: 129–30). For Wallis' treatise *see* Scriba (1966): 21–33.

years ready for press, though in the end it was published only in 1685 as an appendix to Wallis' *Algebra*. Here, Collins tells Leibniz, it is being shown how certain higher equations can be solved by passing from the side of a regular polygon to its diagonals, that is, by sine-functions (he cites equations of the type $x = 2 \sin \theta$, $y = 2 \sin n\theta$). Another English mathematician, versed in this particular field (namely, Pell), had told him that the tan- or sec-functions would serve just as well, for in each case sets of equations are obtained which are solvable step-by-step. To this Leibniz at once responded[76] with a survey of his own treatment of the higher 'Cardan' equations, where it is completely unnecessary to eliminate intermediate terms. At the same time he enquired whether Wallis' treatment of cubics related to reducible equations only or to the rationally irreducible ones also, and whether perhaps one root is to be arbitrarily chosen; in his own method the imaginary disappears after the roots are extracted. When Oldenburg answered this,[77] he enclosed with his letter an excerpt (mentioned above) from a letter by Wallis where he says that he considers the introduction of imaginaries into Cardan's formula as having no essential effect – a belief he had had since 1648, even before he came to be acquainted with Descartes' *Géométrie*. When Collins read Leibniz' description of his solution of generalized Cardan equations, the central idea of which he did not properly grasp, he suspected that this might possibly be the same method that Tschirnhaus and Gregory also possessed. The latter now – apart from his method of eliminating intermediate terms in higher equations by raising the degree – used another procedure in which the roots are expressed by radicals and obtainable from any one of them by appropriate changes of sign. He was now working to establish firm rules (*canones*) for this, but the calculation involved was so extremely laborious that he was willing to put his method and its proof at the free disposal of anyone who would relieve him of all the computations. This was but a short recapitulation on Collins' part of allusions to Gregory's new technique in his letters to him.[78] Gregory's actual

---

[76] Leibniz–Oldenburg, 2 (12) July 1675 (*LBG*: 132). *Compare* note 17 and p. 146.

[77] Oldenburg–Leibniz, 30 Sept. (10 Oct.) 1675 (*LBG*: 141). The subsequent remark on the methods probably employed by Tschirnhaus and Gregory occurs on *LBG*: 142.

[78] We do not know the contents of the now lost letter Gregory–Collins, 28 Mar. (7 Apr.) 1675; *see GT*: 31. The letter of 26 May (5 June) (*GT*: 302–3) deals with the sums of powers of roots of an equation, while in those of 20 (30) Aug. and 11 (21) Sept. 1675 (*GT*: 326, 329) Gregory refers to the difficult calculations required to reduce the degree of an equation.

method, which would seem to have involved some application of group-theoretical concepts, is unfortunately not to be found in those of his manuscripts now extant. Leibniz gained the impression that Gregory's method must be at least closely related to his own if not identical with it;[79] in these circumstances he was unwilling to go into details – and he, too, was frightened off by the lengthy calculations necessitated.

In reply to a remark of his mentioning his algebraic instrument for solving equations, and in connection with his query regarding the logarithmic solution of equations, Leibniz received a brief description of a mechanism of Newton's designed to the same end:[80] in this several parallel (or circular) sliding scales divided logarithmically are used and by means of a movable ruler it is then possible to determine the real roots of all higher numerical equations (with n slides required for an equation of degree n). Leibniz immediately replied[81] that his own procedure was based on quite a different idea: he was however greatly interested in Newton's mechanism, and during his (second) visit to London he made an excerpt from Newton's letter to Collins where it is described.

Already in the spring of 1675 Leibniz had placed an algebraical manuscript in the hands of his mentor Huygens; this was a rather juvenile attempt[82] to replace Cardan's formula for the cubic by a substitution of the form $x = \sqrt{(u + \sqrt{v})}$. During the summer of 1675 Leibniz' new algebraical ideas had become sufficiently clarified in his mind for him to risk again submitting certain of his studies to Huygens with better confidence.[83] The latter amicably promised to peruse the papers critically, showing interest in Leibniz' remarks on the older Italian algebraical literature concerning cubics and bi-quadratics and encouraging him to continue – and perhaps publish – his latest results, but he afterwards relapsed into silence. Leibniz became impatient – we shall later hear for what reason – and by way of a reminder sent Huygens a letter in which he surveys the papers he had previously given to him,[84] accompanying it with a copy of Bombelli's *Algebra*.[85] The latter work had already introduced

[79] Leibniz–Oldenburg, 18 (28) Dec. 1675 (*LBG*: 146).
[80] Oldenburg–Leibniz, 24 June (4 July) 1675 (*LBG*: 130); *see also* Newton–Collins, 20 (30) Aug. 1672 (*NC* 1: 229–30).
[81] Leibniz–Oldenburg, 2 (12) July 1675 (*LBG*: 131).
[82] *C* 937; *compare* ch. 10: note 35.
[83] Leibniz for Huygens, summer 1675; for details *see* note 17.
[84] For the following *compare* Leibniz–Huygens, mid-September 1675 (*HO* 7: 500–4); also the reference in note 22.
[85] Excerpts from this book are found in *C* 1048.

imaginary roots in solving those irreducible cubics for which Cardan's formula yields a positive root, for instance $x^3 = 15x + 4$, where, as Leibniz explicitly quotes, he obtains

$$x = \sqrt[3]{[2 + \sqrt{(-121)}]} + \sqrt[3]{[2 - \sqrt{(-121)}]}$$
$$= 2 + \sqrt{(-1)} + 2 - \sqrt{(-1)} = 4.$$

Bombelli had here promised a geometrical proof, but in reality he only demonstrated the possibility of forming such equations and finding their roots without proceeding to details. In such a case as $x^3 = 12x + 9$ Cardan's formula would lead to a negative root (namely

$$x = \sqrt[3]{\tfrac{1}{2}[9 + \sqrt{(-175)}]} + \sqrt[3]{\tfrac{1}{2}[9 - \sqrt{(-175)}]}$$
$$= \tfrac{1}{2}[-3 + \sqrt{(-7)} - 3 - \sqrt{(-7)}] = -3),$$

and this Bombelli did not admit; so, following Cardan, he found the divisor $x + 3$ of the reordered form $x^3 - 12x - 9$ by testing factors of the absolute term, and after division gave the only positive – that is, for Bombelli, the only admissible – root $x = \tfrac{1}{2}(3 + \sqrt{21})$. Since this could not be obtained by a direct application of Cardan's formula, Bombelli had cast doubt on the general validity of the formula, but wrongly so. Leibniz now claims to be the first to have found that Cardan's rule is valid in every case, quite irrespective of whether the cube roots of the binomial can be extracted in real numbers or not. This he had concluded by analogy with his notable identity[86]

$$\sqrt{[1 + \sqrt{(-3)}]} + \sqrt{[1 - \sqrt{(-3)}]} = \sqrt{6},$$

but similar results held true quite generally for the sums of higher even-degree roots of binomials of this type. Furthermore, he had ascertained that every reducible cubic with rational coefficients possessed a rational root. On this point Descartes' statement in his *Géométrie* was not sufficiently precise; for he spoke simply of quantities, where he ought to have restricted these to being rationals. The solution of equations by radicals could be reduced to two basic geometrical constructions, angle-section and insertion of geometric means; hence in some sense logarithms corresponded to arcs. There exist apparently families of equations which are solvable by generalized Cardan formulae, in such a manner that every equation has to be dealt with by radicals of corresponding degree, as in the instance of the equations of fifth degree here under discussion. Some time ago,

[86] *Compare* note 22 above.

Leibniz concludes, Huygens had suggested to him that he should publish all this and so, since no one could match his judgment thereon, he is now asking for his opinion of his discoveries. Huygens considered his answer very carefully.[87] In his still extant draft reply he unequivocally asserts that the first task must be to establish simple, generally valid rules for finding $\sqrt[3]{(u+\sqrt{v})}$ in order to escape from the awkward period of experimentation that began with Bombelli and was not yet, even with Schooten, come to its close. That a reducible cubic with rational coefficients always had one rational root at least had still to be proved; similarly it needed to be shown that recourse to the imaginary was unavoidable in the irreducible case. The solution of generalized Cardan equations was in itself interesting and useful, but these were special types only, not general ones. As for Leibniz' instrument for solving equations it was in theory well designed but its practical use had still to be proved.

In short: a complete rebuff. But it was not in Huygens, well-mannered and experienced man of the world that he was, to deliver so negative an opinion quite so directly and inconsiderately. The letter he in fact sent to Leibniz begins by apologizing for his overlong delay in replying, caused by his many other obligations and long neglect of practice in algebraical research. The factual points of the draft are taken over into the letter, but they are presented in an amiable if somewhat hollow form. Regarding the algebraic instrument, for instance, Huygens declares that, if he had not seen Leibniz' calculating machine, he would have thought it impossible for anyone to contrive this new piece of apparatus. No mention at all of possible publication of the papers now returned is made. One leaf inadvertently kept by Huygens is sent back to Leibniz a few days later.[88]

Leibniz undoubtedly understood very well what Huygens wanted to express. Huygens' demand for a general method of solving equations, or at least for the general quintic to be conquered, he himself had not, despite his best intentions, accomplished and it would have been ludicrous for him to indulge in any self-deception over this present failure. He strongly felt the inadequacy of his past algebraic studies – a feeling aggravated after he personally met Tschirnhaus who evidently had a far deeper knowledge than he himself in this field – so Leibniz abandoned the unrewarding task of seeking the solution of higher general equations and turned again

---

[87] Huygens–Leibniz, 30 Sept. 1675 (*HO* 7: 504–6).
[88] Huygens–Leibniz, 3 (?) Oct. 1675 (*C* 1104).

with renewed vigour to analysis.[89] And now the insights gained in the realm of algebra were to bear fruit to an extraordinary degree; for without them the Calculus could not have been invented.

Yet this repulse by Huygens – even though their previous close relationships had for some time been loosening – was not at all pleasant. Leibniz' negotiations with Hanover about future employment there had now come to a standstill[90] and so he determined to leave nothing untried to gain if possible a permanent foothold in Paris. He had till now earned a living by taking on political and legal briefs – for instance, that of securing the release of Prince Wilhelm von Fürstenberg and re-establishing peace,[91] or of arranging the divorce of Duke Christian Ludwig of Mecklenburg[92] – and had in fact thereby been able to accumulate considerable savings. Accordingly he could now think of buying a congenial post in Paris and he thought especially of taking up one of the salaried offices at the Académie des Sciences, as Huygens, Römer and Cassini had done. Colbert was the person to arrange this, but he was not to be approached directly, and Leibniz therefore tried to establish a connection by way of the Abbé Gallois, who had great influence with the Duc de Chevreuse, Colbert's son-in-law, and who was willing to support an application by Leibniz.[93] The plan soon assumed a tangible form. Through Dalencé[94] Leibniz was invited to an audience with the Duke at St Germain in order to demonstrate his calculating machine. He evidently made a very good impression there, but caught a cold, had to stay in his rooms for a few days and could not pay his intended formal visit to Gallois. Hence he wrote him a letter conveying his own suggestions on how the matter could most usefully be advanced[95] and in it spoke of his intention of setting down his recent scientific discoveries in the form of letters to well-known experts and also of publishing them should occasion arise. His case seemed to prosper when, by Roberval's death (in late October 1675) not merely a seat in the Académie des Sciences but, moreover, the mathematical

---

[89] The first allusion to 'recently achieved results regarding the treatment of a geometrical problem hitherto almost despaired of' – which doubtless refers to the symbolic representation of infinitesimal procedures – occurs in Leibniz–Oldenburg, 18 (28) Dec. 1675 (*LBG*: 146).

[90] No answer had yet arrived to the letters mentioned in ch. **10**: note 3.

[91] *See* ch. **7**: note 7.

[92] For details *see* Wiedeburg II. 1: 557–603 and *compare LSB* IV. 1: 433–83.

[93] Gallois–Leibniz, 4 Sept. and 3 (?) Oct. 1675 (*C* 1049, 1057).

[94] Dalencé–Leibniz, 29 Oct. 1675 (*C* 1106).

[95] Leibniz–Gallois, 2 Nov. 1675 (*LMG* I: 177–8).

professorship at the Collège Royal and the Ramus chair became vacant. Picard, who was already a member of the Académie des Sciences, was quickly appointed to the professorship at the Collège Royal;[96] the Ramus chair was by its founder's stipulation always conferred for a period of three years at a time only, to the successful competitor in a triennial public disputation (the next one had been planned for March 1676): here Leibniz held high hopes, particularly since it so happened that Tschirnhaus – who had just met Leibniz and was fast becoming an intimate acquaintance – was mathematics tutor to Colbert's son.[97] While all this was going on the Prince Elector Lothar Friedrich von Metternich died, and his successor, Damian Hartard von der Leyen, confirmed Leibniz in his existing post as counsellor in the service of the Elector of Mainz.[98] So far Leibniz had conducted his negotiations with Hanover only halfheartedly[99] but he now saw himself facing a new situation where he could perhaps hope to be able to stay in Paris as political emissary of the ruler in Mainz and of the Duke of Hanover – but his efforts[100] here too remained without success[101] and so he had ultimately to settle for the Hanover post even though the situation there was far from tempting and the Duke insisted that he should take up his abode at his place of office as soon as possible.[102] In the meantime some ill-feeling arose in his relationship with Gallois for a trifling reason,[103] and the extremely

[96] Leibniz–Oldenburg, 18 (28) Dec. 1675 (*LBG*: 144).
[97] Schuller–Spinoza, 14 Nov. 1675 (*SO*: 220–2). Leibniz also attempted to come into direct contact with Colbert: 11 Jan. 1676 (*LSB* I. 1: 457).
[98] Damian von der Leyen–Leibniz, late 1675 (*C* 1266).
[99] Linsingen replied on 22 June 1675 (*LSB* I. 1: 500) to Leibniz' letter of late May (*see* ch. 10: note 7), advising patience on account of the Duke's absence. Hence Leibniz only wrote again to the Duke and to Linsingen on 20 Nov. (*LSB* I. 1: 501, 502); but as now Linsingen was away for some time, a court official, Kahm, took over the negotiations and asked for details: 14 Dec. (*LSB* I. 1: 502). Leibniz recapitulated the contents of his earlier letter and on 11 Jan. 1676 (*LSB* I. 1: 504–5, 505–6), addressed New Year's greetings both to the Duke and to Kahm, adding further letters a few days later (*LSB* I. 1: 506–7, 507–8). Kahm replied on 27 Jan. 1676 (*LSB* I. 1: 506) in a short letter setting out the intended form of the appointment and the composition of the Counsellors' College in Hanover.
[100] Leibniz–Damian von der Leyen, 18 Jan. 1676 (*LSB* I. 1: 398–9).
[101] M. F. Schönborn–Leibniz, 11 Feb. 1676 (*LSB* I. 1: 400–1).
[102] Kahm–Leibniz, 28 Feb. and 19 Mar. 1676 (*LSB* I. 1: 510–11, 513).
[103] Leibniz–Johann Bernoulli, 24 June 1707 (*LMG* III: 816) where Leibniz recalls that during a boring report given by Gallois on peace conditions and negotiations in the Dutch war, he could not refrain from smiling at one point, thereby giving offence. We learn from Leibniz–Johann Bernoulli, 25 May 1698 (*LMG* III: 488) that both Leibniz and La Hire were candidates for the Académie, but since only one could be elected La Hire had agitated against Leibniz – incidentally without knowing him personally. In the end neither was chosen at that time, La Hire's

touchy Abbé dropped his support for Leibniz' case. His hope of a post in Paris was thereafter brought very low; indeed, it would seem that in these changed circumstances Leibniz did not even take part in the disputation for the Ramus professorship, which was gained by Ebert,[104] a fairly insignificant personality. Rather unwillingly Johann Friedrich was persuaded to extend Leibniz' leave to stay in Paris[105] until Whitsun (24 May) 1676. Even then Leibniz did not depart, but in the summer of 1676 made one last attempt to stay in Paris, writing to that end to Huygens,[106] who had again been very ill since the end of 1675 and saw no visitors but was, according to Leibniz' latest information, now on the way to recovery. He replied, immediately before his departure for the Hague,[107] that he had used his influence with both Colbert and Gallois on Leibniz' behalf and was hoping for success; but he could not now, unfortunately, because of the unavoidable preparations for his journey, find time to speak to Leibniz personally.

Now indeed all prospect of staying in Paris had gone and Leibniz reluctantly made his own preparations to travel to Germany. Yet conditions in Hanover turned out to be better than he had originally feared they might be:[108] Leibniz was to assume charge of the Library with his salary backdated to the beginning of the year – but he must come without any more delay, or the post would be given to someone else. A short while afterwards Leibniz received from the Duke an allowance towards the cost of his journey back home,[109] and his passports; he had intended to travel by way of the Spanish Netherlands, but because of passport difficulties entailing a delay of some months this proved impossible to arrange.[110] So with the Duke again urging him to come,[111] Leibniz left Paris in October 1676 – never again to see the metropolis which he had come to love so greatly.

Such is the human background which we must know if we are fully

election came only on 26 Jan. 1678. In a letter to Duke Johann Friedrich, February 1679 (*LSB* I. 2: 124) Leibniz relates that Colbert's son-in-law, the Duc de Chevreuse, had taken him aside and roundly declared that with the Dutch Huygens and the Italian Cassini the Académie already had two salaried foreigners and desired to have no more.

[104] Leibniz–Oldenburg, 2 (12) May 1676 (*LBG*: 168).
[105] Kahm–Leibniz, 12 Apr. 1676 (*LSB* I. 1: 515).
[106] Leibniz–Huygens, mid-June 1676; *see HO* **22**: 696.
[107] Huygens–Leibniz, late June 1676 (*HO* **22**: 696). Huygens left Paris on 22 or 29 June 1676.
[108] Kahm–Leibniz, 2 July 1676 (*LSB* I. 1: 515–16).
[109] Brosseau–Leibniz, 26 July 1676 (*LSB* I. 1: 516).
[110] Brosseau–Leibniz, 13 Sept. 1676 (*LSB* I. 1: 516–17).
[111] Brosseau–Leibniz, 26 Sept. 1676 (*LSB* I. 1: 517).

to understand the increasing tension under which Leibniz lived his day-to-day life from September 1675 until a year later he made the final decision about his future – a tension which would probably have robbed an ordinary person of his power to do profitable work, but which neither made Leibniz lose his strength nor was able to crush him, on the contrary preparing him for his greatest mathematical achievement, his invention of the Calculus.

12

THE MEETING WITH TSCHIRNHAUS

We know the 'public' Leibniz rather accurately from the notes he carefully kept and preserved and the letters he wrote, and we are fairly well informed about his far-reaching plans, his political activity and his scientific studies during his time at Paris. We know far less about his private life. His correspondence with Oldenburg shows us that the young Leibniz of those years must have been a person of great charm, anything but a walking encyclopedia. Wherever it is a matter of direct contact with people he is able to achieve what he wants, and he is powerless only in the face of a cantankerous egotist like Hooke, an unstable depressive like Pell or an unintelligent social climber like Gallois. The unshakeable optimism of his philosophy is the outcome of an equable, irrepressibly optimistic view of life, grown out of a sunny, universally receptive temperament loving work for the sake of the effort it entails, yet he is neither able nor willing to remain alone, but finds his happiness in contact with minds of an equal character. Between them and his various commissions he divides his time.[1] A circle of lively young men had congregated at the Hôtel des Romains in the rue Ste Marguerite where Leibniz had his quarters in the autumn of 1674; here Leibniz got to know the Danish nobleman Walter,[2] who later transmitted two letters to Oldenburg for him. Walter went to Italy in the spring of 1676, whence he sent several exceptionally cordial reports on his travels to Leibniz from Rome;[3] unfortunately Leibniz' replies are not known. From these letters by Walter, enquiring particularly after their host Schütz and

---

[1] Leibniz–Oldenburg, 5 (15) July 1674 (*LBG*: 105).

[2] It is evident from Leibniz–Oldenburg, 6 (16) Oct. 1674 (*LBG*: 106) that Walter brought Leibniz' earlier letter (note 1); on his way back to Paris he similarly carried Oldenburg's letter to Leibniz, 8 (18) Dec. 1674 (*LBG*: 108) and one to Huygens 9 (19) Dec. 1674 (*HO* 7: 395). (In *HO* 7: 399, note 2, and 22: 363 he is erroneously identified as the theologian Michael Walther.) Walter's exchange of letters with Leibniz continued till 1683; this correspondence is still unpublished.

[3] Walter–Leibniz, late April 1676 (*C* 1410); 16 June 1676 (*C* 1443); 22 Sept. 1676 (*C* 774).

other Paris friends, we may obtain some idea of the cheerful spirit pervading the circle of Leibniz' personal friends. Rather more formal, but still suffused by the same warm mood, are two contemporary letters by Schelhammer,[4] who appears also to have met Leibniz in the Hôtel des Romains. Leibniz' strongest and most intimate personal contact, however, was that with Tschirnhaus whom he first met in the autumn of 1675.

The young nobleman from Saxony had gone to the university of Leyden and become a confirmed Cartesian; at the time Heidanus still lectured there, in company with Wittich, the mathematician Pieter van Schooten (the much younger brother of Frans van Schooten) and the physicist Volder. During his years at university Tschirnhaus appears to have acquired the deft facility in formal algebraic calculation that gave him a good start over his contemporaries. For a short time during 1672–3 Tschirnhaus served as a volunteer in the Dutch army, thereafter returning again to extended studies in mathematics and philosophy. During his stay in Amsterdam in 1673 when he probably came into contact with Raey, he also met friends of Spinoza's – in particular the physicians Ludwig Meyer and Schuller – and was introduced by them to the philosopher's doctrines. In his enthusiastic support of Descartes he was at first opposed to Spinoza's view; but he probably came to know him personally at the Hague in 1674 and thereafter carried on a friendly debate with him in a highly interesting correspondence over a long period.[5] In May 1675 Tschirnhaus[6] went to England; he probably brought as an introduction a letter (now lost) from Spinoza to Oldenburg[7] which restarted a correspondence between the two men who had been friends a long time before.

In London Tschirnhaus met with friendly reception; he continued to work on his algebraic researches, in particular the theory of equations; through a letter of introduction from Oldenburg[8] – and perhaps one from Pieter van Schooten also – he came to meet Wallis, who in conversation talked at length with him on various mathe-

---

[4] Schelhammer–Leibniz, early September 1674 (*C* 762); 20 Oct. 1674 (*C* 774₂). Leibniz remained in touch with this excellent medical man, since 1694 a professor at Kiel university, known also as an author, until his death in 1716. Schelhammer's work *De auditu* (1684) is mentioned repeatedly (for instance *LMG* IV: 400 and 243). For Leibniz' efforts concerning the publication of 'Novæ institutiones medicæ' *see LD* v: 338.
[5] For this *compare* the items from the correspondence with Schuller and Tschirnhaus in *SO*.   [6] Collins–Gregory, 3 (13) Aug. 1675 (*GT*: 315).
[7] *Compare SO*: 201–2.
[8] Oldenburg–Wallis, mid-July 1675, *see WO* II: 471.

matical topics:[9] for instance, the determination of quadrable segments of the common cycloid. Tschirnhaus reported on his own recent algebraic results in a spirited manner and made a deep impression on the London mathematicians – so strong indeed that Pell refused to converse with him on scientific questions.[10] He talked with Boyle and Oldenburg about Spinoza and succeeded in overcoming the prejudice which existed in England against the *Tractatus theologico-politicus* as being the work of an 'abominable atheist'.[11] On 30 July (9 August) 1675, immediately before his departure for France, Tschirnhaus met Collins and discussed algebraic topics with him. We have detailed information about this in the correspondence with Gregory,[12] from which the following picture emerges.[13]

Tschirnhaus is an ardent admirer of Cartesian methods; in Paris he intends to write a large work on algebra and geometrical loci, a first draft of which Collins has seen; in it Hudde's rules on reducing and simplifying equations are to be explained. He is occupied in applying Descartes' tangent-method to both geometrical and mechanical curves. He further claims to possess new methods for the quadrature and rectification of curves; to have enlarged the theory of geometrical constructions in several basic ways; and finally to have recently discovered a method of solving all equations, which supersedes Hudde's rules for factorizing equations. Thus, says Collins, we are to acknowledge Tschirnhaus, together with Gregory and Newton, as one of the most important algebraists in all Europe. However, in his admiration for Descartes he would seem to be going a bit far when in all seriousness he claims that, for instance, the new contributions by Sluse and Barrow to the solution of equations and their whole theory of quadratures, rectifications, determinations of centres of gravity and so on are nothing but deductions from Descartes' method. This, comments Collins, is about as grotesque as if one were to assert that Aristotle's remark about the stars being visible in the sky in daytime from the bottom of a deep well[14] embraced Galileo's discoveries with the telescope.

[9] According to *WO* II: 471–2 this conversation took place on 7 (17) July 1675; *compare* also Wallis–Collins, 11 (21) Sept. 1676 (*CR* II: 591).
[10] Collins–Gregory, 21 Sept. (1 Oct.) 1675 (*GT*: 332).
[11] Oldenburg–Spinoza, 8 (18) June 1675 (*SO*: 201–2); Schuller–Spinoza, 25 July 1675 (*ibid.* 203–4).
[12] The date of the conversation is clear from Collins–Gregory, 10 (20) Aug. 1675 (*GT*: 321).
[13] Collins–Gregory, 3 (13) Aug. 1675 (*GT*: 315–16).
[14] *De generatione animalium* v. 1: 780b, 21–2; *compare* Landmann–Fleckenstein (1943).

166

As an example, Tschirnhaus demonstrated – so we are told – how to treat the equation

$$x^4 - px^3 + qx^2 - rx + s = 0$$

when $p^2 s = r^2$; in that case one will obtain

$$x = \frac{p}{4} \pm \sqrt{\left(\frac{p^2}{16} - \frac{q}{4} + \frac{r}{2p}\right)} \pm \sqrt{\left(\frac{p^2}{8} - \frac{q}{4} - \frac{r}{2p} \pm \frac{p}{2}\sqrt{\left[\frac{p^2}{16} - \frac{q}{4} + \frac{r}{2p}\right]}\right)}.$$

Tschirnhaus would not divulge his method, but merely asserted that the example

$$x^4 - 2ax^3 + (2a^2 - c^2)x^2 - 2a^3x + a^4 \ [\equiv (x^2 - ax + a^2)^2 - (a^2 + c^2)x^2] = 0$$

in the *Geometria*[15] was of this type, and that the equation

$$x^4 - 2x^2 + 12x - 18 \ [\equiv [x^2 + (x-3)\sqrt{2}][x^2 - (x-3)\sqrt{2}]] = 0$$

which cannot be solved in real numbers by Descartes' general rule could be treated similarly. Collins determined at once that $p^2 s = r^2$ is not a sufficient general condition for the equation to split up;[16] it is not, for instance, satisfied in the case of

$$x^4 - 8x^3 + 18x^2 - 11x + 2 \equiv (x^2 - 3x + 1)(x^2 - 5x + 2) = 0.$$

Gregory confirmed this by a general calculation[17] and added that evidently Tschirnhaus only employs special methods. This is indeed true, for we readily deduce that Tschirnhaus started with the re-arranged expression

$$\left(x^2 - \frac{px}{2} + \frac{r}{p}\right)^2 = \left(\frac{p^2}{4} - q + \frac{2r}{p}\right)x^2 + \frac{r^2}{p^2} - s$$

which yields the pair of quadratic factors

$$x^2 - \frac{px}{2} + \frac{r}{p} \pm \sqrt{\left(\frac{p^2}{4} - q + \frac{2r}{p}\right)} x$$

when $p^2 s = r^2$. We also hear that Tschirnhaus has deposited a manuscript with Oldenburg in which he resolves all equations whose roots are in arithmetical progression.[18] Let the equation be

$$x^n + px^{n-1} + qx^{n-2} + \ldots = 0,$$

[15] The same equation with its solutions also appears in Tschirnhaus–Gent, early August 1675 (MS: Amsterdam, Wisk. G. 49d).
[16] *Geometria* (1659): 82–4; the reference is to a *neusis*-problem out of Pappus, mentioned on p. 42 above; it also occurs in Tschirnhaus–Gent, early August 1675 (MS: note 15). [17] Gregory–Collins, 20 (30) Aug. 1675 (*GT*: 324–5).
[18] Collins for Gregory, 3 (13) Aug. 1675 (*GT*: 318–19).

then the n solutions would be found to be

$$x_k = -p/n - (\tfrac{1}{2}[n+1] - k)\sqrt{w},$$

where $\quad w = \left(\dfrac{n-1}{2n}p^2 - q\right)\bigg/\left(\dfrac{n[n^2-1]}{24}\right), \quad k = 1, 2, ..., n,$

observing that $\quad \Sigma x_k = -p, \quad \underset{i \neq k}{\Sigma} x_i x_k = q.$

The copy we have is distorted by many mistranscriptions and contains the solutions only for the cases from 2 to 9 (distinguished into even and odd n), with the added remark that the procedure is general. But it is manifest that Tschirnhaus has understood the matter fully. Gregory sees nothing very remarkable in it,[19] because it is easy to write down the conditions between the coefficients for the roots to be in arithmetical progression. In the case of the cubic equation $x^3 + px^2 + qx + r = 0$, for instance, the condition to be satisfied is found to be $27r = 9pq - 2p^3$.

To find extreme values of polynomials of the form

$$f(x) = a_0 x^n + a_1 x^{n-1} + ... + a_n$$

Hudde had developed a rule which amounts to equating the derivative

$$f'(x) \equiv na_0 x^{n-1} + (n-1)a_1 x^{n-2} + ... + a_{n-1}$$

to zero; whence if x_0 is a common root of $f(x) = 0$ and $f'(x) = 0$, then $f(x) = 0$ has a double root $x = x_0$. Hudde expresses this slightly differently:[20] for a double root the original equation and the sum which arises when its terms are multiplied by an arbitrary arithmetical progression $pf(x) + kxf'(x)$ must be equal to zero; for a triple root the last expression must again be multiplied term by term by an arithmetical progression and the sum equated to zero, and so on. Tschirnhaus maintains[21] that Hudde did not fully comprehend the matter; otherwise he would have pointed out that in the case of a triple root the sum of the terms multiplied by a progression of triangular numbers also vanishes, and correspondingly for a quadruple root the sum multiplied by pyramidal numbers also vanishes, and so forth. We see from this that, along with $f_1(x) \equiv f(x) = 0$ and

$$f_2(x) \equiv nf(x) - xf'(x) \equiv 0.a_0 x^n + 1.a_1 x^{n-1} + ... + n.a_n = 0,$$

[19] Gregory–Collins, 20 (30) Aug. 1675 (*GT*: 325).
[20] *Geometria* (1659): 507–16.
[21] Collins–Gregory, 3 (13) Aug. 1675 (*GT*: 315).

he also employed the conditions

$$f_3(x) \equiv \binom{n+1}{2} f(x) - \binom{n}{1} x f'(x) + \tfrac{1}{2} x^2 f''(x)$$

$$\equiv 0 . a_0 x^n + 1 . a_1 x^{n-1} + 3 . a_2 x^{n-2} + \ldots + \binom{n+1}{2} a_n = 0,$$

$$f_4(x) \equiv \binom{n+2}{3} f(x) - \binom{n+1}{2} x f'(x) + \tfrac{1}{2} n x^2 f''(x) - \tfrac{1}{6} x^3 f'''(x)$$

$$\equiv 0 . a_0 x^n + 1 . a_1 x^{n-1} + 4 . a_2 x^{n-2} + \ldots + \binom{n+2}{3} a_n = 0,$$

and so on. Gregory prefers Hudde's procedure to that of Tschirnhaus.[22]

Continuing his report to Gregory,[23] Collins produces his own objection that a biquadratic is easily solvable by square roots only when it has two pairs of opposite roots. Tschirnhaus had replied that his method could be applied also to equations with four arbitrary roots: he possessed a wealth of auxiliary procedures but intended not so much to exhibit his general rule as produce some easily intelligible special cases. Thus, where the additional condition $p^3 + 8r = 4pq$ is fufilled, then

$$x = \frac{p}{4} \pm \sqrt{\left(\frac{p^2}{16} - \frac{r}{p} \pm \sqrt{\left[\frac{r^2}{p^2} - s\right]}\right)}.$$

To this class belong biquadratics whose roots are in arithmetical progression; those, again, which arise as the product of two quadratic factors with the same middle term; furthermore, the equation which resolves the problem of inscribing in a circle-quadrant

$$y^2 = x(2c - x), \quad 0 \leqslant x \leqslant c,$$

a rectangle of area ab (as may be seen on setting $(c - x)y = ab$); and, finally, the equation $x^4 - ux^2 + v = 0$ with roots

$$x = \pm \sqrt{\left(\frac{u}{4} - \sqrt{\frac{v}{4}}\right)} \pm \sqrt{\left(\frac{u}{4} + \sqrt{\frac{v}{4}}\right)}$$

all belong to this class. Gregory[24] sees in all this merely an application of a well known rule for eliminating simultaneously the second and fourth terms which we may readily deduce from rearranging

$$x^4 - px^3 + qx^2 - rx + s = 0$$

[22] Gregory–Collins, 20 (30) Aug. 1675 (*GT*: 325).
[23] Collins–Gregory, 10 (20) Aug. 1675 (*GT*: 321); contained also in Tschirnhaus–Gent, early August 1675 (MS: note 15).
[24] Gregory–Collins, 20 (30) Aug. 1675 (*GT*: 325).

as $\quad [x(x-\tfrac{1}{2}p)]^2+2rx(x-\tfrac{1}{2}p)/p+s \ = \ x^2(p^2/4+2r/p-q).$

Collins ends his account[25] with a report of a construction for cubics which, perhaps because he was in a hurry, he seems not to have fully understood: at an end-point of the diameter $\tfrac{1}{2}t\sqrt{(2+\sqrt{5})}$ of a semi-circle erect a perpendicular of length $t/4$, join its end-point to the centre of the circle, and from the other end-point of the diameter project the point of intersection back on to the perpendicular; in this way we obtain the x required. (We have here, in fact, to do with a biquadratic: from

$$x:\tfrac{1}{2}t\sqrt{(2+\sqrt{5})} \ = \ \tfrac{1}{4}t:[\tfrac{1}{4}t\sqrt{(2+\sqrt{5})}+\tfrac{1}{4}t\sqrt{(3+\sqrt{5})}]$$

there follows $\qquad x \ = \ t/2[1+\sqrt{(\sqrt{5}-1)}],$

that is, $\qquad t^4-8xt^3+32x^2t^2-64x^3t-16x^4 \ = \ 0\,;$

whence, if we now take x as given, and t unknown, the condition $p^3+8r \ = \ 4pq$ is satisfied, and thus the way to the solution by means of the factorization into quadratic pairs $t^2-4xt+(8\pm4\sqrt{5})x^2$ indicated.)

Tschirnhaus left on Collins the impression that these instances represented only a very simple application of his general rule, and that in other cases things would prove to be far more complicated: from all he was told Collins was led to conclude that Tschirnhaus really did possess comprehensive methods for the solution of higher equations. But in his reply to Collins' letter, which contains a number of interesting indications of his own general method for the rational solution of equations, Gregory speaks rather deprecatingly about what has been passed on to him:[26] he cannot think very highly in scientific matters of a person who so grossly overvalued Descartes; whatever had been said was entirely confined to special cases; and he failed to see any connection between his own recent research and the results of Tschirnhaus.

In the meantime Collins had somewhat recovered from his first surprise contact with the German nobleman and now had time for more sober second thoughts about all he had heard. He had, after all, not been told anything tangible about the new general method other than that in London Tschirnhaus had been working on the rules for

[25] Collins–Gregory, 10 (20) Aug. 1675 (*GT*: 322); contained also in Tschirnhaus–Gent, early August 1675 (MS: note 15).
[26] Gregory–Collins, 20 (30) Aug. 1675 (*GT*: 325).

THE MEETING WITH TSCHIRNHAUS

solving equations up to the eighth degree;[27] now about to depart and
with his papers all packed, he had not even shown him his new
method for solving cubics, which involved cubic roots of binomials
but in a way different from Cardan's.[28] This was presently to be
given to the Dane Georg Mohr, who not long afterwards became a
close friend of Tschirnhaus. It is evident from Tschirnhaus' communications to van Gent at this time[29] that he started with the solutions
and from there went back to form an equation. We find for instance
substitutions like $x = \sqrt[p]{[a + \sqrt[q]{b}]}$ (where p and q are simple integers),
or $x = a + \sqrt{[b + \sqrt{(c + \ldots)}]}$ which later recur in conversation and
collaboration with Leibniz. By some unlucky chance the symmetrical
transformation $x = \sqrt{a} + \sqrt{b} + \sqrt{c}$ was not followed through in a full
evaluation; it would have led to the equation

$$x^4 - 2(a + b + c)x^2 - 8\sqrt{(abc)}\, x + (a + b + c)^2 - 4(bc + ca + ab) = 0$$

and so would have permitted immediately to construct the cubic
resolvent $(t - a)(t - b)(t - c) = 0$ from the reduced quartic.[30] His
crucial meeting with Collins occurred on 9 August 1675; the whole
tenor of Collins' account makes it appear certain that he had not
been in contact with Tschirnhaus before that date. This is an important point to remember, because in the early summer of 1676 there
took place an interesting, still unpublished exchange of letters on the
Cartesian method between Oldenburg (briefed by Collins) and
Tschirnhaus which the *Commercium epistolicum* later erroneously
antedated by a whole year.[31] On this occasion Tschirnhaus was to
receive information about English infinitesimal methods and, in
particular, Newton's tangent-rule. If 1675 were the correct date, then
indeed Tschirnhaus on coming to know Leibniz could possibly have
transmitted to him significant details of Newton's method. In
addition – though Collins says nothing of this to Gregory – Oldenburg reports[32] that Gregory's rectification of a circular arc (probably
an excerpt from Gregory[33] which we shall take up later on) had been

---

[27] Collins–Gregory, 21 Sept. (1 Oct.) 1675 (*GT*: 332).
[28] Collins–Gregory, 4 (14) Sept. 1675 (*GT*: 327).
[29] Tschirnhaus–Gent, early August 1675 (MS: note 15).
[30] This is Euler's approach in his *Algebra* (1770) = *Opera* I. 1 (1911): 309.
[31] We find there a Latin excerpt from the draft letter Collins–Oldenburg for Tschirnhaus, late May 1676 (*CE (1712)*: 43) and a reference to Tschirnhaus' reply of
mid-June 1676 (*CE (1712)*: 43). Oldenburg has noted the date of arrival on the
manuscript as '8 June 1675' instead of 1676; it was incorporated in *CE (1722)*: 97.
[32] Oldenburg–Leibniz, 30 Sept. (10 Oct.) 1675 (*LBG*: 142).
[33] Gregory–Collins, 15 (25) Feb. 1668/9 (*GT*: 68–9); *see* pp. 205–6 and p. 215. This
concerns a notable, if fallacious, attempted approximation for a circular arc by

171

given to Tschirnhaus and Leibniz might have obtained the details from him. In fact the document in question (of which we possess only Collins' draft) was composed not in 1675 but beyond doubt in 1676, for it contains a remark by Gregory on Tschirnhaus' excessive overvaluation of the Cartesian method,[34] which was written only in the autumn of 1675 and was altogether improper to be passed on to those not directly involved. Just this remark was left out when the piece was copied in an important letter to Leibniz.[35] The English draft composed by Collins is thus later in date than May 1675 and was certainly not transmitted to Tschirnhaus in its original form. All conclusions hitherto drawn from the existence of this piece are untenable since they are based on a faulty premiss. We shall return to this later. There remains the question as to whether Leibniz could have received from Tschirnhaus any essential notion of the results obtained by the English in analysis. As we have seen, of the more important English mathematicians, Tschirnhaus talked in fact only to two, Wallis and Collins. The conversation with Wallis was concerned mainly with the rational quadrature of a cycloidal zone and probably also algebraical details, and it is doubtful whether Wallis on this occasion was himself very forthcoming; furthermore, he was at that time still ignorant of the new methods that Newton and Gregory had thought out. Of the conversation with Collins we have the latter's own report. This shows us distinctly that Collins' first reaction to Tschirnhaus was one of unqualified admiration. He tried, it is true, to raise objections to Tschirnhaus' assertions, hoping to elicit thereby more exact information about his general method for resolving equations, yet on the whole it was not he who led the conversation but Tschirnhaus, and he took care to talk only of his methods of quadrature and rectification, stressing their origin in the Cartesian *Géométrie*. It was in no way in Tschirnhaus' interest to learn overmuch of the English methods since this might well in their eyes have imperilled his own claim to originality: certainly he received no more than general hints known already to Leibniz and the one item on Gregory's circle-rectification. He went back to Paris bursting with the desire to be recognized as one of Europe's leading mathematicians and believing he had no further need to sustain himself on the fruits

linear combinations of chords which had aroused Leibniz' strongest interest; *compare* ch. 6: note 15. In the surviving notes of Leibniz' and Tschirnhaus' cooperation there is no record of these approximations.
[34] Gregory–Collins, 20 (30) Aug. 1675 (*GT*: 325).
[35] Oldenburg–Leibniz and Tschirnhaus, 26 July (5 Aug.) 1676 (*LBG*: 173); *see also* ch. 15: note 57.

of anyone else's thought. We know that during his London visit he
bought Barrow's *Lectiones opticæ et geometricæ*, possibly in the
edition with the 1674 title page.[36] Tschirnhaus had studied these with
care; he is the true author of the erroneous suggestion[37] that Leibniz
received the decisive impulse for his discovery of the higher analysis
from Barrow's *Lectiones geometricæ*.[38]

Tschirnhaus seems to have left London[39] about the middle of
August 1675. He took with him a letter of recommendation from
Oldenburg[40] and further letters for Huygens,[41] also one for Leibniz;[42]
he first met Leibniz towards the end of September[43] and a few days
later he presented himself to Huygens,[44] who received him amicably
and enquired eagerly after common friends and acquaintances in the
Hague. Spinoza, too, and his *Tractatus theologico-politicus*, which
Huygens knew already,[45] were discussed, though when asked about
further writings of the Dutch philosopher Tschirnhaus replied, as
arranged, rather guardedly according to Spinoza's own wish.[46]
Tschirnhaus then knew no French and made great efforts to learn it
quickly;[47] he also taught mathematics to Colbert's son using Latin
for the purpose[48] (which was most welcome to Colbert *père* for
educational reasons). The collection and elaboration of his earlier
algebraic studies he no longer worked at; this elegant young twenty-
five-year-old who had, through his family connections, entry into

---

[36] Collins–Gregory, 3 (13) Aug. 1675 (*GT*: 315).
[37] Collins–Gregory, 19 (29) Oct. 1675 (*GT*: 343).
[38] Tschirnhaus–Leibniz, March 1679 (*LBG*: 389); *compare* ch. 6: note 61.
[39] The last preserved letter from London is Tschirnhaus–Spinoza, dated 12 Aug. 1675
(*SO*: 206–7), obviously an N.S. date to make sense of Collins' remark in a letter to
Gregory 10 (20) Aug. 1675 (*GT*: 321) where he refers to Tschirnhaus' visit 'shortly
before his departure'; *compare* also the dates of letters to Huygens which Tschirnhaus was to take along.
[40] Oldenburg–Huygens, 30 July (9 Aug.) 1675 (*HO* 7: 486).
[41] Smethwick–Huygens, 30 June (10 July) 1675 (*HO* 7: 487–8); Papin–Huygens,
10 Aug. 1675 (*HO* 7: 490–1).
[42] Oldenburg–Leibniz, 30 (?) July (9 Aug.) 1675; *see LBG*: 142.
[43] We may presume that the first conversation with Leibniz of which Tschirnhaus
made his notes, 1 Oct. 1675 (*C* 1055) was preceded by a formal call where he
presented his letter of recommendation.
[44] Huygens–Smethwick, 29 Sept. (9 Oct.) 1675 (*HO* 7: 512). Evidently Huygens
already knew Tschirnhaus at that time for in a letter to Leibniz, 3 (?) Oct. 1675
(*C* 1104) he asks for his address.
[45] Schuller–Spinoza, 14 Nov. 1675 (*SO*: 220–1).
[46] Spinoza-Schuller, 18 Nov. 1675 (*SO*: 223–4).
[47] Leibniz–Oldenburg, 18 (28) Dec. 1675; we refer to a phrase in the introduction
which was crossed out by Oldenburg himself and thus does not appear in the
copies nor in printed versions.
[48] Schuller–Spinoza, 14 Nov. 1675 (*SO*: 220–1).

every salon in Paris, was presumably only too glad to be sidetracked by the pleasures of high society. Yet he had to think of the future and he would, thinking of the future, have preferred above all to find a position as mentor and travel companion to some German prince,[49] but his efforts towards this end proved fruitless. The close scientific relationship between Leibniz and Tschirnhaus is documented by the numerous notes preserved of their conversations and joint studies. At first, Tschirnhaus reported on his newest results; here occurs the remark[50] that Kersey's *Algebra* was quite valueless. The talk turning to Spinoza, Leibniz, who had exchanged letters with Spinoza in connection with a question in optics,[51] expressed interest in his *Tractatus theologico-politicus* and Tschirnhaus had to debate with himself whether to reveal further details about the excerpts he had made from Spinoza's manuscripts,[52] but in deference to the wishes of the hermit in the Hague he refrained from doing so.[53] In a further conversation in which Justel and Mathion also took part, Tschirnhaus reported amongst other things on rare books in Holland.[54] Discussion of mathematical topics only began, as far as we can judge from surviving evidence, towards the end of November 1675, that is to say at a time when Leibniz already possessed his notation for the infinitesimal symbolism. An early note from Tschirnhaus' hand[55] begins with the equation

$$y = (3a - 2x) \sqrt{(2ax - x^2)/(4a - 2x)}$$

for which the curve in Cartesian coordinates is constructed through the proportionality of certain line-segments, a curve later known as Maclaurin's trisectrix.[56] Tschirnhaus next mentions an integral transformation employed by Roberval[57] which he has seen in Gregory's *Geometriæ pars universalis*;[58] he goes on to determine the extreme

[49] Leibniz–Habbeus, 14 Feb. and 22 Mar. 1676 (*LSB* I. 1: 445 and 447–8); also Leibniz–Kahm, 14 Feb. 1676 (*LSB* I. 1: 509).
[50] *C* 1055.
[51] Leibniz–Spinoza, 5 Oct. 1671 (*LSB* II. 1: 155); Spinoza–Leibniz, 9 Nov. 1671 (*LSB* II. 1: 184–5).
[52] Schuller–Spinoza, 14 Nov. 1675 (*SO*: 220–2).
[53] Spinoza–Schuller, 18 Nov. 1675 (*SO*: 223–4).
[54] *C* 1056.   [55] *C* 1127.
[56] *Treatise of fluxions* I (1742): 261–3, fig. 124. Tschirnhaus did not notice the angle-trisection appended there.
[57] *See* also p. 57, fig. 11.
[58] Tschirnhaus in fact mentions Gregory's *Geometriæ pars universalis* where the theorem appears as prop. 11 (pp. 27–9) in his letter to Gent of early August 1675, referring in particular to prop. 69 (pp. 128–30) for some deductions on circle chords, of which there is no note in the papers from his time in Paris. For the

value of $ax - x^2$ by Fermat's method,[59] and ends by finding the sub-tangent for the curve $y^3 = x^3 + ax^2 + bx + c$.[60] To this sheet is attached another probably used to write further notes during the same conversation.[61] Here we get the subnormal of a common parabola, almost exactly in the manner described in Schooten's commentary.[62] On the remaining parts of these two sheets Leibniz himself wrote down a few details of infinitesimal problems arising in differential geometry. On the second sheet we find, for instance,[63] the rule for differentiating a product in the form $d\bar{x}y \, \Pi \, d\bar{x}\bar{y} - xd\bar{y}$ with the comment that this is a remarkable theorem, valid 'for all curves'.

Further calculations and figures,[64] only roughly jotted down by Tschirnhaus, deal with quadratic and biquadratic equations, with the geometrical interpretation of the solutions – perhaps suggested by De Beaune[65] – and with the imaginary. On this topic we find in a piece of work by Leibniz[66] the interpretation of $\sqrt{(-x^2)}$ as the ordinate of length x, understood as geometric mean of $-x$ and $+x$ on the axis of x; Tschirnhaus is expressly named as the inventor,

subject matter, see Hofmann (1971): 100–3. Roberval's theorem is expounded in the letter to Gent, 19 Dec. 1675 (MS: note 15; 49g). Tschirnhaus (in C 1127) draws three diagrams in one of which the original curve resembles a parabola $y^2 = 2px$, in another a circle-quadrant $y^2 = 2ax - x^2$, $0 \leqslant x \leqslant a$; there are no calculations. Tschirnhaus here comes very near to what Leibniz had done shortly before (in C 1076) in extension of his transmutation theorem; for the contents of this manuscript see Hofmann (1970): 97–101. The coordinated curve is called *curva sylloga* by Tschirnhaus as we know from Leibniz' notes in C 1208 where the curve however is wrongly drawn so that Leibniz misses the connection with C 1076.

[59] Tschirnhaus became acquainted with the method for finding extreme values (he never mentions Fermat's name in the context) either during his studies at Leyden or by his reading of the *Supplementum* to Hérigone's *Cursus* (1642, ₂1644): prop. 26. Of course an example also comes in the lengthy insert into the first version (1649) of Schooten's commentary in the *Geometria* (1659): 253–64, particularly 263, the example namely of finding the extremes of $x^2(a-x)$ by determining

$$x^2(a-x) - y^2(a-y) \equiv (x-y)[a(x+y) - (x^2+xy+y^2)] = 0$$

when $y = x$. The author of this method is again Fermat, see *FO* I: 149.

[60] The result is wrong, presumably written down from memory. The procedure already occurs in the letter to Oldenburg, 6 (16) July 1675 (MS: note 15; 49a), and consists in determining not the subtangent t as usual from the equation of the curve $f(x, y) = 0$, but of $u = t - x$. Tschirnhaus has explained this method by further examples in the *AE* (December 1682): 391–3. Responding to this opening Leibniz at once demonstrated his own procedure in the case of the equation of a general conic, using Sluse's tangent-rule. For this see p. 73.

[61] C 1132.

[62] *Geometria* (1659): 246; 246–7 (addition to first version of 1649).

[63] C 1131.

[64] C 1135–7.

[65] 'Notæ breves' in *Geometria* (1659): 112–14 as explanation to the text pp. 6–7.

[66] C 1144.

though we may doubt if it was an original idea of his and did not have its source in a conversation with Wallis[67] in whose *Algebra*, largely complete in manuscript by 1676, the same remark occurs.[68] Studies in the solution of equations loom large in further conversations between Leibniz and Tschirnhaus, concerning at first cubics,[69] biquadratics[70] and quintics.[71] We find here again Tschirnhaus' formulae which obviously are contrived by working backwards from solutions. But there is nothing here about attempts at solving the general equation of nth degree. The (fallacious) method which Tschirnhaus published in 1683[72] is of later origin than this discussion, dating from just before his departure for the South of France and Italy.[73] According to Oldenburg's account[74] Tschirnhaus had received from Collins some remarks on rectification of circle arcs, but nothing of this being preserved in our manuscripts, we can only point

[67] *Compare* pp. 165–6.
[68] *Algebra* (1685): 764–73; also *WO* II: 286–95.
[69] Only one of several recorded attempts, *C* 1161, is successful where using

$$(x-2)(x-3)(x-5) \equiv x^3 - 10x^2 + 31x - 30 = 0$$

as a pattern, Tschirnhaus solves the cubic

$$x^3 - px^2 + qx - r = 0$$

after reduction by $x = y + p/3$ in the usual way by Cardan's formula. Abbreviating the constituent parts of x by $a = p/3$, $b = \sqrt[3]{[M + \sqrt{N}]}$, $c = \sqrt[3]{[M - \sqrt{N}]}$ he can write the three solutions as

$$a+b+c, \quad a - \tfrac{1}{2}(b+c) \pm \tfrac{1}{2}\sqrt{(-3b^2 + 6bc - 3c^2)}$$

(where a slip of the pen is corrected in *C* 1413); Leibniz has added the simplified expression $\frac{1}{2}(c-b)\sqrt{(-1)}$ (where $\sqrt{(-3)}$ is intended) for this last root term.
[70] In *C* 1296 those equations only are considered which are lacking the terms in x^3 and x; they are tackled in *C* 1198–9 by setting $x = \sqrt{a} + \sqrt{b}$; also the special cases of biquadratics referred to above on p. 169 are here dealt with.
[71] In *C* 1296 Tschirnhaus tentatively sets $x = a + b + c$ for the equation

$$x^5 + px^3 + qx^2 + rx + s = 0$$

hoping to eliminate thereby the terms in x^3 and x^2 but he does not pursue the matter.
[72] *AE* (May 1683): 204–7.
[73] This dating is indicated by Tschirnhaus in a letter to Leibniz, 17 Apr. 1677 (*LBG*: 332–3) where he refers to a preceding letter evidently now lost, which must have been despatched on 20 Nov. at the latest, immediately before Tschirnhaus' departure for the South of France. Leibniz refers to Tschirnhaus' proposed solution on 18 Sept. 1679 (*HO* 8: 215) without actually communicating its basic idea which Tschirnhaus had shown him in the instance of the cubic. In Tschirnhaus' published account in *AE* (May 1683): 204–7 again this example alone is worked out. Leibniz had remarked early in 1680 (*LBG*: 402–3) that this gave a way for transforming equations without solving them, and this he believed to be able to demonstrate already for the quintic case; he expresses himself similarly again in a note of summer 1683 (*LBG*: 315–16).
[74] *See* note 32 above.

in this connection to a joint memorandum where areas of inscribed regular polygons of 4.2^n sides are calculated step-by-step from the relation $2r_{2n}/R = \sqrt{(2 + 2r_n/R)}$.[75] The joint efforts to find partial sums of the harmonic progression proved unsuccessful;[76] Leibniz had not been satisfied[77] with Collins' report on this subject[78] though he conceded the value of a near approximation in default of anything better.[79] More interesting is perhaps Leibniz' attempt, albeit abortive, to reach a law for the distribution of primes by focusing attention on the gaps between primes in the natural sequence of numbers; the only result of this search was a pretty geometrical illustration of the sieve-process.[80] There is also evidence to show that during these discussions Leibniz told Tschirnhaus of his results on sums of reciprocal figurate numbers[81] and the technique of difference arrays;[82] but whether he also heard of the discovery by the (unknown) ingenious pupil of

[75] C 1362 and 1411 obviously influenced by Viète, *Variorum...liber* VIII (1593): ch. 18 = *Opera* (1646): 398–400. Leibniz' remark in the 'Bisectio laterum', C 1138 on Tschirnhaus' treatment of angular sections (*see* also ch. 11: note 17) perhaps refers to this, or to the instrument invented for the purpose by Tschirnhaus and mentioned in his letter to Gent, 6 Nov. 1675 (MS: note 15; 49f), *see* Hofmann (1971): 105.

[76] For instance in the notes written down on C 1310, 1340 and 1513.

[77] Leibniz–Oldenburg, 10 (20) May 1675 (*LBG*: 123).

[78] Oldenburg–Leibniz, 12 (22) Apr. 1675 (*LBG*: 121–2); *compare* p. 138.

[79] At a later stage Leibniz believed to have found an exact formula for partial sums of the harmonic progression as is evidenced in *AE* (February 1682): 45 (*LMG* V: 121). To Jakob Bernoulli's question, 19 Oct. 1695 (*LMG* III: 22) he replied, 12 Dec. 1695 (*LMG* III: 23–7) merely by indicating his line of thought which is but a variant of Collins' information of 1675. When Johann Bernoulli, 22 Sept. 1696 (*LMG* III: 327–8) drew his attention to a note made by Huygens in his personal copy of the *AE* (1682) (*HO* 22: 787) to the effect that 'he knew of no such formula', Leibniz acknowledged in his letters 16 Oct. and 16 Nov. 1696 (*LMG* III: 331; 336), that he had deceived himself.

[80] C 1310 and similarly in LH 35. IV: 17; *see* also Mahnke (1913): 29–61. In the paper published in *JS* No. 7 of 28 Feb. 1678 (*LMG* VII: 119–20) Leibniz only remarked that every prime greater than 3 is of the form $6n \pm 1$. Amongst his examples occurs the composite number $510\,511 = 6 \times 85\,085 + 1$ (which factorizes into

$$19 \times 97 \times 277).$$

Since $510510 = 2 \times 3 \times 5 \times 7 \times 11 \times 13 \times 17$ this faulty example shows that Leibniz seems to have believed at the time that the number $p_1 \times p_2 \ldots \times p_n + 1$, where the p_k are the first n primes, is itself a prime. (In fact

$$2 \times 3 \times 5 \times 7 \times 11 \times 13 + 1 = 30031 = 59 \times 509$$

represents the smallest counter-example to Leibniz' presumed rule, which had been published already in 1599 in P. Bongo, *Numerorum mysteria*).

[81] In C 1181 Tschirnhaus tries, without success, to generalize the scheme mentioned on p. 81.

[82] In C 1340 we find the beginning of the difference array for the sequence of $n!$.

Dechales concerning Σn^{-k} we cannot now tell.[83] Tschirnhaus also then got to know the basic idea of Leibniz' circle quadrature,[84] later indeed he was shown the full manuscript of the 'Quadratura circuli' before it went to the printer.[85]

An analysis of these notes of Leibniz' suggests that Tschirnhaus, having made Leibniz' acquaintance towards the end of September, probably had two conversations with him, on 1 and 3 October 1675, mainly general in tone, and came only about 26 November 1675 to discuss mathematical topics with him in any detail. These referred to Tschirnhaus' own independent results and the use of imaginary quantities in the solution of geometrical problems whose algebraical solution Tschirnhaus produced. We have no indication at all that during these weeks Leibniz learnt from him anything significant about the latest work of the English mathematicians. It is certain only that Tschirnhaus, at the behest of Oldenburg and Collins, enquired after the whereabouts and possible publication of the writings of several French mathematicians – notably Desargues' 'Leçons des tenebres', Lalouvère's and Fermat's remains, Roberval's papers and especially Pascal's manuscripts – to which Leibniz' attention also had earlier been drawn.[86] Hardly anything by Roberval

[83] Leibniz has re-written what he was told in C 1210: where

$$A = \sum_{n=1}^{\infty} 1/(2n-1)^k, \quad B = \sum_{n=1}^{\infty} 1/(2n)^k,$$

then $\quad C = \sum_{n=1}^{\infty} 1/n^k = A+B = 2^k.B,$ and so $\quad A-B = (2^k-2).C/2^k.$

Leibniz shows this explicitly for $k = 2$ and indicates the procedure for $k = 3$. Dechales' anonymous disciple had remarked that for $k = 2$ the value of $A - B$ lies between 0·820 and 0·821: more correctly we find in fact 0·822 and 0·823 for these bounds. Twenty years later Leibniz recalled in a letter to L'Hospital, March 1695 (*LMG* II: 276–7) that during his time in Paris he had met an extraordinarily gifted young mathematician whose name he had now forgotten; remembering only that he came from Lyons and intended to enter a business career. In his reply, 25 Apr. 1695 (*LMG* II: 280), L'Hospital had no information to give; Dechales had died long ago, in 1678. The unknown young man is again mentioned in Leibniz–Johann Bernoulli, 29 Jan. 1697 (*LMG* III: 362–3).
[84] In C 1471 Tschirnhaus checks the expansion of $ay^2/(a^2+y^2)$ and draws some related diagrams.
[85] C 1232 (the preface: *LMG* V: 93–8) and 1233; for the excerpt available in print *see* ch. 6: note 58. The manuscript has never fully been published in printed form; on its fate *see* ch. 17: note 78. Tschirnhaus reports from Rome, 27 Jan. 1678 (*LBG*: 341, 350) that he had informed the mathematicians there of Leibniz' circle quadrature and had found them highly interested in it.
[86] Oldenburg's requests of 24 June (4 July) 1675 (*LBG*: 130). Collins had enumerated the titles of several unpublished writings by Lalouvère and Fermat in a letter to Gregory, 14 (24) Mar. 1671/2 (*GT*: 225).

had then been printed, but some of his treatises and lectures were circulated in manuscript amongst his friends. Collins had wished Leibniz to secure copies[87] but was informed that this would be unnecessary since Roberval intended to publish his papers himself;[88] however, in the midst of his preparations for such an edition Roberval had died leaving, so we are told, what was presumably the most important portion of his papers to the Académie Royale des Sciences, naming Blondel, Buot and Picard as his executors.[89] Leibniz was able to inspect his 'Elementa geometrica' and also communicated the manuscript to Tschirnhaus: both expressed the opinion that it would be hardly worth while to have this single piece printed.[90]

As for Pascal, Leibniz had already in the summer of 1674 written that he was expecting a report from his heir, Périer, on his manuscripts.[91] At last, almost a whole year later, he received from Étienne's two younger brothers, who at that time were being taught by Arnauld and Nicole in Paris, a number of Pascal's papers,[92] hardly anything more than fragments,[93] amongst which were unfinished studies in elementary geometry – these later stimulated Arnauld to write his famous *Nouveaux elemens de geometrie* – and an address to Montmor's Paris Académie. This last had been designed to appear at the head of his tract on the *Triangle arithmétique* which, though printed in 1654, was only collectively published in 1665. Also a fragment from an 'Introduction to geometry' was in the possession of the Abbé des Billettes, where Leibniz saw it and made excerpts from it.[94] Leibniz copied the whole of the address to the Paris Académie in which Pascal gives a survey of his already completed and also his projected mathe-

---

[87] Oldenburg–Leibniz, 24 June (4 July) 1675 (*LBG*: 130).
[88] Leibniz–Oldenburg, 2 (12) July 1675 (*LBG*: 132).
[89] Roberval's papers were prepared for the press in 1686–8 by La Hire, but were published only in 1693.
[90] Leibniz–Oldenburg, 18 (28) Dec. 1675 (*LBG*: 144). Leibniz later mentions the still rather imperfect axiomatic considerations contained in this letter; they have not hitherto appeared in print; according to Auger (1926): 138, note 1, they survive in the Bibliothèque Nationale, MS fds. fr. 9119–20. *Compare* Leibniz' critical remarks on Descartes' *Principia*: *LPG* IV: 355 and the references in the 'Nouveaux essais' (1703–5) I. 3 §24 and IV. 7 §10 (*LSB* VI. 6: 107–8 and 414).
[91] *Compare* ch. 7: note 5, and p. 79.
[92] The receipt of such papers from the Périer brothers is acknowledged on 4 June 1675 (*C* 978).
[93] Shortly before he received a first portion of Pascal's papers Leibniz writes to Oldenburg, 10 (20) May 1675 (*LBG*: 123) his impression from a conversation with Étienne Périer in mid-June 1674 (*compare* ch. 7: note 2) that nothing but fragments have survived. Evidently this was true only for the fragments concerning elementary geometry which Leibniz later mentions in the *AE* (November 1684): 537–42 (*LPG* IV: 426).   [94] *C* 1501 = *PO* IX: 291–4.

matical writings.[95] The fragments on conics which were of particular interest to Collins[96] had then not yet come into Leibniz' hands; they were probably communicated to him a few months later in January 1676.[97] A fairly large excerpt[98] made by Leibniz himself is preceded by two smaller ones[99] containing additions by both Leibniz and Tschirnhaus, while a note on *hyperbolæ oppositæ* (that is the two complementary branches of a hyperbola) is by Tschirnhaus alone.[100] All in all, Leibniz and Tschirnhaus were highly pleased with what they found in Pascal's papers. In the accompanying letter which Leibniz addressed to Périer when he returned them[101] – and which was perhaps intended to serve as an introduction to the publication of these manuscripts which Leibniz strongly recommended – he pronounced the treatise on conics as being in a fit state to be printed.[102] Périer presumably passed the bundle of papers on to his publisher, Desprez, but the manuscripts appear thereafter to have perished[103] and Leibniz' summary alone now allows us a unique insight into their content. Leibniz suggests beginning the printed version with a piece on the 'Generatio conisectionum', in which conics are generated from the circle by central projection; to this could be joined Pascal's disquisition on his hexagon and on the configuration (sometimes variously ascribed to Maclaurin or to Simson) of four tangents to a

---

95 *C* 1500 = *PO* III: 305–8.
96 Collins had already heard in Vernon's letter 1 (11) Mar. 1671/2 (*CR* I: 186–7) of the whereabouts of Pascal's unpublished remains and had forwarded this information to Leibniz through Oldenburg 12 (22) Apr. 1675 (*LBG*: 121). For Collins' particular interest in Pascal's *Conica* see p. 37.
97 Leibniz in the letters 2 (12) July and 18 (28) Dec. 1675 (*LBG*: 132 and 144) does not yet mention the *Conica* among Pascal's posthumous papers, only referring to them in a note of January 1676 (*C* 1292); it consists of explanatory additions by Leibniz to rough diagrams with short notes by Tschirnhaus. Pascal's interesting sketch, *PO* IX: 232–3 has been edited by Costabel (1962): 259–60, 267–8 in facsimile with French translation: it shows that Pascal arrived at his theorem on the *hexagrammum mysticum* by an easily verifiable consideration on the ellipse which he then elegantly generalized.
98 *C* 1499 (*LBG*: 134–40 = *PO* IX: 234–43).
99 *C* 1498 (*PO* IX: 227–8) and *C* 1496 (*PO* IX: 229–31); see Costabel (1962): 257–9, 260–5.
100 *C* 1377 (*PO* IX: 233); see Costabel (1962): 260, 268.
101 Leibniz–É. Périer, 30 Aug. 1676 (*LBG*: 133–5 = *PO* IX: 220–6).
102 Collins–Bernard, 17 (27) Nov. 1676 (MS: Copenhagen, Boll. Ud 1). Leibniz–Remond, 14 Mar. 1714 (*LPG* III: 612–13). Since Pascal's *Conica* are not mentioned in the correspondence with Oldenburg we may assume that Leibniz reported on them only during his second visit to London.
103 Thus Leibniz writes to Vagetius, 7 Dec. 1686 (*LD* VI: 34). Later, in letters to Fabricius, 7 July 1707 (*LD* V: 421), and to Pinsson, 14 May 1700 (*LD* V: 470), he talks of extensive excerpts out of Pascal's *Conica*, adding that he intends to publish them if an opportunity arises.

conic and their points of contact and, furthermore, his studies on the
proportions existing between chords, secants and tangents, on the
determination of conics from five given elements, points or tangents,
and on solid loci (this last had apparently been initially planned as
an independent tract). At the end of the volume should come – in
addition to his *Essay pour les coniques* of 1640 – some of the fragments,
notably one (of which a draft only was found) on the *problema
magnum*, namely, to construct the right circular cone which shall
have a given conic as its base. To publish this collection would, writes
Leibniz, be a meritorious deed; but its printing should not be much
longer delayed since a similar treatise was known to be in prepara-
tion – a probable allusion to La Hire's *Sectiones conicæ* which did
not however appear till 1685.

It seems that Tschirnhaus had already returned the papers of
Roberval and Pascal lent him by Leibniz by 17 December 1675, for
a few days later Leibniz declared in a letter to Oldenburg that to
send him a young man of such intelligence and high promise was
indeed an act of friendship;[104] Tschirnhaus, he writes, had shown him
some of his notes on geometry and analysis and these were of great
elegance; Tschirnhaus was now fully occupied in learning French and
hence unable in the foreseeable future to keep his promise of sending
any contribution to the *Philosophical Transactions*, in particular the
one he had agreed to write on the solution of equations in irrationals,
and had asked him some time ago to apologize to Oldenburg on his
behalf for his protracted silence. As desired, he had taken some
trouble to seek out the papers left by Roberval, Pascal and Frénicle,[105]
but with only partial success; Frénicle, too, had died in the autumn
of 1675, leaving behind, ready for printing, a manuscript on rational
rightangled triangles which was to be edited with additions by
Mariotte. We shall revert to this later in a different context.

In his letter Leibniz goes on to discuss, in sequel to a remark by
Boyle in his *Excellency of theology*,[106] the question of how far his
scientia de mente can be dealt with by geometrical methods.[107] In his
own view the metaphysical constructions of the Cartesians are far
too strongly based on the concept of 'idea': the difference between
matter and mind cannot be exclusively represented by separating
extension and consciousness. In proof of the eternity of mind Des-

104 Leibniz–Oldenburg, 18 (28) Dec. 1675 (*LBG*: 143); *see* also above, pp. 173–4.
105 Leibniz has written 'Fermat' instead of Frénicle.
106 Excerpts from this essay, with a few notes, are found in *C* 1172.
107 Details are given in *C* 1173 and 1175.

cartes and others had adduced a pure concept of being whose fundamental quality was existence itself; yet, says Leibniz, it is not at all certain whether what is thinkable has also real existence: of this type are such concepts as the number of all numbers, infinity, smallest, largest, most perfect, totality and other notions of that sort, which are not by their nature self-evident, and become fit to use only when clear and unambiguous criteria for their existence have been established. It all amounts to our making a truth mechanically, as it were, reliable, precise and so irrefutable: that this should at all be possible, is an all but incomprehensible sign of grace. Here Leibniz reproduces views which he had formulated in writing in discussing Spinoza's ethics with Tschirnhaus.[108] There follows a reference to one of Leibniz' favourite notions, namely his *characteristica universalis*, or conceptual script, by whose aid he hoped to reduce the processes of thought to a more or less mechanical procedure so that even the most difficult concepts like God, or Mind, might be handled as readily as numbers and diagrams. Contemporary algebra, he felt, afforded only a weak foretaste of the possibilities; the essential result of his symbolic logic would be one's inability to err even if one so wished.

This communication crossed in transit with a short letter from Oldenburg[109] – now become impatient since he was still without reply to his enquiry of 10 October 1675 – complaining that from Tschirnhaus he had as yet received no news at all: had he quite forgotten the promises he had earlier given in London? Leibniz' crossing letter remained at his request without immediate answer since he intended to be away for some weeks and was not due to return to Paris before the end of January 1676. He had in fact been invited by Abbé Gravel, with whom he had ably collaborated at Mainz, to take part in a conference at Marchienne about the neutrality of Liège, but he did not get from Hanover the political appointment he expected for this purpose.[110] It was during these highly charged opening days of 1676, with his whole future at stake, that he looked over Pascal's papers. He also frequently met Tschirnhaus and with him visited Clerselier,[111] the executor of Descartes' literary remains; together, the

108 Related studies occur in *C* 1216–17.
109 Oldenburg–Leibniz, 20 (30) Dec. 1675 (*LBG*: 143).
110 Leibniz–Duke Johann Friedrich, mid-January 1676 (*LSB* I. 1: 506–7). We can see from Leibniz–Kahm, mid-January 1676 (*LSB* I. 1: 507–8) that Leibniz did not accept Gravel's invitation as he feared it might be embarrassing to the Duke if someone in his employ had a close connection with the French diplomat.
111 Leibniz and Tschirnhaus probably called on Clerselier around 24 Feb. 1676 (*C* 1321). For Leibniz' excerpts *see C* 1322–5.

two inspected many of Descartes' then unpublished papers and made excerpts from them; Leibniz, as his notes make clear, was particularly interested in medical, meteorological and mathematical topics, and at a later date he obtained, through Tschirnhaus, copies of still other unpublished manuscripts of Descartes.[112]

We can readily understand why Leibniz did not write anything further to Oldenburg during the following weeks. About himself he had nothing to report. With a heavy heart he had accepted the office in Hanover after all his efforts to obtain a secure position in Paris had failed. Yet another bitter disappointment came when it proved impossible to remain longer in Paris even for a brief time on behalf of his new master. He was aware, it is true, that the conception of the new infinitesimal analysis was of the greatest importance for the widening of mathematical knowledge altogether, yet it all still was *in statu nascendi* and not ready to be communicated. In algebra, too, which had supplied the main topics of the preceding dialogue with London, Leibniz believed himself to have achieved a decisive break-through, but felt baffled by the immense labour of computation involved. What he learnt from Tschirnhaus about the latter's own attempts in this field appeared quite unsatisfactory to Leibniz. He clearly saw the shortcomings of the various procedures they developed and pointed them out to Tschirnhaus, so preventing his highly imaginative but insufficiently critical younger friend from passing on their current results before they might reconsider their effectiveness. Tschirnhaus was one of those people – no rarity among mathematicians – who are so strongly wrapped up in their own private world of thought that they can react creatively to suggestive outside stimuli only indirectly and after a period of incubation. This emerges very clearly from the later correspondence between Leibniz and Tschirnhaus. There we again meet with the circumstance, previously observed by us, that Tschirnhaus was during his stay in Paris strongly urged to employ the substitution $x = u + v$ in the treatment of generalized Cardan equations and certain extensions of these but, only half attending to it, rejected the method without testing its validity,[113] only to take it up again later – independently, he imagined[114] – without getting any further ahead with it than Leibniz

---

[112] The 'Recherche de la vérité' came as an enclosure with the letter Tschirnhaus–Leibniz, 16 Nov. 1676 (*LSB* II. 1: 277–86); the 'Tractatus [*sic*] de methodo' with Tschirnhaus–Leibniz, 24 Apr. 1683 (*LBG*: 446).

[113] Leibniz–Tschirnhaus, mid-May 1678 (preliminary draft) (*LBG*: 521).

[114] Tschirnhaus for Leibniz, 10 Apr. 1678 (*LBG*: 366–7); similar to Leibniz' approach in *C* 1042 though not of equal scope: *see* p. 146.

had years before. In the meantime the latter believed he had established that this substitution leads, for an equation of degree 4, to an auxiliary equation of degree 3×4, and again for an equation of degree 5 to one of degree 4×5, and thus is of no practical use.[115] The two friends had also already discussed in Paris the effectiveness of the transform $x = \sqrt[3]{u} + \sqrt[3]{v}$ for reducing the cubic[116] (and similar variants in the case of higher equations), but by now Leibniz knew that this substitution was identical with his previous one.

Far more interesting is Tschirnhaus' attitude to Leibniz' infinitesimal methods. In Paris Leibniz told him in general terms about his new calculus and showed him in detail the transmutation by which he had obtained his arithmetical circle quadrature. In their later correspondence Tschirnhaus remembered this second bit accurately, but the first not at all clearly,[117] recalling only that, whereas Leibniz had wanted to treat the theory of combinations as a separate science, he himself at best viewed it as an appendix to algebra and could not rightly understand what Leibniz meant when he said that the doctrine of combinations can be considered as a theory of forms[118] (of the symmetric functions). To a remark by Leibniz on his projected Universal Language he replied that he really could not see any use in it;[119] Leibniz thereupon wrote back discussing his *characteristica universalis* more explicitly[120] but without winning Tschirnhaus over.[121] It is during this discussion that Leibniz insisted that we must on no account be satisfied to introduce new conceptual structures by pure definition, but need to add a proof of their existence.[122]

In fact Tschirnhaus had heard in Paris of Leibniz' infinitesimal calculus, too; so, for instance, he still has the vague recollection that Leibniz had substantially extended Guldin's theorems on centres of

---

[115] Leibniz–Tschirnhaus, mid-May 1678 (preliminary draft) (*LBG*: 522).

[116] *C* 1296; *see* above p. 176.

[117] This report is already referred to in Leibniz–Tschirnhaus, early March 1678 (*LBG*: 353). Tschirnhaus immediately could only remember the transmutation theorem: *see* the letter to Leibniz, 10 Apr. 1678 (*LBG*: 358). The further text however shows that he must also have heard of the creation of a new notation, for he comments that the introduction of new symbols while useful as an abbreviation made understanding more difficult and was not in fact a proper method to use. *See* also note 125.

[118] Tschirnhaus–Leibniz, March 1679 (*LBG*: 392) referring back to Leibniz–Tschirnhaus, late May 1678 (*LBG*: 379–80).

[119] Tschirnhaus–Leibniz, 10 Apr. 1678 (*LBG*: 370). Tschirnhaus here appears to refer to a not preserved letter, dated perhaps in February 1678.

[120] Leibniz–Tschirnhaus, late May 1678 (*LBG*: 379–80).

[121] Tschirnhaus–Leibniz, March 1679 (*LBG*: 393).

[122] Leibniz–Tschirnhaus, late May 1678 (*LBG*: 381).

gravity by means of his characteristic triangle.[123] Leibniz now seeks to remind Tschirnhaus of how he had wanted thoroughly to explain all this to him, but how Tschirnhaus had seen in his new symbols nothing but useless signs which would merely serve to obscure; he had not even been willing to be shown an example.[124] How little Tschirnhaus had understood of it is shown by his remark that Leibniz' procedure was at best a *compendium methodi*, but not at all as yet the method itself.[125]

Where it was a simple matter of algorithms, Tschirnhaus felt at once in his element; thus the method of quadrature which he contrived is entirely algorithmic.[126] Though he computes with indivisibles, he thinks the introduction of infinitely narrow rectangles – considered by Leibniz as unavoidable – to be unnecessary. By pursuing Cavalieri's procedure and elaborating Cartesian lines of thought he achieved his aim. In modern terms, Tschirnhaus' method is based on the following geometrical consideration: the volume of a solid placed between the three mutually perpendicular coordinate-planes and the cylinders erected on the bases $\int_0^a y \,.\, \mathrm{d}x$ and $\int_0^a z \,.\, \mathrm{d}x$ is given by $\int_0^a yz \,.\, \mathrm{d}x$ and also by $\iint z \,.\, \mathrm{d}x \,.\, \mathrm{d}y$. Tschirnhaus uses this equivalence to evaluate by an integration by parts $\int_0^1 y^m \,.\, \mathrm{d}x$ where $y^n = x$; $\int_0^1 x^p y^q \,.\, \mathrm{d}x$ where $x + y = 1$; and $\int_0^1 x^{pq-1} y \,.\, \mathrm{d}x$ where $y^2 = 1 - x^p$. But he demonstrates only the simplest cases having integer indices explicitly and leaves the reader to do the rest by extending his particular calculations. He also transforms $\int_0^{2a} xy \,.\, \mathrm{d}x$, where $(y/a)^n = (2a - x)/x$, into $\int_0^\infty 2a^{2n+2} \,.\, \mathrm{d}y/(a^n + y^n)^2$ by integration

---

[123] Tschirnhaus–Leibniz, late November 1677 (*LBG*: 400).

[124] Leibniz–Tschirnhaus, late May 1678 (*LBG*: 375).

[125] Tschirnhaus' comments in his *Medicina mentis* (1687; ₂1695: 184) on the undesirability of newly coined concepts, technical terms or symbols are obviously directed against Leibniz. When he met Wolff at the Leipzig Easter Fair in 1705, Tschirnhaus repeated what he had written on 10 Apr. 1678 (*see* note 117), that Leibniz had invented an abbreviating procedure, not a really comprehensive method. *See* Wolff's autobiography (1841): 125–6.

[126] He communicated his first example (squaring the parabola) orally to Leibniz: *C* 1090. Further references in Tschirnhaus–Leibniz, April–May 1677 (*LBG*: 338). A more detailed presentation comes in Tschirnhaus–Leibniz, 27 Jan. 1678 (*LBG*: 341–9). See Hofmann (1941): 55–69.

by parts, again only considering the simplest cases $n = 1, 2, 3, 4$. In his reply[127] Leibniz pointed out the close relationship of these procedures and the notions underlying them to those elaborated by Grégoire de Saint-Vincent, Pascal and Fabri, and suggested that one might prove the theorems in Wallis' *Arithmetica infinitorum* similarly, or perhaps even generalize them; but he could see no connection with Descartes. Tschirnhaus now declared[128] that the procedures of Grégoire and Pascal were not as general as his own. Fabri's book had come to his knowledge only through Leibniz; in it, as he had now ascertained, everything was expressed in a more roundabout and incomplete way. He himself valued three differing methods of quadratures most: Heuraet's rectification of curves together with Barrow's applications in his *Lectiones geometricæ*; Leibniz' method of transmutation; and his own method. Thereupon Leibniz once again stressed the significance of his general method,[129] regretting deeply that Tschirnhaus, with his prejudice against symbols, had shown so little enthusiasm for the new analysis.

Though they belong to a later date we have touched upon these details here because they reveal that Tschirnhaus had not really penetrated to any fundamental level of mathematical understanding. He remained trapped on the surface and in his algorithms, not recognizing the value of creating a symbolic language truly adequate to a topic. Apart from this he was by reason of his peculiar mental make-up totally unable quickly and profitably to accept an external stimulus or to comprehend its significance. He would thus never have been capable of transmitting accurate reports of English mathematical methods to Leibniz; not even if he had in fact been told anything in detail. Leibniz could learn from an association with him, but only because he gained in him a fellow-student of an extraordinarily strong mathematical and scholarly bent, with whom he came quickly to develop an intimate understanding; but to learn anything from Tschirnhaus himself was, as matters stood in early 1676, quite impossible for Leibniz.

127 Leibniz–Tschirnhaus, early March 1678 (*LBG*: 352–4).
128 Tschirnhaus–Leibniz, 10 Apr. 1678 (*LBG*: 357–8).
129 Leibniz–Tschirnhaus, late May 1678 (*LBG*: 375); *see* note 124.

13

THE INVENTION OF THE CALCULUS

After the large group of documents sent on 12 (22) April 1675, we hear little about infinitesimal problems in the letters exchanged by Leibniz with Oldenburg. Of interest is a renewed enquiry by Leibniz the following June concerning the rectification of the ellipse and hyperbola;[1] he was told in reply that no one in England could perform this geometrically (that is, by means of a definite algebraic formula) but only by approximation[2] (that is, by means of series). At this time Leibniz still believed that he could reduce both rectifications to the quadrature of circle and hyperbola,[3] and he communicated this 'result' to Gallois as one of his most important recent discoveries.[4] We still find the same assertion in a later letter to Oldenburg.[5] Leibniz' error arose from a mistake in computation and is not one of principle. He now established the equality of the strip $\int_{x_1}^{x_2} (ab/x) \, dx$ of a hyperbola with the corresponding sector[6] – this is an old theorem to be found already in the *Opus geometricum* of Grégoire de Saint-Vincent – and measured its area by logarithms;[7] he also investigated similar curves and 'parallel' curves[8] (families of involutes). More generally, he considered into what divisions the

---

[1] Leibniz–Oldenburg, 2 (12) July 1675 (*LBG*: 132).
[2] Oldenburg–Leibniz, 30 Sept. (10 Oct.) 1675 (*LBG*: 142).
[3] *C* 1085, 1089, 1091, 1165–6, 1277, 1458–9.
[4] Leibniz–Gallois, 2 Nov. 1675 (*LMG* I: 177–8).
[5] Leibniz–Oldenburg, 21 June (1 July) and 12 (22) July 1677 (*NC* 2: 216, 232); Leibniz–Bertet (?), 1677 (*LSB* II. 1: 383). It was probably Newton himself who having seen Leibniz' observation of 12 (22) July 1677 in a copy of the letter made by Collins, first noticed the error in Leibniz' rectification and consequently urged him in the letter 16 (26) Oct. 1693 (*NC* 3: 286) to publish his assertion – implying the request for a proof. Leibniz evidently meanwhile had become uncertain, asking Johann Bernoulli in a letter, 17 June 1694 (*LMG* III: 142) to check whether the rectification of ellipse and hyperbola can really be reduced to their quadrature since he himself had no time to do it. Bernoulli replied, 12 Sept. 1694 (*LMG* III: 150), that this reduction was not feasible and thereafter Leibniz did not return to the problem.
[6] *C* 1076.
[7] *C* 1087.
[8] *C* 1071–3; 1831.

ars inveniendi is to be grouped:[9] the 'combinatory' part is to contain the general method, while the 'analytical' one shall furnish the solutions in individual cases.

Next he resumes his earlier researches into the general determination of centres of gravity,[10] pursuing them at greater depth.[11] At the outset[12] we find the transformation

$$\int_0^a xy \,.\, \mathrm{d}x = \tfrac{1}{2}a^2 b - \int_0^b \tfrac{1}{2}x^2 \,.\, \mathrm{d}y$$

where $\int_0^a x(y \,.\, \mathrm{d}x)$ is regarded as the moment of the area $\int_0^a y \,.\, \mathrm{d}x$ about the y-axis and $\int_0^b \tfrac{1}{2}x(x \,.\, \mathrm{d}y)$ as that of $\int_0^b x \,.\, \mathrm{d}y$. To symbolize integration Leibniz here, following Cavalieri, Mengoli and Fabri, uses the abbreviation '*omn.*' for *omnes*; for example, he writes '*omn.* $\overline{xy\ ad\ x}$' for $\int_0^a xy \,.\, \mathrm{d}x$ the horizontal bar serving as a vinculum (corresponding to our brackets). Leibniz now sets $xy = z$ and $x^2 = 2t$, and so there follows

$$\int_0^a z \,.\, \mathrm{d}x = \tfrac{1}{2}a^2 b - \int_0^b t \,.\, \mathrm{d}y.$$

If a particular implicit equation $f(x, y) = 0$ is now given, it can be rewritten as either $U(x, z) = 0$ or as $V(t, y) = 0$: the above relationship between integrals then tells us that the corresponding strips of area cut out from the two curves are equal. The instance of a general conic is considered in detail.

For geometrical reasons it is clear that a centre of gravity is determined as soon as we know the area and its moments about two intersecting axes. Where only the moments about two parallel axes are known, the area can be determined (from

$$a \int_0^c y \,.\, \mathrm{d}x = \int_0^c (a+x)y \,.\, \mathrm{d}x - \int_0^c xy \,.\, \mathrm{d}x)$$

but not the centre of gravity; when however the moments about three non-concurrent axes are known, then both the area and the position of the centroid can be found. This can, Leibniz goes on to explain, be extended also to the arcs of curves. In general[13] we have

$$\int_0^a xy \,.\, \mathrm{d}x = a \int_0^a y \,.\, \mathrm{d}x - \int_0^a \left(\int_0^x y \,.\, \mathrm{d}x \right) . \mathrm{d}x;$$

[9] *C* 1078-9, 1097-8.   [10] *C* 547, 551, 564.   [11] *C* 1086.
[12] C 1089 (*LBG*: 147-9).   [13] *C* 1090 (*LBG*: 149-51).

here the terminal abscissa which we have denoted by the letter a is called $ult(ima)$ x by Leibniz, while the double integral is denoted by $\overline{omn.\ omn.\ y}$. From this he develops the further integral-transforms

$$\int_0^a y.\,\mathrm{d}x = a\int_0^a \frac{y}{x}.\,\mathrm{d}x - \int_0^a\int_0^x \frac{y}{x}.\,\mathrm{d}x.\,\mathrm{d}x = a^2\int_0^a \frac{y}{x^2}.\,\mathrm{d}x - a\int_0^a\int_0^x \frac{y}{x^2}.\,\mathrm{d}x.\,\mathrm{d}x$$

$$- \int_0^a\left(x\int_0^x \frac{y}{x^2}.\,\mathrm{d}x - \int_0^x\int_0^x \frac{y}{x^2}.\,\mathrm{d}x.\,\mathrm{d}x\right).\,\mathrm{d}x = \dots,$$

and so on. Leibniz knows that

$$\int_0^a \log x.\,\mathrm{d}x = a \log a - a$$

can be found in this way by setting $y = 1$ in the above formula (tacitly assuming $\lim_{x\to 0}(x \log x) = 0$), and that, by setting $cz = \int_0^a y.\,\mathrm{d}x$ we can attain the arc-length

$$s = \int_0^a \sqrt{[1 + (\mathrm{d}z/\mathrm{d}x)^2]}.\,\mathrm{d}x = \int_0^a \sqrt{[1 + (y/c)^2]}.\,\mathrm{d}x$$

of the quadratrix $z = \phi(x)$. Next, the moment of an area $\int_0^a y.\,\mathrm{d}x$ about an arbitrary straight line is required to be determined.[14] This Leibniz fixes by the segment e on the x-axis, the distance f of the line from the origin and by the length g of the segment on the line which makes with e and f a rightangled triangle, whence in modern notation the line has the equation $(fx + gy - ef)/e = 0$ and so, since the centroid of the area-element $y.\,\mathrm{d}x$ is at $(x, \tfrac{1}{2}y)$, the moment of the area $\int_0^a y.\,\mathrm{d}x$ about the line will be $\int_0^a (f[x - e] + \tfrac{1}{2}gy)(y/e).\,\mathrm{d}x$; again the centroid of the arc-element $\mathrm{d}s$ is at (x, y), hence the moment of the arc $\int_0^c \mathrm{d}s$ about the line is $\int_0^c ((f[x - e] + gy)/e).\,\mathrm{d}s$. Leibniz finds these results through a direct calculation, corresponding to the modern transformation to a 'standard' form for the equation of a straight line in analytical geometry; he is fully conversant with the change to a new axis and is aware of the generality and the importance of the result.[15] By this means, he believes, numerous separate theorems by Cavalieri, Grégoire de Saint-Vincent, Wallis, Gregory and Barrow can be presented in a uniform manner. Cognate

14 C 1106 (LBG: 157–60).
15 The calculation is performed in detail for the circle in C 1106, and subsequently for the other conics in C 1165–7.

to these deductions, he claims, is Grégoire's *ductus plani in planum*, his method of integration for the solids $\int_0^a uv.dx$ contained between the coordinate-planes and the cylinders $\int_0^a u.dx$ and $\int_0^a v.dx$. If, for instance, $uv = ay = a\sqrt{(a^2-x^2)}$, then we achieve the quadrature of the circle by a *ductus* of two *subaltern* (auxiliary) parabolas

$$u^2 = a(a+x), \quad v^2 = a(a-x),$$

one onto the other.[16]

Later on, Leibniz investigated Guldin's theorem on centres of gravity more closely[17] and found that the area swept over by a thread unwinding from a curve equals the length of the thread multiplied into the arc described by the centroid of the total thread[18] (that is, the unwinding + unwound portion). He also noticed the connection with the problems of finding surface-areas and volumes of solids of revolution and 'mouldings'. During his following stay at Lyons, Tschirnhaus discussed the topic with Regnauld who showed him the place in Commandino's translation of the *Mathematical Collection*[19] from which it is evident that Pappus already in classical times knew Guldin's theorem – though it is not easily recognized there because of lacunae in the Greek codex used in that translation.

For Leibniz the main issue in the problem of quadratures was to determine in what circumstances the algebraic curve defined by some polynomial $f(x, y) = 0$ gives rise to an 'analytical' integral

$$z = (1/a)\int_0^x y.dx,$$

that is, one which can itself be defined by a polynomial $F(x, z) = 0$. The question had long occupied him;[20] now at last he believed he had reached his goal.[21] In order to determine the desired connection, Leibniz finds by a direct calculation using Sluse's tangent-rule

[16] Hence Grégoire de Saint-Vincent had by a faulty integration deduced his abortive quadrature of the circle in book x of the *Opus geometricum*. See further Hofmann (1941): 69–72.   [17] *C* 1286; 1419–21.
[18] *AE* (November 1695): 493–5 (*LMG* vii, 337–9).
[19] Tschirnhaus–Leibniz, 17 Apr. 1677 (*LBG*: 328); Leibniz–Johann Bernoulli, 28 July 1705 (*LMG* iii: 772). The reference is to the end of the introduction to book vii of Pappus' *Collectiones*. The imperfect text of the first edition (1588 = 1602) has been slightly emended in the light of Guldin's *Centrobaryca* (1635–41) in Manolessi's revised edition of 1660.
[20] *C* 820, 882, 906.
[21] *C* 1090 (*LBG*: 149–51).

$tF_x + zF_z = 0$ the subtangent t at a general point (x, z) and on eliminating t by means of $z/t = y/a$ he obtains $aF_x + yF_z = 0$. He then imagines z to be eliminated by combining this equation with $F(x, z) = 0$ and then has to check by comparing coefficients whether this result can be adapted to represent the integral or not. In the latter case, since as he is aware, the degree of the equation cannot be predetermined, we will need[22] – in analogy with the rationalized representation of irrational quantities – to add certain terms to the original equation which, though insignificantly altered thereby, will assume a form in which it is quadrable. (Leibniz did not yet reach the essential kernel of the modern theory, namely that $\int_0^x y \, . \, dx$ can be given as $R(x, y) + T(x)$ where $R(x, y)$ is a polynomial in x and y and $T(x)$ a transcendental function which itself can be written as a sum of certain characteristic integrals of first, second or third species.)[23] The elaboration of this procedure which he showed to Tschirnhaus at the time was to occupy him for many years thereafter.[24] In this way, by combining it with his rationalizing transformation,[25] he learnt how to characterize all problems dependent solely on the quadrature of the circle or the hyperbola.[26] Tschirnhaus, too, had investigated the feasibility of performing quadratures analytically[27] without conclusive results.

We return to Leibniz' studies[28] of October 1675; these again have for their aim the investigation of an important topic in which Leibniz had long interested himself: the solution of 'inverse-tangent' problems. Already in the autumn of 1674 he had been able to determine a curve, given its subnormal;[29] but by the following spring

---

[22] C 1074.
[23] The complete distinction was only determined in 1833–8 by Liouville.
[24] Leibniz–Tschirnhaus, early 1680 (LBG: 405); this is preceded (on p. 404) by a refutation of Tschirnhaus' assertions, based perhaps on the notes on C 1092 (LBG: 151–2) where Leibniz' own approach appears at the beginning of the manuscript.
[25] See note 20.
[26] Earlier examples can be found in C 560–1 and 823, later ones in C 1237–8, 1242, 1244, 1482, 1519; see also Leibniz–La Roque, late 1675 (C 1228).
[27] Tschirnhaus–Oldenburg, 22 Aug. (1 Sept.) 1676 (MS: London, RS: LXXXI, 33); of this extensive letter CE (1712): 66 only quotes a short extract. Tschirnhaus later, evincing an amazing ignorance of the facts, published in AE (October 1683): 433–7 a method for deciding the algebraic integrability of algebraic curves – without any explicit reference to Leibniz whose approach he had changed for the worse and marred by inaccurate additions. The controversy between Leibniz and Tschirnhaus mentioned in ch. 6: note 6 took its start from here; see further Hofmann (1965): 288–313.
[28] C 1092 (LBG: 151–6).   [29] C 840.

he had not really advanced much further.[30] Now the same question is taken up afresh. Leibniz starts from the relationship

$$\int_0^x p \,.\, dx = \int_0^y y \,.\, dy = \tfrac{1}{2}y^2$$

where he thinks of y as written in the form of the integral $\int_0^y dy$; and so of $\tfrac{1}{2}\left(\int_0^y dy\right)^2$ as not equivalent to the double integral $\int_0^y \int_0^y dy \,.\, dy$ in that the double integral differs from the product of the two integrals. In the middle of this paper Leibniz replaces the abbreviation *omn.* by the sign $\int$ (a 'long s', the initial letter of the word *summa* whose place it takes), at first writing $\int y$ where we would set $\int_0^x y \,.\, dx$: all integrals are understood to be definite, but no special notation for the limits is used. It is particularly noticed by Leibniz that the operation $\int$ raises the dimension by one degree. Where $\int y = z$, he puts, conversely,

$$y = \frac{z}{d}$$

and again emphasizes that the d-operation lowers the dimension by one degree. Immediately afterwards he presents some simple examples, from which it becomes evident that Leibniz takes a constant factor before the integral sign, and that he treats the sum of integrals as the integral of the sum. His final remark concerns his 'transmutation': let the tangent at the point $P(x, y)$ of the curve cut the y-axis in a point $T(0, y - x \,dy/dx)$ and on the ordinate through P lay off the length $\tfrac{1}{2}OT$ to obtain a point Q $(x, (y\,dx - x\,dy)/2\,dx)$, then the area

$$\int_0^x \left(y - \frac{y\,dx - x\,dy}{2\,dx} \right) . \, dx$$

generated *ordinatim* from PQ will be equal to the triangle $\tfrac{1}{2}xy$, as can be deduced immediately from the resulting integral

$$\int_0^x \frac{x\,dy + y\,dx}{2\,dx} . \, dx.$$

Thus his general transmutation theorem is built directly into the developing structure of his new analysis.

A few days later[31] his thoughts turned to curves with a given subnormal. From $y \,.\, (y\,dy/dx) = a^2$ Leibniz finds $a^2x = y^3/3$, not noticing that a constant of integration ought to be added since he makes the integral-curve always to start from the origin; there follows a check by means of Sluse's tangent-rule. If $x(y\,dy/dx) = a^2$, then $\frac{1}{2}y^2 = a^2\!\int dx/x$, a curve which, says Leibniz, is transcendental and which can be constructed by means of the logarithmic line. If Leibniz had gone into the fine details here, he would of necessity have been led to consider the limits of the integral. Again, if $x + y\,dy/dx = a^2/y$ then $x^2 + y^2 = 2a^2\!\int dx/y$; here Leibniz mistakenly regards the last integral function as a logarithm, failing to distinguish integrand and base variable. During the calculation he changes his notation, replacing

$$\frac{`x\,'}{d}$$

by the equivalent $`dx\,'$

which now appears more appropriate to him and which he retains in all his future work: he is already able to conceive such an 'indefinite' integration as (in modern style) $\int y\,.\,dy = \frac{1}{2}y^2$. Leibniz has to gain familiarity with his newly introduced notation and at first, through carelessness and misunderstanding, commits a number of errors but he is on the right lines. The example $x + y\,dy/dx = a^2/x$ is readily tackled by him: setting $a^2\!\int dx/x = \log z$, he obtains $x^2 + y^2 = 2\log z$. From a final example $\sqrt{(x^2+y^2)} = y\,dy/dx$, he derives

$$\tfrac{1}{2}y^2 = \int_0^x \sqrt{(x^2+y^2)}\,.\,dx.$$

Leibniz clearly feels that this is only a first step, not yet an acceptable solution, and so he tries an approximation: from the particular solution $x = y = 0$ he advances by a first infinitesimal unit $dx = h$, determines the corresponding dy from

$$(dy)^2 \approx h\sqrt{(h^2+[dy]^2)} \quad \text{or} \quad h^2(h^2+[dy]^2) \approx (dy)^4,$$

in general, he aims step-by-step to construct an approximating polygon whose sides are to be fitted between adjacent pairs y_k of equidistant ordinates constructed in the base points

$$(x_k, 0), \quad x_k = h, 2h, 3h, \ldots$$

successively at a slope $\sqrt{(1+[x_{k-1}/y_{k-1}]^2)}$. Leibniz had already used this idea more than two years before[32] when he sought to solve the differential equation $dy/dx = y/2x$. He hoped to convert this into a

[32] *C* 575; *compare* Mahnke (1926): 55–6 and Hofmann–Wieleitner (1931): 287–8.

useful method of obtaining a solution for a given differential equation through a limit procedure, but this goal he reached neither at his first attempt nor now.

In sequel Leibniz returns to a methodical investigation of his newly contrived calculus. Where x and y are functions of t, he remarks that neither

$$\frac{\mathrm{d}x}{\mathrm{d}t}\cdot\frac{\mathrm{d}y}{\mathrm{d}t} = \frac{\mathrm{d}(xy)}{\mathrm{d}t}$$

nor

$$\frac{\mathrm{d}(y/x)}{\mathrm{d}t} = \frac{\mathrm{d}y/\mathrm{d}t}{\mathrm{d}x/\mathrm{d}t}$$

is valid. What he actually intended to show we may gather only with difficulty from a number of his casual jottings and incompletely recorded notes where correct statements are intermixed with bad mistakes. In the end he calculates y from the general implicit Cartesian equation of a conic as a function of x and $\mathrm{d}y/\mathrm{d}x$. Characteristically, Leibniz has first to obtain the subtangent t according to Sluse's rule and then to eliminate it by $t = y\,\mathrm{d}x/\mathrm{d}y$. Leibniz asserts that in this way all similarly formed expressions can be squared, obviously with an eye on the solution of simple linear differential equations of the form $\mathrm{d}y/\mathrm{d}x = -(ax+by+d)/(bx+cy+e)$.

These, then, are Leibniz' famous notes whose composition led him, struggling to attain the simplest and most obvious way of presentation, to invent the Calculus. At first there can be no question on his part of consciously creating something new; it was simply a matter of suitably and formally abbreviating various integral transforms which he had formulated in the reduction of certain inverse-tangent problems, and in particular of eliminating the opaque and long-winded verbal descriptions which barred the way to a general view-point. Once this first, crucial step towards the 'algebraization' of infinitesimal problems had been taken, a new vision disclosed itself to a man experienced in identifying general, characteristic elements in a medley of similar things.

Much was still wanting which only a later age contributed, as for instance, any distinction between the definite and the indefinite integral. And much needed to be done, such as superseding Sluse's rule, unsuitable as a formalism, however useful as a calculating device. But the creator of this new and ingenious tool was a man blessed with an abundant store of viable ideas and the will to give them shape and form. He had a clear sense of much of what was still lacking in his calculus, but he knew that its defects could be cured, and that the way into new country was now open. In this ebullient mood of certainty he writes to Gallois[33] of his intention of publishing in the form of letters to well known persons of repute, what he had so far discovered and what he further envisaged; while to Oldenburg he writes[34] that he will, on an appropriate occasion, communicate (together with a description of his instrument for the solution of algebraic equations, and the arithmetical circle-quadrature which he had given more than two years before to his friends in Paris) his investigations of a different geometrical problem whose solution he had almost given up in despair – in all probability an allusion to his renewed studies of the inverse-tangent problems,[35] considerably advanced through his extended calculus formalisms. It is interesting to note that from this letter the *Commercium epistolicum*[36] cites only the references to Tschirnhaus mentioned previously and to his algebraic instrument and arithmetical circle-quadrature, but omits this final remark, favourable to his claim of independent discovery.

That Leibniz by this time could achieve many results virtually effortlessly is shown in a note concerning his meeting with Dechales, who in 1673 had come from Lyons to succeed Pardies in Paris and had a high contemporary reputation as the author of a much praised *Cursus mathematicus*.[37] Dechales set Leibniz the task of finding the portion of a circular cone cut off between its base and a plane parallel to the axis of the cone. Leibniz found the answer the very same evening and stated that it depends on the quadrature of the circle and that of a hyperbola, as is indeed immediately evident.[38]

Of more interest is Leibniz' discussion of 'Bertet's Curve', to which Ozanam had drawn his attention some years earlier.[39] This curve is generated from a circle of radius a by adding to this the arc-lengths (measured from a certain point on the circumference). The equation of its radius vector is therefore $r = a+s$. To find its

[33] Leibniz–Gallois, 2 Nov. 1675, unpublished draft: *C* 1107B.
[34] Leibniz–Oldenburg, 18 (28) Dec. 1675 (*LBG*: 146).
[35] *C* 1131, 1165–6.
[36] *CE* (*1712*): 45; *see* also p. 181.
[37] *Compare* Leibniz–Oldenburg, 14 (24) May 1673 (*LBG*: 96) also the reviews in *JS* No. 2 of 14 Jan. 1675 and in *PT* 9 No. 110 of 25 Jan. (4 Feb.) 1674/5.
[38] *C* 1063.
[39] Leibniz–Bertet, 3 (?) Nov. 1675 (*C* 1111). We do not know on what occasion Ozanam drew Leibniz' attention to 'Bertet's curve'; it figures in Ozanam's *Dictionnaire* (1690 = 1691): 97 where Leibniz' construction of the tangent is given though without quoting his name. The curve is in fact nothing but an Archimedean spiral $r = a+a\theta$.

general tangent Leibniz draws the corresponding tangent to the generating circle and on it lays off a length rs/a (in the opposite direction to that in which the curve is generated), and then joins the end-point to the point on the Bertet curve; the area of a sector of the curve is expressed[40] by $\frac{1}{2}rs + s^3/6a$. To find the point on the curve whose tangent shall be at right angles to the initial radius is considered an easy problem by Leibniz; but we should take this comment to apply only to the immediate derivability of the condition

$$\frac{a+s}{a} \tan \frac{s}{a} = 1$$

and not to its solution. Leibniz at once indicates the generalization to a *curva protensa* in which the circle is replaced by an arbitrary generating curve. In a note obviously belonging in his context[41] the curve is cleverly generated by a string-construction. The curve can in fact most easily be characterized as having a polar subnormal of constant length, as Roberval certainly had been well aware.[42]

Simultaneously, Leibniz was working on a new summary account of his arithmetical circle quadrature intended to be published in the *Journal des Sçavans*.[43] We possess a draft in French, its recast and changed form and later additional marginal notes in Latin. The presentation of its argument is far more fluent than that originally formulated for Huygens,[44] but it is marred by surprising errors of memory and other mistakes which show that Leibniz wrote this paper (which in the end he did not submit) in considerable haste. He had first intended to mention Gregory's quadrature of the cissoid but later again abandoned the plan: this establishes its fairly close relationship to the earlier paper seen by Huygens; notable is his present remark that it is possible by means of his general transmutation rigorously to prove the quadratures of general parabolas and hyperbolas which Wallis had plausibly derived by 'induction' from their tangent properties, and also to determine the rational segment of the cycloid. In his added notes a method is indicated for finding the rational quadrature of the equilateral hyperbola: taking the equation of the curve to be $y^2 = x(2a + x)$ Leibniz joins the general point (x, y)

[40] *C* 1110, 1112. [41] *C* 1112.
[42] Roberval's tract on 'Composition des mouvemens', expounding his mechanical tangent-construction was hardly known before it appeared in the *Divers ouvrages* (1693): particularly 92-6.
[43] Leibniz–La Roque, late 1675 (*C* 1228).
[44] Leibniz for Huygens, October 1674 (*C* 773, 797).

with the opposing vertex $(-2a, 0)$, so constructing on the vertical axis $x = -a$ the length

$$z = ay/(2a+x) = ax/y = a\sqrt{[x/(2a+x)]}.$$

From this we deduce at once that

$$x = 2az^2/(a^2 - z^2), \quad y = 2a^2z/(a^2 - z^2)$$

and so

$$\int_0^x y \, . \, dx = \int_0^z 8a^5z^2 \, . \, dz/(a^2 - z^2)^3,$$

an integral which may be evaluated by rendering the integrand into an infinite series.

The same topic is further explored in a draft version in French which was possibly intended for Gallois.[45] From its introduction we single out Leibniz' general remarks about the essence of the arithmetical quadrature: the infinite series $1 - \frac{1}{3} + \frac{1}{5} - \ldots$ represents a quantity which we can readily imagine by the help of a circle, just as one may adduce the diagonal of a square as an illustration for the sum of the series[46]

$$1 + \frac{1}{2} - \frac{1}{4 \, . \, 3} - \frac{1}{8 \, . \, 3 \, . \, 17} - \ldots = \sqrt{2}.$$

We detect here a decisive change in concept: a quantity is now defined by a converging infinite series – a considerable advance in the new analytic mode. The arithmetization of higher geometrical problems, says Leibniz, is a fundamental step forward which takes us far beyond Viète and Descartes on the one hand, and Cavalieri, Fermat and Wallis on the other; the greatest obstacle to the development of that older vision was its exclusive appeal to infinitesimal rectangles, whereas by the introduction also of infinitesimal triangles following Desargues and Pascal the field of view is enlarged beyond expectation. In two separate notes Leibniz deals explicitly with his transmutation theorem, performing the quadrature of higher parabolas by its aid as he had long ago explained.

It appears that Leibniz talked to Bertet of his quadratures of general parabolas and hyperbolas in the spring of 1676, thereupon receiving in return more details about Bertet's method of quadrature employing geometric progressions.[47] We now possess only a diagram,

[45] Leibniz for Gallois (?), late 1675 (C 1227, 1230–1).

[46] See ch. 10: note 32.

[47] Bertet for Leibniz, 9 (?) Feb. 1676 (C 1113, 1304).

where a parabolic segment, a triangle and a hyperbolic strip are drawn side by side; in each the altitude is divided by a continued section into portions in geometric progression, starting from the baseline, and a corresponding set of inscribed rectangles is drawn. In a further diagram the same is done for an arc of an equilateral hyperbola whose mirror-image with respect to an asymptote is also drawn: this probably relates to the solid of revolution of the hyperbola about the asymptote. No accompanying text exists; we may surmise that the related calculations were made on a separate sheet of paper which has now disappeared.

Evidently, we have here to do with Fermat's method for the quadrature of the higher parabolas and hyperbolas as developed in a paper of his 'De æquationum localium transmutatione' which had not yet at that time appeared in print. The first part of this essay, containing the quadrature of the higher parabolas, had been already communicated in 1629 to certain of Fermat's friends (to d'Espagnet, Carcavy and Beaugrand at least for sure), and further copies were taken to Italy by Mersenne during the winter of 1644–5, one going to Santini in Genoa, another to Torricelli,[48] who greatly admired its method and confessed that he himself had succeeded so far only in squaring the parabolas $y/b = (x/a)^n$ for n a positive integer,[49] whereas Fermat was more generally able to deal with the case where n had a fractional value.[50] Torricelli had, a few years earlier, independently determined the infinitely extended solid of revolution[51]

$$\pi \int_a^\infty y^2 . dx$$ generated by the rotation of the hyperbola $xy = ab$ round

$y = 0$, using for this purpose the equivalent of an integration by parts, and then proceeded along the same way as Fermat to the quadrature of the higher hyperbolas[52] $(x/a)^m . (y/b)^n = 1$, $(m, n$ integers). Fermat appears to have subsequently forwarded his own proof of this result to Torricelli.[53] Bertet, who later tried to apply the method using a division by points in geometric progression, may have heard something of it in France or in Italy but evidently (since he

[48] Ricci–Torricelli, 28 Jan. 1645 (*TO* III: 258).
[49] Torricelli–Ricci, 27 Aug. 1644 (*TO* III: 222); Ricci–Torricelli, 10 Sept. 1644 (*TO* III: 225). [50] Torricelli–Ricci, 6 Feb. 1645 (*TO* III: 271).
[51] Cavalieri–Torricelli, 17 Dec. 1641 (*TO* III: 65); Torricelli–Nicéron, June (?) 1643 (*TO* III: 129); *De solido hyperbolico acuto* was available in print in the *Opera* (1644). Roberval had reached the same result as is evident from his letter to Mersenne for Torricelli, summer 1643 (*TO* III: 135–8).
[52] In particular: Torricelli–Cavalieri, 5 May 1646 (*TO* III: 373).
[53] Fermat–Torricelli, late 1646 (lost); *compare FO* II: 338.

does not even name its author), if he did, this was only in a rather distorted form or perhaps simply a general verbal description.

Fermat's technique may be illustrated as follows:[54] in the case of the general parabolas $(y/b)^n = (x/a)^m$ $(m, n$ positive coprime integers), where $0 < t < 1$, we construct a sequence of points $x_k = at^{kn}$ in the base and a corresponding set of ordinates $y_k = bt^{km}$ perpendicular to it; in each parabolic strip $\int_{x_k}^{x_{k-1}} y \, . \, dx$ we inscribe a rectangle

$$(x_{k-1} - x_k) y_k = ab(1 - t^n) t^{m((k-1)(m+n))}$$

until in the whole parabola we have inscribed a step-figure of area

$$\frac{ab(1 - t^n) t^m}{1 - t^{m+n}} = ab \frac{(1 + t + \ldots + t^{n-1}) t^m}{1 + t + \ldots + t^{m+n-1}} \, ;$$

this tends to the parabola's area $\int_0^a y \, . \, dx$ in the limit as $t \to 1$, whence we obtain

$$\int_0^a y \, . \, dx = abn/(m+n).$$

At first Fermat treated only the instance $m = 1$, but later on mastered the general case also. For the corresponding hyperbolas $(x/a)^m (y/b)^n = 1$ he started not with inscribed rectangles, as Bertet indicates, but with circumscribed ones, and there are certain other discrepancies in his version: thus, Bertet is seemingly not content with computing the area $\int y \, . \, dx$ but considers also its mirror-image in the x-axis, while in his diagram – as we have remarked above – only the case of the progression for $t = \frac{1}{2}$ is sketched, and the reader has to supply the rest himself. That Bertet has not fully understood the method (and so probably not invented it himself independently) is indicated by his unsuccessful attempt to apply the technique in rectifying the arc of the parabola $ay = x^2$. Here he sets $x_k = at^k$, $y_k = at^{2k}$ and forms the element

$$\sqrt{[(x_{k-1} - x_k)^2 + (y_{k-1} - y_k)^2]} = a(1-t)t^{k-1} \sqrt{\{1 + [(1+t)t^{k-1}]^2\}},$$

but the proportionality of this to the ordinates of a hyperbola now asserted is true neither in general nor in the particular example $t = \frac{1}{2}$ adduced by him.

[54] We give here a modern generalized transcription of the treatise *FO* I: 255–66 mentioned in ch. 5: note 39.

THE INVENTION OF THE CALCULUS

Unfortunately, we know of no remark by Leibniz on Fermat's method. He might quite conceivably have refashioned it into some infinitesimal equivalent. The transformation

$$(x/a)^m = (y/b)^n = \exp(mnt)$$

was, indeed, not yet at his disposal, but this device might have been replaced by a transition to logarithmic form[55] by means of the substitution $dx/x = n \, dt$ (and hence $dy/y = m \, dt$), when he would deduce that

$$d(xy) = (m+n)xy \, dt$$

and so

$$\int_0^a y \, . \, dx = n \int_0^1 xy \, . \, dt = (n/[m+n]) \int_0^{ab} d(xy) = abn/(m+n).$$

Torricelli's refined method in his later treatises on parabolas and hyperbolas[56] (which remained unpublished in Leibniz' day) may similarly be translated into infinitesimals; in the case of the parabolas, for example, we conclude from their tangent property that

$$(1/n) \int_\alpha^a y \, . \, dx = (1/m) \int_\beta^b x \, . \, dy = (1/[m+n]) \int_{\alpha\beta}^{ab} d(xy).$$

Only in one area did Leibniz remain at this time still a novice, namely number theory. He did indeed occupy himself with that topic too, but without succeeding in attaining any fruitful method. During a meeting of his with Arnauld the question was raised as to how one might sensibly classify all integral rightangled triangles or represent them by a general formula. Arnauld, who knew his Euclid well, was of course aware that in *Elements* x, 28, Lemma 1 the rational solution of the equation $x^2+y^2 = z^2$ is given in the form

$$x = \tfrac{1}{2}(u^2-v^2)w, \quad y = uvw, \quad z = \tfrac{1}{2}(u^2+v^2)w.$$

The problem he now proposed was probably that of formulating a rule which would give all such primitive triangles without exception. Leibniz at once began to tabulate low numbers of the form $r = p^2+q^2$ and then pick out the resulting squares.[57] He came to believe he could show that the hypotenuse of each such triangle has to be set as the sum of two consecutive squares, or as one half of the square of an odd number increased by unity, but as soon as he handed over a paper containing this statement to Arnauld he noticed his error and he

[55] It is interesting to note that Fermat himself already speaks of a logarithmic method (*FO* I: 265). [56] *See* ch. 5: note 40.
[57] Leibniz for Arnauld, 11–12 Dec. 1675 (MS: Hanover, LH 35 III B8).

hastened at once to supply the Euclidean rule, which, it would appear, he had not known before and now deduced unaided from considerations of divisibility. He added that 2 can never be a side of a rightangled triangle with integral sides, and, furthermore, that one at most of the shorter sides can be a prime, and that in a primitive triangle the sum of the hypotenuse and a side is always double a square number. He also dealt with the question whether the area of a rational rightangled triangle can be a square number or its double and developed a proof of the impossibility of this which he afterwards remodelled.[58] This last problem was posed to him by Mariotte who, having taken charge of the edition of Frénicle's *Traité des triangles rectangles en nombres*, discussed certain details of it with Leibniz.[59] During their conversation Leibniz gave, basing his account on some earlier notes,[60] a new proof of the formulae for square numbers of the form $3n$ or $3n + 1$ and mentioned the lesser Fermat theorem;[61] Mariotte in turn communicated his step-by-step method for determining all integral rightangled triangles whose two sides differ by unity.[62] When Leibniz met Ozanam again, the conversation turned to the determination of two numbers such that, when the difference of their squares is either added to the smaller of the two, or subtracted from the larger one, the result will be the same square number;[63] the problem of six squares was also discussed,[64] and in connection with this the attempt at getting an expression of higher degree, like $x^4 + 3y^4$, to become a square number.[65]

[58] *C* 1150–1; *see* also Leibniz–Oldenburg, 2 (12) May 1676 (*LBG*: 168) and the note of 8 Jan. 1679 (*LMG* VII: 120–5); further the subsequent fruitless discussion with Molanus and Eckhard (*LPG* I: 272–314), and the reference in Leibniz–Huygens, 18 Sept. 1679 (*HO* 8: 215).

[59] *C* 1305.

[60] *C* 727, 1253, 1282; also Leibniz–Oldenburg, 2 (12) May 1676 (*LBG*: 168).

[61] Leibniz could hardly have known of Fermat's communication to Frénicle, 18 Oct. 1640 (*FO* I: 209) so that his was an independent discovery; *see* Mahnke (1913): 57. For the later remark in the 'Nova algebræ promotio' (*c.* 1694) (*LMG* VII: 180–1) Fermat's text might have been accessible in the *Varia Opera* (1679): 163.

[62] Mariotte for Leibniz, February (?) 1676 (*C* 1306), takes up Frénicle, *Traité* (1676), book I, prop. 19 to deal with a problem following on representing the sides of Pythagorean triangles by means of coprime natural numbers p, q (of which one is even) in the form $a = p^2 - q^2$, $b = 2pq$, $c = p^2 + q^2$: if now $a - b = 1$, $P = 2p + q$, $Q = p$, then $A = P^2 - Q^2$, $B = 2PQ$, $C = P^2 + Q^2$ will give $A - B = -1$.

[63] *C* 1201.

[64] *C* 1201, 1211, 1368, 1376, 1382; *see* also p. 90.

[65] *C* 1385.

14

DISPUTE ABOUT DESCARTES' METHOD

We have earlier mentioned that at the beginning of May 1676 Tschirnhaus addressed a letter to Oldenburg in which he repeated an assertion, originally made in conversation with Collins, to the effect that Descartes was the true founder of the new mathematical method, and that what his successors and rivals had added was really only the continuation and elaboration of his own ideas. This letter was passed on to Collins, who himself at once began to compose a lengthy reply to it.[1] We now know only its English draft, not Oldenburg's final Latin version which evidently contained a number of changes and omissions. We can gather something, at least, of its contents, from Tschirnhaus' answer,[2] though this, unfortunately, is known only in a copyist's excerpt, and that disfigured by a number of mistranscriptions. The clash of opinion over Descartes' method is especially important because of its close connection with the correspondence exchanged at this same period between Collins, Newton, Oldenburg and Leibniz; but it also commands our interest in its own right for the light it sheds on many previously unknown details, even given its lack – typical of Collins' writings – of a firm structure and despite the carelessness shown by Tschirnhaus in framing his reply.

Collins begins with the observation that current opinion about the mathematical achievements of the present generation is divided. Isaac Vossius (who had lived in London since 1670 and been made a canon of Windsor in 1673) had gone furthest in the one direction, for according to him the achievements of the Ancients had yet nowhere near been matched; others, again, ascribed to the moderns, and in particular to Descartes, far more than was in fairness their due. True, he had been very clever in thinking out his ovals,[3] but he expounded

---

[1] Collins–Oldenburg for Tschirnhaus, late May 1676 (MS: London, RS: LXXXI, 25), partially printed in *CE* (*1712*): 43.
[2] Tschirnhaus–Oldenburg for Collins, mid-June 1676 (MS: note 1).
[3] Descartes treats of ovals in the *Geometria* (1659): 50–66; the method employed there is much improved in Newton's *Methodus* of 1671 (*NP* **3**: 136–9).

their theory so subtly that an eminent geometer (Barrow? or Newton whose *Lectiones opticæ* Collins knew in manuscript?) had tried in vain to elicit the idea underlying his method of construction. Collins here alludes to the oval interfaces between two homogeneous media of differing density which refract rays emanating from a point source in the one medium to a point in the other: this involves the solution of a simple differential equation treated most elegantly by using bipolar coordinates to define the curve – but Descartes gives only veiled indications of his method.

On the whole, Descartes' *Géométrie* certainly is, in Collins' eyes, a work of great merit, though it should not be forgotten that much of its detail was taken over from other sources. Its treatment of quadratic equations is a rather superfluous triviality,[4] but its discussion of the '*locus ad tres et quattuor lineas*' (in modern terms, conics defined by linear forms tied by the conditions $g_1g_2 = g_3^2$ and $g_1g_2 = g_3g_4$ and their generalizations) is well done[5] – though Collins prefers Newton's method of generating conics as the meet of two rigid angles rotating about fixed poles with their second meet constrained to be in a straight line.[6] The distinction of geometrical problems from mechanical ones Collins considers well founded;[7] in the interim, however, it had by means of infinite series been possible to achieve decided progress in the study of mechanical problems, such as, for instance, Kepler's problem of dividing a semicircle in a given ratio, in whose solution by Wren and Gregory the prolate cycloid had proved most useful.[8] Collins regards as unsatisfactory Descartes' determination of several geometric means by his moving angle construction;[9] Gregory had since done this by calculation (Collins refers to his use of the binomial theorem[10] in this context) in connection with his investigation of the particular logarithmic spiral ($r = ae^\theta$) whose stereographic projection is the 45° spherical loxodrome[11] – and this was more valuable than anything to be found in Descartes on that topic.

Concerning solid problems, constructed by Descartes with the aid of a fixed parabola and a varying circle,[12] we can, Collins reminds us, find something not only in Werner's and Barozzi's books on conics

4 *Geometria* (1659): 5–7.   5 *Geometria* (1659): 7–16.
6 *See* ch. 4: note 45.   7 *Geometria* (1659): 17–18.
8 *See* pp. 217–18.
9 *Geometria* (1659): 20–1, 67–9 enlarging on Pappus IV: prop. 5.
10 *See* p. 134.
11 This seems to be based on direct verbal communication by Gregory.
12 *Geometria* (1659): 85–96.

but already even in Eutocius' commentary on Archimedes; Sluse had considerably improved the method (in his *Mesolabum*)[13] and had further refined it in his correspondence with Huygens on Alhazen's problem[14] (to find the points on a circle at which rays of light from one given point are reflected to a second). In substitute for such awkward constructions as those where the intersections occur at a very narrow angle Barrow, Newton and Gregory had contrived excellent graphical approximations;[15] they had also investigated the degree of the algebraic equations defined by the meets of various curves. By means of Newton's construction of conics (whose intersections yield the solution of 'solid' problems) it was possible to find any point on them without drawing the curve itself, while by the use of 'spherick' projection every conic can be transformed into a circle on a sphere.[16] For problems higher than 'solid', Collins went on, Descartes needs a different curve in every case, and so a graphical method would here seem preferable. Newton solves all equations from the fifth to the ninth degree by means of intersections of a single fixed cubic parabola with a conic, for the equation of ninth degree he needs two cubic parabolas or one of fourth degree and a conic.[17] Descartes has introduced a quartic equation[18] in his solution of the (classical Apollonian) verging problem where a line is to be inclined through a corner of a square so that its intercept between the two opposing sides is of given length: quite needlessly so, since the problem is in fact a 'plane' (quadratic) one, as Pell has remarked long ago.

The dramatist, William Joyner, had, Collins proceeds to tell us, brought back from France a rumour that Roberval had, in a public lecture, once accused Descartes of having taken essential hints on algebra from Harriot,[19] and in regard to his tangent-method, from

---

13 Collins knew Sluse's *Mesolabum* (1668) in detail having reviewed it in *PT* 4, No. 45 of 25 Mar. (4 Apr.) 1669: 903–9.
14 Oldenburg has given a good selection of letters addressed to him on Alhazen's problem by Sluse and Huygens in the *PT* 8, Nos. 97: 6119–26 and 98: 6140–6.
15 *See* ch. 4: notes 43–4.   16 *See* p. 137.
17 *See* ch. 10: note 74.
18 *See* ch. 11: note 10, and p. 145.
19 The clash with Roberval apparently happened in the summer of 1648 during Descartes' last stay in Paris, for which *see* Roberval's note (partly) reproduced in *DO* xi: 687–90 and Leibniz' text in his 'Notæ quædam' (1693) (*LPG* iv: 312). Roberval's statement was repeated by Wallis in ch. 53 of his *Algebra* (1685): 198, and this in its turn is the source for Leibniz' account in 'De ortu...algebræ' (1685–6) (*LMG* vii: 213) and in 'Nova algebræ promotio' (1694?) (*LMG* vii: 157). Concerning the sign-rule *see* also Leibniz–Hermann, 18 Jan. 1707 (*LMG* iv: 308). Johann Bernoulli too considered the possibility of Descartes' dependence on Harriot, as in his letter to Leibniz, 1 Sept. 1708 (*LMG* iii: 837).

Hérigone's *Cursus mathematicus* of 1631; and no doubt, too, Ricci's allusion in the preface to his *Exercitationes geometricæ*, where he announces an impending publication of his concerning the errors committed by certain famous persons, referred to Descartes. The assertions here made by Collins are not at all well-founded: there is no proof whatever of the allegation that Descartes was dependent on Harriot, nor is its truth likely; as for the story about the tangent method, this is certainly incorrect since Hérigone's short account of Fermat's method was given only in the book of his *Cursus* published in 1642, five years after the *Géométrie* had appeared. Ricci's algebraical manuscript remains unpublished to this day: whether the reference in the preface is to Descartes, is wholly uncertain. What further Collins has to say regarding the alleged incorrectness of other specific assertions in Descartes' *Géométrie* is either immaterial, illusory or mere self-deception.

Far better founded is his remark (later to be repeated in the *Commercium epistolicum*) that, when Descartes asserted that of all imaginable tangent-methods his was the best, he was boasting without real justification.[20] That Sluse contrived his tangent-method from studying Descartes (as Tschirnhaus had evidently previously suggested) is not Collins' opinion. Gregory and Newton had made decisive progress in this field. The latter had some time before heard from Collins[21] that Sluse and Gregory had come upon new tangent-methods and that Barrow's *Lectiones geometricæ*, where the topic was also treated, enjoyed a high reputation with all scholars:[22] this it was that had induced him to communicate his own techniques. There now follows the text of Newton's December 1672 letter on tangents,[23] though in the extant draft in fact it is not part of Collins' main text but inserted with a caret at a later place. Very possibly it was not this extended version that was sent to Tschirnhaus, but only a brief résumé of Newton's procedure, and the same presumably holds true for the ensuing excerpts from Gregory's letters, in which Barrow's method is accorded highest praise[24] and where, incidentally, a remark

---

[20] Descartes–Mersenne, 18 (?) Jan. 1638 (*DO* III: 300–5 = *MC* **7**: 17–18). A Latin version of the passage in question was inserted in *CE* (*1712*): 43.
[21] Collins–Newton, mid-December 1672 (*NC* **1**: 247). See also Hofmann (1943): 83–7, 89–94.   [22] *Lectiones geometricæ* (1670): 80–1; see ch. **6**: note 48.
[23] Newton–Collins, 10 (20) Dec. 1672 (*NC* **1**; 247–8). Evidently no more than an allusion was given, for in his reply of mid-June 1676 where he otherwise takes up Collins' letter point by point, Tschirnhaus makes no mention of Newton's tangent-method at all.
[24] Gregory–Collins, 5 (15) Sept. and 23 Nov. (3 Dec.) 1670 (*GT*: 103, 121).

on Gregory's discussion of the *spiralis arcuum rectificatrix*[25] is inserted. At the end of this present paragraph comes the remark mentioned above where Gregory registers his dissent from Tschirn-haus' over-estimation of Cartesian methods in general[26] – certainly this was wholly unsuitable to be straightforwardly transmitted.

Now follows Collins' survey of the results achieved by Newton with his method of infinite series[27] – that is in the rectification and quadra-ture of curves, the mensuration of volumes and surfaces of solids of revolution and their segments and the determination of their corresponding centres of gravity, and also apart from all this, in the solution of equations. If you like, Collins continues, you might well say that Newton found his inspiration for all this in Descartes, but you might just as well cite Plato as his master: in fact, Descartes had explicitly denied that it was possible to rectify a geometrical curve; Barrow, however, had given a number of examples of rectification, and, before him, Neil had reduced the rectification of the Apollonian parabola to the quadrature of a hyperbola, while Wren had done the same for the cycloid 'by means of the circle quadrature'. We at once see that Collins here gives a completely mistaken picture of develop-ments in the rectification dispute. Instead of Plato he should have cited Aristotle, while Neil ought to be linked with the rectification of the semi-cubic parabola, and determining the arc-length of the cycloid has nothing directly to do with squaring the circle. The whole para-graph, in the version communicated, was completely rewritten by Oldenburg, who was careful to exercise the utmost restraint after all the recent bitter dispute on this topic between Huygens and other Fellows of the Royal Society. From Tschirnhaus' reply we gather, for instance, that, in the revise, Heuraet too was mentioned in appreciative terms. Some time later Wallis saw Collins' English draft[28] and in all friendliness pointed out the deep inaccuracy of his account.[29]

Collins continues with a survey of Newton's successful treatment of trigonometrical functions: thus from the tangent he can find the arc, and from the arc the log sin, log tan and log sec and their corresponding inverses;[30] further, he can solve not only numerical

---

[25] For this *see* below ch. **15**: note 54.
[26] Gregory–Collins, 20 (30) Aug. 1675 (*GT*: 325). *See* p. 172.
[27] *Compare* the similar wording in Collins–Borelli, December 1671 and Collins–Vernon, 26 Dec. 1671 (5 Jan. 1672), mentioned in ch. **4**: note 54.
[28] Collins–Wallis, 20 (30) Sept. 1677 (*CR* II: 608).
[29] Wallis–Collins, 8 (18) Oct. 1677 (*CR* II: 608).
[30] *See* note 27.

equations but also quite general literal equations, taking as his basis an adaptation of Viète's method, wholly independently of Descartes. The portion of Collins' draft to which Tschirnhaus refers in his reply terminates with a brief reference to Dulaurens, who had pointed out in his *Specimina* how little this branch of algebra had hitherto been cultivated.

The remainder of the English draft consists in several hasty notes, among which we may mention the citation of a long letter by Wallis on cubic equations,[31] together with references to Pell's and Gregory's studies and to finding limits for roots of equations – of this, again, nothing had been said by Descartes – and a caret indicating that an insertion dealing with Pell's methods was to be made. This last is missing, and we shall soon see why. It is further worth noting that Bombelli's name occurs in the context of a note on cubics. This earliest citation by Collins of this Italian mathematician is probably a recollection of what he had heard from Leibniz himself[32] – fresh confirmation that the present draft letter was not written in May 1675 but only in May 1676. In sequel, Collins has noted down some details of his conversation with Tschirnhaus on 30 July (9 August) 1675 concerning quartics, $x^4 + px^3 + qx^2 + rx + s = 0$ for which $p^2s = r^2$ (*compare* p. 167), and made a collection of unsolved problems and other mathematical desiderata: here he sets first the solution of Davenant's problem[33] mentioned above on p. 138; next, the computation of really extensive tables of logarithms, of square and cube roots and the areas of circle segments (on which, we are told, Newton has already done a great deal of preliminary work);[34] thereafter that of determining a geometrical locus from a number of equidistant given ordinates, and of treating expressions of the form

$$\sqrt[3]{(a+\sqrt{b})} + \sqrt[3]{(a-\sqrt{b})};$$

and lastly, that of determining sufficient conditions for a quartic to be reducible. The inclusion in this enumeration of the problem of interpolation appears as something of an inconsistency on Collins' part, since on this very question Collins had long ago received all the information he desired from Gregory.[35]

[31] Wallis–Collins, 12 (22) Apr. 1673 (*CR* II: 564–76).
[32] Leibniz–Oldenburg, 2 (12) July 1675 (*LBG*: 131).
[33] *See* below pp. 255–6.   [34] *See* ch. **10**: note 70.
[35] Gregory–Collins, 23 Nov. (3 Dec.) 1670; *see* p. 32. Collins by then was probably aware that Newton could manage interpolations for unequal intervals of the argument (*see NP* **4**: 14–21 for notes of May 1675), but felt not at liberty to point this out without specific permission from the author.

This whole paper of Collins' is, indeed, not of much importance; his criticism of Descartes is essentially petty and misses the main points at issue, while what is being noted about far-reaching new developments in Britain is confined wholly to Collins' immediate circle of friends. We have dealt somewhat thoroughly with the details only because they show that all Collins could produce for Tschirnhaus, too, were near-irrelevancies or obscure indications of a general nature from which nothing precise about the methods employed could have been deduced. If Tschirnhaus had had the eyes to see it, Collins would have been small fry, but he was after all, still very young, believing himself adequately qualified to pass a definitive judgment in the matter – which he certainly was not – and determined to present his own hero, Descartes, as an unsurpassed model to all others.

In his reply,[36] accordingly, Tschirnhaus begins with a long enumeration of mathematical achievements whose discovery was undeniably the property of Descartes: for instance, his comprehensive discussion of all classical 'solid' loci by means of the defining equation $y = ax + b \pm \sqrt{(cx^2 + dx + e)}$, and the indication of its extension to higher 'linear' curves: his 'simplest possible' mechanical generation of higher curves by movable straight lines; his methods for lowering the degree of an equation and for its graphical solution which were later further developed by Hudde and Kinckhuysen; and, lastly, his principle of comparing coefficients. The performance of all quadratures and cubatures so far known and the determination of centroids by a uniform procedure had appeared so easy a task to Descartes that he forbore to give a detailed account of them;[37] yet he had widened the scope of elementary geometry – till then restricted to constructions by ruler and compass – by including all 'geometrical' (algebraic) curves; and, lastly, he had – as Tschirnhaus is ready to demonstrate – contrived the best possible method for constructing tangents to geometrical curves, though this could, of course, be varied in several ways. Not even Fermat who had engaged Descartes by letter in a dispute over this last method, had denied his originality; so, if Collins asserted the contrary, he was obviously in error.

Tschirnhaus is ready to concede that Descartes might have found one or two things, particularly algebraic items, in the older authors, but it was well established that he had studied hardly any mathe-

---

[36] Tschirnhaus–Oldenburg for Collins, mid-June 1676 (MS: note 1).
[37] Descartes–Mersenne, 13 July 1638 (*DC* II No. 89 = *MC* 7: 344).

matical books, thinking out almost everything for himself. His aim had always been to attain the most general viewpoint even though he allowed himself the occasional numerical example. It was only by restricting mathematics in this way to the realm of pure algebraical problems that his fundamental, rigorous clarity of treatment had been achieved, whereby distinction might be made between truly mathematical questions capable of a complete solution and inexact approximations; accordingly the chief merit of men like Sluse, Wallis, Gregory, Hudde and Barrow consisted in extending and refining existing mathematical methods. Descartes himself had quite deliberately given indications only, gladly leaving their detailed execution to others. Tschirnhaus, of course, admits that in the single case of the geometrical rectification of geometrical curves Descartes had indeed been in error; but even Heuraet, who had solved this problem so felicitously and generally, confessed that this had been possible only on the basis of the principles earlier laid down by Descartes. Certainly, he admits, Newton's mechanical construction of conics is very interesting, but it is no more than a geometrical detail, while he suspects that his solution of equations of seventh or eighth degree, as mentioned by Collins, with the aid of an intersecting cubic parabola and conic, or by two cubic parabolas, does not relate to the most general equations of this type, but only to certain special ones. With regard to the Apollonian *neusis*-problem Descartes had passed only some very brief comments, but in fact, Tschirnhaus claims to have elucidated this affair now; it reduced to the solution of an equation, namely $x^4 - 2ax^3 + (2a^2 - c^2)x^2 - 2a^3x + a^4 = 0$, which Collins had some time ago passed on to Gregory.[38] Referring to Collins' further account of Newton's and Gregory's recent results Tschirnhaus was ready to concede that these were all very praiseworthy; however since it did not pertain to geometrical problems and methods, but rather to mechanical ones, they did not belong to the realm of Cartesian mathematics. Tschirnhaus ends by emphasizing that he deliberately bases his comments only on what is accessible in print, and that he knows nothing in detail of the latest unpublished mathematical researches in England.

These statements by Tschirnhaus allow us a good deal of insight into his mind, showing clearly that he was interested only in the geometrical (algebraically expressible) problems on which he concen-

---

[38] *See* note 18. Tschirnhaus had shown the example to Collins during their conversation of 30 Aug. (9 Sept.) 1675 (*GT*: 316); Collins passed it on to Gregory, 3 (13) Aug. 1675 (*GT*: 316).

trated all his skill and energy, while everything else he heard of other types of questions he tended to consider more or less as unimportant mechanical approximations. For this reason he listened only with diminished attention to Leibniz' explanations of his own discoveries in the realm of infinitesimals, since all this, in his opinion, belonged in no way to pure mathematics. Whether Leibniz himself encountered any novelties in this exchange of letters remains uncertain. We can identify only a single note by him that can be related to Collins' letter; in it he mentions the discussion between Huygens and Sluse on the Alhazen problem,[39] alluding to Barrow's mechanical solution of it in lecture IX of his *Lectiones opticæ*. There can here be no question of any narrow dependence, however, for Leibniz may well have been led to study this same problem for quite a different reason. Certainly, Tschirnhaus' final observation would not have been approved by Leibniz who again and again urged that by restricting mathematics to the purely algebraical Descartes had quite un- naturally cramped his freedom of vision and thereby rendered deeper insight on his part into relationships impossible. Tschirnhaus was well aware of this attitude of Leibniz', but refused to take notice of it. Tschirnhaus' own testimony that he was inadequately informed about the current English studies in mathematics is of importance, for if this was indeed the case he cannot have passed on anything essential to Leibniz. In conjunction with his phrase about Descartes' tangent method this remark of Tschirnhaus' makes it completely certain that he did not receive from Oldenburg the full text of Newton's December 1672 letter on tangents. Collins evidently did not feel entitled to divulge details of this letter without the direct consent of Newton himself, who alone could decide how much it was fitting to communicate.

[39] *C* 1451.

15

THE REPORT ON GREGORY'S
RESULTS AND PELL'S METHODS

Leibniz came to be directly acquainted with Mohr, who had left London[1] in the late autumn of 1675, apparently only in the spring of 1676.[2] The Dane had received the sine and arcsine series from Collins and passed both on to Leibniz, who, particularly surprised by the elegance of the sine-series, wrote to Oldenburg asking that Collins should send the relevant proof.[3] He was himself engaged, he said, in giving the finishing touches to the proof of his arithmetical circle quadrature, a result he had already shown some years ago to his friends in Paris; its complete version would be sent to England as a present in return for the demonstration desired. Not a word is said to indicate that Leibniz had already received these very same series in the spring of 1675,[4] and the 'editor' of the *Commercium epistolicum* is, at the point where this passage is quoted, rightly astonished by the omission.[5] The subject is taken up in a letter of Newton's in the spring of 1716, at a time when the priority dispute was at its height;[6] there the reproach is made that Leibniz' enquiry showed clearly that he did not then know the series, and that it was now indeed obvious why he had responded to the first transmission only in vague terms.[7] To English eyes, the appearances were against Leibniz; his excuse of simply forgetting through the pressure of his Paris affairs,[8] and his reiteration that what he especially wanted was (as he also noted in the margin[9] of his copy of the *Commercium epistolicum*) not the result nor

[1] Oldenburg–Leibniz, 30 Sept. (10 Oct.) 1675 (*LBG*: 140); Collins–Gregory, 19 (29) Oct. 1675 (*GT*: 338).
[2] In his letter to Oldenburg 18 (28) Dec. 1675 (*LBG*: 148), referring to Mohr's essay mentioned in ch. 11: note 48, Leibniz writes of 'a treatise whose author he does not know'.
[3] Leibniz–Oldenburg, 2 (12) May 1676 (*LBG*: 167).
[4] Oldenburg–Leibniz, 12 (22) Apr. 1675 (*LBG*: 114). Compare p. 131.
[5] *CE* (*1712*): 45: *Quasi ante annum easdem non acciperet ab Oldenburgo.*
[6] Newton–Conti, 26 Feb. (7 Mar.) 1715/16 (*LBG*: 272).
[7] Leibniz–Oldenburg, 10 (20) May 1675 (*LBG*: 122).
[8] Leibniz–Conti, 9 Apr. 1716 (*LBG*: 279–80).
[9] *CE* (*1712*): 45: *Sed volebat modum, quo eo fuisset perventum.*

the law by which the series progressed but its proof – all this was looked upon as mere evasion by his critics even though it was historically true. We must not ignore the difficult decision which faced Leibniz at the time: whether or not to submit to the wishes of a new master who pressed him to take up his office immediately and was unwilling to extend Leibniz' stay in Paris beyond Whitsun (24 May 1676).[10] Of these complications the English knew nothing; Leibniz had always been careful not to reveal details of his impending journey home.

The remainder of his letter of 12 May 1676 is written in haste; we have already mentioned most of its content.[11] In a few words Ebert's appointment to the Ramus professorship is noted, Tschirnhaus' latest studies on angular sections and regular polygons briefly mentioned, and Hardy's enquiry about manuscripts of Apollonius in Arabic passed on. Frénicle's *Traité des triangles* has been published and it has presumably meanwhile arrived in London too; Leibniz now says that he is able to give new proofs for most of the theorems in it, and indeed extend them considerably, with the exception of Frénicle's excellent proof that the area of a rightangled triangle with integral sides cannot be a square number. On the back of Leibniz' letter Oldenburg has noted that Wallis also had already proved this.[12] Towards the end of his letter Leibniz asks that great care be taken to see to it that Gregory's papers are adequately preserved and in particular his invaluable studies in number theory; he himself had once more interested himself in solving equations by infinite series, and in applying trigonometrical and logarithmical tables to the solution of affected equations, because he was dissatisfied with his own results in this field.[13] Pardies, indeed, had promised the solution of affected equations by means of the logarithmic curve – no notes by Pardies relevant to this have survived – but Leibniz thinks this improbable. For his own part he could extract every root of a number or binomial either directly or at least approximately but only in a rather clumsy fashion;[14] no doubt the English had gone farther in this field.

Part of this letter by Leibniz was read to the Royal Society,[15] many of whose members were at the time much occupied in assessing

10 Kahm–Leibniz, 12 Apr. 1676 (*LSB* I. 1: 515); *compare* p. 162.
11 Leibniz–Oldenburg, 2 (12) May 1676 (*LBG*: 168–9).
12 Wallis–Digby, 20 (30) June 1658 (*WCE*: 181). The attempted proof is fallacious.
13 *C* 985–90, 1119.   14 *C* 1030.
15 RS meeting, 18 (28) May 1676 (*BH* III: 315).

Linus' and Gascoine's objections to Newton's theory of colours. The tone of the discussion – we cannot here follow it in detail[16] – had become steadily more acrid; between the autumn of 1675 and the next spring Newton was subjected to a stream of criticism. Matters came to a head during a meeting of the Society on 16 (26) December 1675 when Hooke declared[17] in all seriousness that the essence of Newton's 'Hypothesis of Light'[18] was lifted from his own *Micrographia*. The subsequent successful performance of Newton's experiments in front of the Society[19] did not bring the end of the debate since a new adversary, Lucas, now appeared on the field; obstinate and unteachable, he held stubbornly to the objections of Linus, his predecessor at Louvain, and to his own criticism, demanded the publication by the Society of all arguments brought forward by himself[20] and eventually forced Newton to break off what had obviously become by now a fruitless exchange of opinions.[21]

Oldenburg intended to fall in with Leibniz' wishes and send him the more important items from Gregory's correspondence and as much of Newton's latest results as he should think appropriate. He had, after conversation with Collins,[22] began to make some preparatory notes for his reply on the back of Leibniz' letter, but not finding these satisfactory he asked Collins for further information. Collins thereupon gathered all he considered in any way important from Gregory's letters to him – some had already been included in the draft he had given to Oldenburg for his reply to Tschirnhaus – and planned also to add an account of Pell's studies. In addition Oldenburg sent Newton an excerpt from Leibniz' letter, asking him for a contribution to his intended reply to Paris.[23] Newton was at first reluctant to comply and merely gave Collins a slightly altered version of his earlier letter on tangents[24] – one which was in fact already included in Collins' collection. We may perhaps conclude from this that Newton, following his usual custom, had asked Collins for the original of his letter of 10 (20) December 1672 to be returned to him – and now

[16] *Compare* Rosenfeld (1927).
[17] RS meeting, 16 (26) Dec. 1675 (*BH* III: 262–9).
[18] Newton to Oldenburg, 7 (17) Dec. 1675 (*NC* 1: 362–86).
[19] RS meeting, 27 Apr. (7 May) 1676 (*BH* III: 313–14).
[20] Lucas–Oldenburg, 23 Jan. (2 Feb.) 1676/7 (*NC* 2: 192).
[21] Oldenburg–Leibniz, 2 (12) May 1677 (*LBG*: 238).
[22] An extract from Leibniz–Oldenburg, 2 (12) May 1676 in *NC* 2: 3–4 is accompanied in note 1 on p. 5 by a reference to Oldenburg's notes the last of which is reproduced there.
[23] Oldenburg–Newton, 15 (25) May 1676 (*NC* 2: 7).
[24] Newton–Collins, 27 May (6 June) 1676 (*NC* 2: 14–15).

acknowledged that he had no objection to Collins making notes on the letter for his own private purposes, but at the same time indicated that he did not wish the text of the document passed on in an official letter without his direct consent.

The elaborate document which Collins wrote on this occasion – the so-called 'Historiola'[25] – was in Oldenburg's opinion far too prolix and detailed. He wanted a shorter survey, and attached particular value to a direct contribution from Newton, especially with the change in the situation arising from the receipt of Tschirnhaus' letter.[26] It was now desirable to produce more accurate texts to demonstrate that in mathematics the English had gone beyond the results of Descartes in every direction. Collins meanwhile had completed his account of Pell's work, originally planned as an appendix to the letter for Tschirnhaus and then as a supplement to the 'Historiola', and had sent this together with a summary of the 'Historiola' (the so-called 'Abridgment') to Oldenburg,[27] who thought it proper to ask for Newton's opinion on this 'Abridgment' because of the remarks regarding himself contained in it. He did this when he replied[28] to the long letter which Newton intended to be transmitted to Leibniz and Tschirnhaus[29] and which Oldenburg had officially read at a meeting of the Royal Society;[30] at the same time Collins sent an account of the 'Abridgment' to Baker, referring to his survey of Pell's methods and Newton's own letter.[31] For some reason which we do not know Newton's reply was delayed, so that Oldenburg (as we can see from his much altered first draft) first of all began with his translation of Collins' report to Pell and only thereafter worked on the 'Abridgment'. Newton asked to change two passages in the 'Abridgment',[32] one of the alterations is carried out in Collins' own hand, while the other was introduced into the translation. On account of the importance of the whole affair Oldenburg thought it

[25] Collins–Oldenburg, May/June 1676, extract in *CE* (*1712*): 46–7; larger extract *NC* 2: 18–19.

[26] Tschirnhaus–Oldenburg, mid-June 1676, extracts passed on to Newton probably late June, compare *BH* III, 318.

[27] Collins–Oldenburg for Leibniz and Tschirnhaus, 14 (24) June 1676: extract in *CE* (*1712*): 46; another extract in *NC* 2: 50–1.

[28] Oldenburg–Newton, 17 (27) June 1676 (lost); its existence follows from Oldenburg's remark on Newton–Oldenburg, 13 (23) June 1676 (*NC* 2: 41, note 1).

[29] Newton–Oldenburg, 13 (23) June 1676 (*NC* 2: 20–32).

[30] RS meeting 15 (25) June 1676 (*BH* III, 319).

[31] Collins–Baker, 22 June (2 July) 1676, inferred from the reply Baker–Collins, 20 (30) July 1676 (*CR* II: 1, 4).

[32] Newton–Oldenburg, late July 1676, inferred from Newton–Collins, 5 (15) Sept. 1676 (*NC* 2: 95).

proper to retain Newton's autograph letter and his own autograph translation of Collins' draft in his own files, and to forward to Leibniz copies made by his secretary.[33] He himself entered only a few corrections, adding a date and greeting and a postscript to Newton's letter in which he apologizes for employing a copyist because of the extraordinary amount of extra work involved. He had not even found enough time to check the complete text with the care it deserved and was indeed afraid that in a few places some errors distorting the sense might have gone undetected, but a man of Leibniz' experience would readily correct any such corrupt passages. Both letters were expressly intended jointly for Leibniz and Tschirnhaus, and an answer from each of the two recipients was expected. Oldenburg ends with his hope of an early reply from Leibniz and that in it he would also redeem his promise to describe his calculating machine.

We shall now, first of all, take a closer look at Oldenburg's letter – not the 'Historiola' (to which we shall return in a different context) – with only a glance at the 'Abridgment' which, but for some inessential omissions and transpositions, was now incorporated practically without alteration in the translation.

In his opening words Oldenburg expresses his pleasure that Leibniz has written again after so long a silence; both Collins and Newton were much pre-occupied and so the reply came after some delay. Collins gives the law by which the coefficients $\frac{1}{6}$, $\frac{3}{40}$, $\frac{5}{112}$, ... of the inverse sine series are formed as follows: $\frac{1}{6} = \frac{1}{2} \cdot \frac{1}{3}$, $\frac{3}{40} = \frac{1}{6} \cdot \frac{3}{4} \cdot \frac{3}{5}$, $\frac{5}{112} = \frac{3}{40} \cdot \frac{5}{6} \cdot \frac{5}{7}$, ...; this is no less elegant than that for the coefficients of the sine-series which Leibniz had praised, he says, and assures Leibniz that there is lively interest in the results he has announced, particularly since he had indicated that they were of a different nature altogether. Now it appears that by means of the theory of series every difficulty could be overcome, so that Gregory had said that the power of all previous methods had the same ratio to that of series as the glimmer of dawn to the splendour of the noonday sun[34] – this although he had himself previously developed a differing excellent method for the circle, as the following example would show (fig. 26).[35] Let $AB = s_0$ be a chord, $AP_1B = S$ its arc (less than a semicircle), $AT_0 = t_0$ the length of the tangent as far as its inter-

[33] *See* Oldenburg's postscript, 26 July (5 Aug.) 1676 (*LBG*: 192) to Newton–Oldenburg for Leibniz, 13 (23) June 1676.
[34] Oldenburg–Leibniz, 26 July (5 Aug.) 1676 (*LBG*: 171).
[35] *Ibid.* reproducing the contents of Gregory–Collins, 15 (25) Feb. 1668/9 (*GT*: 68–9).

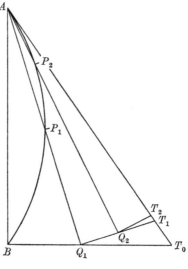

Fig. 26

section with BT_0 perpendicular to the chord; then evidently $s_0 < S < t_0$. Now bisect the arc AB in P_1, produce AP_1 to meet BT_0 in Q_1, at Q_1 erect the perpendicular to AQ_1 meeting AT_0 in T_1; then one obtains correspondingly $s_1 = AQ_1 < S < AT_1 = t_1$. If now P_2 bisects the arc AP_1 and the construction is continued analogously, Gregory derives $s_2 = AQ_2 < S < AT_2 = t_2$; and so on. It can easily be seen that this construction does no more than line up sequences of chords and tangents that have been formed, as in Archimedes' procedure, by repeated angle bisection. Gregory had then, according to Collins, established the inequalities

$$(-22s_0 + t_0 + 96s_1)/75 > (-3s_0 + 16s_1 + 2t_1)/15 > S$$

and $\qquad (-56s_0 - t_0 + 320s_1 + 52t_1)/315 < S$

and further

$$\frac{64s_2 - 20s_1 + s_0}{45} < \frac{4096s_3 - 1344s_2 + 84s_1 - s_0}{2835}$$

$$< \frac{1\,048\,576s_4 - 348\,160s_3 + 22\,848s_2 - 340s_1 + s_0}{722\,925} < S.$$

How these approximations are obtained – they also occur in the form of statements about areas in Gregory's *Exercitationes geometricæ* – we have already explained.[36] Better rational approximations can hardly,

[36] *See* p. 68.

216

in Collins' opinion, be found for the arc of a circle; Gregory had, he tells Leibniz, subsequently extended this procedure to rectifying arcs of other curves. Unfortunately no pertinent details have been preserved, so we cannot determine whether Gregory really knew the extensions of his approximations which hold for a smooth arc without inflexion when the process of angle-bisection is replaced by construction of the sequence of points at which the tangent is parallel to the related chord. Of this type is the extremely close approximation in four terms $\frac{1}{15}(16[2s_1+t_1]/3-[2s_0+t_0]/3)$. This is the place in the 'Abridgment' where there was a reference to Newton which Oldenburg, at his desire, later eliminated from the translation. The approximations quoted were known to Leibniz from reading the *Exercitationes geometricæ*, and in his reply he emphasized how important it would be for him to know the proof of the approximations;[37] he believed he himself possessed still better approximations for a circular arc, without bisecting the arcs.

As a further example of the power of Gregory's methods Collins adduces his treatment of Kepler's problem.[38] (He had misplaced the sheet that was to be included in the 'Abridgment' and so this particular passage did not reach Newton.) The problem occurs at the end of chapter 60 in the *Astronomia nova*: to draw (fig. 27) a line PQ from a given point P on the diameter AB of a semicircle to meet the circumference at Q in such a way that the area is divided in a given ratio $p:q$. For its geometrical solution Wallis had in his *Tractatus de cycloide*,[39] presenting Wren's method, employed a prolate cycloid. Let the radius of the circle be r and, where $OP = d$ is the distance of the given point P from its centre O construct $R = r^2/d$; now let a circle of radius R roll along a line perpendicular to the base OP at D where $OD = R$, and the point A will then describe the cycloid. Its ordinate is $(BT =)\ z = r\sin\theta + R\theta$. Furthermore, where $y = r\sin\theta$ is the ordinate of the circle,

$$\text{area } APQ = \tfrac{1}{2}d.y+\tfrac{1}{2}r^2\theta = \tfrac{1}{2}r^2\,(r\sin\theta+R\theta)/R = \tfrac{1}{2}(r^2/R)z,$$

and $$\text{area } BPQ = \tfrac{1}{2}\pi r^2 - \tfrac{1}{2}(r^2/R)z = \tfrac{1}{2}(r^2/R)(\pi R-z);$$

and so, on setting off $BT = z$, the point T divides the length $BE = \pi R$ in the ratio of the areas $APQ:BPQ$. Gregory had taken up this problem in proposition 35 of his *Vera quadratura*: at the end

[37] Leibniz–Oldenburg, 17 (27) Aug. 1676 (*LBG*: 198). The subsequent allusion is probably to the series for the inverse tangent.
[38] Oldenburg–Leibniz, 26 July (5 Aug.) 1676 (*LBG*: 171).
[39] Wallis, *Tractatus duo* (1659), 80–1.

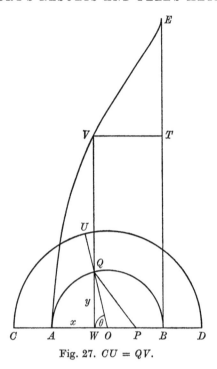

Fig. 27. $CU = QV$.

of 1670 he mentions in a letter to Collins[40] that he can give a solution by development in series, and this before unravelling the puzzle of Newton's series for a circular zone. From his re-discovered worksheets it is now clear how he proceeded.[41] Gregory takes the equation of the circle in the form $y^2 = x(2r-x)$ so deriving the relation

$$\mathrm{d}x/y = \mathrm{d}y/(r-x) = \mathrm{d}s/r,$$

and then introduces $t = s+y$ (the ordinate of a common cycloid). He then obtains

$$\mathrm{d}x/\mathrm{d}t = y/(2r-x) = x/y$$

and

$$\mathrm{d}y/\mathrm{d}t = (r-x)/(2r-x) = (rx-x^2)/y^2,$$

whence by repeated differentiation follows

$$\frac{\mathrm{d}^2x}{\mathrm{d}t^2} = \frac{rx^2}{y^4}, \quad \frac{\mathrm{d}^3x}{\mathrm{d}t^3} = \frac{2rx^4}{y^7},$$

$$\frac{\mathrm{d}^4x}{\mathrm{d}t^4} = \frac{2r^2x^5 + 6rx^6}{y^{10}}, \quad \frac{\mathrm{d}^5x}{\mathrm{d}t^5} = \frac{22r^2x^7 + 24rx^8}{y^{13}}, \dots.$$

[40] Gregory–Collins, 23 Nov. (3 Dec.) 1670 (GT: 120).
[41] GT: 360–1, 363.

218

With $x_0 = r$, $y_0 = r$ this yields a Taylor expansion in the form

$$-x = \frac{t}{r}x_0' - \frac{t^2}{r^2}\frac{x_0''}{2!} + \frac{t^3}{r^3}\frac{x_0'''}{3!} - \ldots = t - \frac{t^2}{2r} + \frac{t^3}{3r^2} - \frac{t^4}{3r^3} + \frac{23t^5}{60r^4} - \ldots$$

Gregory now proceeds to the prolate cycloid by substituting

$$z = Rs/r + y$$

and finds by similar reasoning that

$$\frac{dx}{dz} = \frac{y}{R+r-x}, \quad \frac{dy}{dz} = \frac{r-x}{R+r-x}, \quad \frac{d^2x}{dz^2} = \frac{r^2+(r-x)R}{(R+r-x)^3}, \ldots$$

With $x_0 = r$, $y_0 = r$ this now leads to

$$r - x = \frac{rz}{R} - \frac{r^2z^2}{2!R^3} + \frac{rz^3(3r^2-R^2)}{3!R^5} - \frac{r^2z^4(15r^2-7R^2)}{4!R^7}$$
$$+ \frac{rz^5(105r^4-60r^2R^2+R^4)}{5!R^9} - \ldots$$

In the account sent by him to Collins[42] only the bare result is given with the remark that the expansion is of use when P is near to B, that is when $R \approx r$, otherwise the following expansion, where $R+r = e$, is to be preferred (it arises when $x_0 = 0$, $y_0 = 0$):

$$x = \frac{rz^2}{2!e^2} + \frac{rz^4(4r-e)}{4!e^5} + \frac{rz^6(70r^2-26er+e^2)}{6!e^8} + \ldots$$

Now all that remains is to set $z = \pi pR/(p+q)$. Gregory gives as an instance the rapidly converging series which results in the case $r^2/R = r/9$, that is for $e = 10r$. As Gregory sent simply the final expansion as a ready-made formula, Collins was surprised and dissatisfied that the general method for forming the terms was missing. In consequence the publication originally planned did not take place[43] – a decision about which Gregory later rightly complained.[44] After Gregory's death his elder brother David asked for[45] and obtained[46] a copy of this piece; new plans were made for publication[47] but once more they came to nothing. Leibniz made a note

[42] Gregory–Collins, 9 (19) Apr. 1672 (*GT*: 227–8).
[43] Collins–Gregory, 28 May (7 June) 1672 (*GT*: 233); 26 Dec. 1672 (5 Jan. 1673) (*GT*: 248).
[44] Gregory–Collins, 11 (21) Sept. 1675 (*GT*: 329); Collins–Gregory, 21 Sept. (1 Oct.) (*GT*: 330); Gregory–Collins, 2 (12) Oct. (*GT*: 334–5).
[45] David Gregory–Collins, 10 (20) June 1676 (*CR* I: 225).
[46] Collins–David Gregory, 11 (21) Aug. 1676 (*GT*: 344).
[47] Collins–Wallis, 16 (26) Feb. 1676/7 (*see CR* II: 604–6).

about the problem[48] but did not, as far as we know, afterwards return to the subject.

Following his report on Kepler's problem Collins remarks[49] that series also find an effective use in the treatment of equations though, on account of the great number of indeterminate quantities involved, their application tends to be lengthy and will prove reasonably convenient only in some special cases. He then adds, thereby supplementing certain earlier communications of his, the series[50]

$$r \tan s/r \, = \, s + \frac{s^3}{3r^2} + \frac{2s^5}{15r^4} + \dots$$

and

$$r \log \tan \left(\frac{\pi}{4} + \frac{s}{2r} \right) \, = \, s + \frac{s^3}{6r^2} + \frac{5s^5}{24r^4} + \dots$$

which, as we have already remarked, were determined by repeated differentiation. Gregory came upon his method, according to Collins, after he had been shown the series for a circle-zone,[51] yet this was nothing but a special case of Gregory's binomial expansion.[52] We have seen above that Gregory apparently hit upon his general method somewhat earlier, but at the time Collins was unable to assess all these details accurately. He had, he goes on, after Gregory's death collected extensive excerpts from their exchange of letters (namely the 'Historiola') in which the history of the development of the theory of series was also given. On this occasion (he had originally intended to say nothing on this subject) Collins makes a mention of Newton's tangent-method found by him some five years previously; at Newton's suggestion Collins gave the date of the letter on tangents[53] but avoided quoting the explanatory example without which the method he had used was of course quite unidentifiable; he cites only Newton's final general remark from the letter to the effect that the new procedure is to be regarded as merely a corollary to a more general method by which, without a great deal of labour, tangents to all geometrical and mechanical curves can be found and several other

---

[48] C 1454.

[49] Oldenburg–Leibniz, 26 July (5 Aug.) 1676 (*LBG*: 171) reproduces the contents of Gregory–Collins, 9 (19) Apr. 1672 (*GT*: 227–8).

[50] Oldenburg–Leibniz, 26 July (5 Aug.) 1676 (*LBG*: 171–2), taken from Gregory–Collins, 15 (25) Feb. 1670/1 (*GT*: 170); this supplements the record in Oldenburg–Leibniz, 12 (22) Apr. 1675 (*LBG*: 115).

[51] Oldenburg–Leibniz, 26 July (5 Aug.) 1676 (*LBG*: 172), referring to Collins–Gregory, 24 Mar. (3 Apr.) 1669/70 (*GT*: 89); *see* also ch. 7: note 78 and the reports mentioned in ch. 7: note 77 sent out by Collins concerning Newton's results.

[52] This corresponds to what is written in Gregory–Collins, 19 (29) Dec. 1670 (*GT*: 148).   [53] Oldenburg–Leibniz, 26 July (5 Aug.) 1676 (*LBG*: 172).

related questions – such as curvature, areas, length of arcs, centres of gravity and the like – dealt with; it was not, like Hudde's method for extreme values or Sluse's new tangent-method, restricted to equations free from radicals. Newton had, Collins goes on to say, linked it with his series-expansions and had given some hint of it to Barrow when the latter was preparing to publish his *Lectiones geometricæ* (1670), but had then omitted to communicate a fuller account through other preoccupations. This is perhaps merely a polite way of saying that Newton would have found it tactless to show his new method to Barrow at the very moment when the latter was developing his own variant of it in the *Lectiones*. In consequence of his reading of Barrow, Collins tells us, Gregory had also arrived at a general tangent-method[54] that he was able to apply to such complicated cases as the *spiralis arcuum rectificatrix*. A revised text of Gregory's *Vera quadratura* was nearly completed.[55] To avoid the inconveniences of series-expansions Gregory had adopted a certain type of symbolic method by whose means he could remove all intermediate terms while raising the degree of the equation;[56] yet, as Collins relates, Gregory did not believe that Tschirnhaus' special methods had anything to do with his own general one.[57] Gregory's unfavourable opinion of Tschirnhaus, which (as we have mentioned several times above) derived from his overestimation of the Cartesian method, was of course left out in the final version of the letter. Gregory was in fact, Collins continues, now able to reduce any equation[58] where from any one solution all others can be found rationally to a linear equation; where the solutions depend rationally on any two, a quadratic equation will be needed, and so on. For his general procedure he used a representation by radicals with variations of signs.[59] In some of his letters, moreover, Gregory had spoken of

[54] Gregory–Collins, 5 (15) Sept. 1670 (*GT*: 103). The details concerning the construction of the tangent of the *spiralis arcuum rectificatrix* $r^2 - 2ar \cos \theta + a^2 = a^2\theta^2$ are found in Gregory for Collins, 23 Nov. (3 Dec.) 1670 (*GT*: 134–7).

[55] This remark is based on Gregory–Collins, 23 Nov. (3 Dec.) 1670 (*GT*: 118); nothing has hitherto been found amongst Gregory's papers indicating work on a new edition of the *Vera quadratura*.

[56] After recapitulating a relevant remark from his letter to Collins, 17 (27) May 1671 (*GT*: 187–8), Gregory writes that in order to reduce a quintic equation into a solvable form he has to employ an auxiliary equation of degree 20 = 5 × 4: Gregory–Collins, 17 (27) Jan. 1671/2 (*GT*: 211). For the method used *see* p. 153.

[57] Oldenburg–Leibniz, 26 July (5 Aug.) 1676 (*LBG*: 173) referring to Gregory–Collins, 20 (30) Aug. 1675 (*GT*: 325). *See* also p. 172.

[58] Collins here refers to Gregory's letter of 11 (21) Sept. 1675 (*GT*: 328).

[59] Oldenburg–Leibniz, 26 July (5 Aug.) 1676 (*LBG*: 174), referring to Gregory–Collins, 2 (12) Oct. 1675 (*GT*: 335).

new successes in dealing with Diophantine problems,[60] and so too, in fact, had Pell. Gregory had also worked on the graphical solution of equations[61] and tried his hand at Davenant's problem,[62] for which he needed an equation of thirtieth degree, though it was Collins' own belief that equations of much lower degree would be sufficient.

The 'Abridgment' conveys, as this summary shows, by a few aptly chosen examples quite an imposing impression of Gregory's achievements in the fields of power series and algebra. No proofs are given, but Collins is not to blame for that; he himself had received from Gregory no more than certain individual results with some indication of their significance, but had never been told anything concrete about the methods employed. This first part of the letter was meant to satisfy Leibniz' curiosity and to demonstrate to him and to Tschirnhaus the scope of the highly important achievements which stood to the gifted Scotsman's credit. Would Tschirnhaus still wish to contend that all this was no more than the working out of Cartesian ideas and that the Cartesian tangent-method was the best possible one?

In the second part of his letter[63] Collins takes up Leibniz' remark about Tschirnhaus' work on angular sections and regular polygons. He reports on a theorem in Gregory's *Geometriæ pars universalis* (prop. 69)[64] bearing on the topic and its more extended version in Wallis' *Sectiones angulares*,[65] the manuscript of which was at that moment in his custody in London, and takes a somewhat closer look at Pell's contributions to algebra. Collins reminds us how Pell had already in 1650 in his 'Idea of mathematicks' demanded that everything that earlier mathematicians had none too systematically, but indeed by chance, discovered ought to be uniformly collected in a clear, lucid and organic order so that all conceivable problems would be classified and the aids required for their solution categorized or, where appropriate, the impossibility of a solution demonstrated. This might be said to be the content of his 'Cribrum Eratosthenis' of which Boyle had already given a survey (where, is now no longer

[60] For an example we may point to Gregory's treatment of the problem of six squares; *see* p. 90.
[61] An instance is preserved in the 'Historiola'.
[62] *See* below, pp. 255–6.
[63] Collins' draft, 14 (24) June 1676 is printed in *CR* I: 243–8; in translating it Oldenburg has made slight changes, *see* his letter to Leibniz, 26 July (5 Aug.) 1676 (*LBG*: 176–9).
[64] $2n+1$ equidistant points A_k on the circumference are joined to a further point P on the circle, then $\sum_{k=1}^{2n+1} (-1)^k A_k P = 0$; *see* Hofmann (1969c): 143–5.
[65] On the contents of Wallis' treatise on *Angular sections* see Scriba (1966): 21–3.

known). Pell had no very high opinion of Descartes; Collins reports him as saying that previous algebraists had not worked carefully enough and thus had left a number of ripe fruits to be harvested by the epigones (this sentence is left out by Oldenburg in his translation). Descartes had spoken unfavourably of Pell in a letter of which Theodor Haak possessed a copy, but he wished not to divulge it so as not to hurt his friend Pell (this passage too is suppressed by Oldenburg). Perhaps, Collins suggests, Clerselier has found something about this amongst Descartes' papers and could make it public?[66]

Pell, he tells us,[67] had from time to time shown a large (bundle of) manuscript containing his notes on equations of sixth degree with some 400 examples arranged in order of magnitude of the coefficients, together with their solution by the aid of sine-tables. He avoided Viète's rather cumbersome approximation procedure and had need of no lemmas on angular sections, but only of very accurate tables. Having first determined the truly lowest degree of a given problem he then solved it by a step-by-step use of limits for the roots, a technique whose theory had so far only incompletely been developed: the details were set out in his 'Canon mathematicus'. With the use of series he had evidently not occupied himself very thoroughly – he fully acknowledged their effectiveness but thought his own approach better. Hitherto, Collins confirms yet again, he could not be induced to divulge his discoveries; he would communicate them only after those of Gregory and Newton had been published when they could be judged in their full extent. Newton, we hear, had given lectures on his methods of series and on algebra, and had deposited their text in the University Library. By this phrase Collins signifies – though the uninitiated reader will scarcely guess as much from this indication – that Newton had foresworn any intention of printing his lectures on mathematics.[68] (These subsequently published as his *Arithmetica universalis*, were said to be delivered between October

[66] Oldenburg–Leibniz, 26 July (5 Aug.) 1676 (*LBG*: 178). Descartes' opinion is contained in his letter to van Hogelande, 8 Feb. 1640 (*DO* supplément: 2–4 = *MC* 9: 308–9); it is not at all wholly unfavourable. Descartes bases himself on Pell's pamphlet *Idea matheseos* (1638). It was reprinted as 'Idea of Mathematicks' in 1650: *see* P. J. Wallis (1967). Eventually with Haak's consent, Hooke published both the *Idea* and Descartes' letter together with some supplementary pieces in the *Philosophical Collections* 5 of February 1682/3: 127–37.
[67] Oldenburg–Leibniz, 26 July (5 Aug.) 1676 (*LBG*: 177–8).
[68] The manuscript deposited by Newton in the University Library, Cambridge has now been edited in *NP* 5: 54–491, with additional matter on pp. 492–517. The unauthorized *editio princeps* by Whiston (1707) has been thoroughly reviewed by Leibniz in the *AE* (November 1708): 519–26 = *NP* 5: 23–31.

1673 and October 1683 and do not touch on infinitesimal problems at all.) Collins' remark is somewhat erroneous. According to the founder's statutes for the Lucasian professorship[69] which Newton held in Cambridge he was obliged to give a public lecture once every week in term time and be available twice a week for two hours for private tuition. He was at liberty either to have his lectures printed or to deposit the texts of ten lectures each year at the University Library for perusal by members of the University.

At the end of his letter Collins expresses the hope that he has satisfied the curiosity of the Paris mathematicians with his account, and in return now asks for some account of their own plans and results. In the English draft Leibniz is advised of the impending visit to Paris of an English musician of the name of J. Smith, who was on his way back from Rome, bringing new books which he had procured with Borelli's assistance; Leibniz' friends are asked to help him acquire Frénicle's *Traité des triangles en nombres* and La Hire's latest work on conics (*Nouvelle Methode*, 1673). This passage is omitted in Oldenburg's translation – the transmission of the letter had been delayed for weeks and the commission had presumably become outdated.

[69] The statutes concerning the professorship are reproduced in *NP* **3**: xx–xxvii.

16

NEWTON'S FIRST LETTER
FOR LEIBNIZ

We have already drawn attention to the highly charged atmosphere in which the decisive debate between Leibniz and Newton was now to take place. A variety of causes had contributed to make the English suspicious. During his first stay in London Leibniz had quite seriously compromised his reputation and this had not been forgotten by his opponents. He had made a lot of promises and kept very few: what substance lay behind his words and hints? Had he in fact penetrated to new insights or did he only pretend to their knowledge so that he could appropriate the finest results of the English? What provoked him to revert once more to the sine and arcsine series which he had been previously given? Did he really believe that Newton would divulge his method, carefully kept back and anxiously guarded as it was? You could not call him unadroit, this German; a man who could say such clever things about the nature of the new sciences might well be expected to have reached equivalent results or indeed – and who was to judge – something even better. Well, one had to take the risk: Oldenburg, the faithful mediator in the uneasy disputes over the theory of colours, was to have his way and be given a summary of the most important results – only results, it was understood, not methods or proofs, particularly since Newton himself was not yet quite clear about the best and most sensible form in which to present them: the *De Analysi*, of which Collins had previously made a copy, was now no longer a satisfactory account of his series method and the *Methodus* augmenting its scope was still unfinished. But even in solely transmitting details the greatest caution had to be observed. A charlatan would be unable to do much with them; yet to someone who knew enough to divine the rest, the slightest indiscretion might reveal too much. On the other hand there was little hope of producing a coherent exposition in print within a foreseeable space of time; a carefully thought out letter gave one the chance of securing one's priority for the most important results and the general principles which could there be stated in an impenetrably disguised form. The utmost reserve seemed necessary; had not Descartes all too rashly prided himself that his methods were the best and most effective and had he not now been far outdistanced by Newton's researches? And Leibniz was no Tschirnhaus, no blind and credulous campfollower of the Cartesians, but a profound and independent thinker. This was evident from the few details that Newton had so far gleaned from his letters to Collins and Oldenburg. So Newton took some of his earlier papers up again and attained a number of new and interesting insights in reworking his older notes.

Right at the beginning[1] Newton emphasizes that he intends to give only a selection from his work on series; Leibniz had most likely contrived a similar procedure and presumably obtained comparable, if not better, individual results. To obtain a series-expansion for fractions we are to use long division, and for the solution of equations the well known procedure of Viète in his *Resolutio numerosa* or of Oughtred in his *Clavis*, remodelled in letters, but in a shorter and improved form. For ordinary radicals[2] the formula

$$(P+PQ)^{m/n} = \underbrace{P^{m/n}}_{A} + \underbrace{\frac{m}{n}AQ}_{B} + \underbrace{\frac{m-n}{2n}BQ}_{C} + \underbrace{\frac{m-2n}{3n}CQ}_{D} + \dots$$

where m/n can be any positive or negative fraction, is of great value, and division too can be dealt with by this expression. This he illustrates by examples of varying degrees of difficulty.

He next reproduces his scheme for solving the equation

$$x^3 - 2x - 5 = 0$$

[1] Newton–Oldenburg for Leibniz and Tschirnhaus, 13 (23) June 1676: (*NC* 2: 20–32), transcript passed on by Oldenburg on 26 July (5 Aug.) 1676. Extracts from this extensive letter were included, in translation, in Wallis' *Algebra* (1685): 330–4, 338–9, 341–6; these extracts are somewhat enlarged in the Latin edition of the *Algebra* (1693): 368–72, 376–7, 381–9. The full text in *WO* III (1699): 622–9 is based on a slightly variant version probably sent to Wallis by Newton in May 1695. Wallis returned his own transcript for checking, 30 May (9 June) 1695 (*NC* 4: 129) and after further prompting by him, 3 (13) July 1695 (*NC* 4: 139–40), Newton proposed a few alterations, probably in July 1695 (*NC* 4: 140–1), relating however not to the first, but only to the second letter of 24 Oct. (3 Nov.) 1676. The copy sent to Leibniz contained the errors of the original version as well as a number of mistakes made by the scribe which Oldenburg had not found time to expunge by collation with the original. In the following we shall, based on *NP* 4: 667–70, give references simply to *NC* 2 without details of date etc.

[2] *NC* 2: 21–3. On the invention of the binomial expansion *see* the remarks to Newton–Oldenburg for Leibniz, 24 Oct. (3 Nov.) 1676: p. 261.

in numbers, and correspondingly $y^3 + axy + a^2y - x^3 - 2a^3 = 0$ in letters but the explanation he gives is wholly inadequate.[3] He instances its application in deriving the series[4] for $\sin x$ and $\sin^{-1} z$ followed by the expression[5] of $y = \sin nt$ as a function of $x = \sin t$. Kepler's problem for the ellipse $(y/b)^2 = x(2a - x)/a^2$ is treated: Newton sets[6] the distance of the point of division E from the origin B to be t and expresses the ordinate y of a point G as a function of

$$z = 2 \text{ area } BEG = 2\int_0^x y \,.\, \mathrm{d}x + (t - x)y = ty + \int_0^y y^2 \,.\, \mathrm{d}(x/y).$$

To do so we need to expand

$$\frac{x}{y} = \frac{a}{y}\left\{1 - \sqrt{\left[1 - \left(\frac{y}{b}\right)^2\right]}\right\}$$

and then, having substituted it, to determine z. To evaluate the length of arc $s = \int_0^x \sqrt{[1 + (\mathrm{d}y/\mathrm{d}x)^2]} \,.\, \mathrm{d}x$ of an ellipse Newton[7] takes its equation to be $y^2 = b^2 - kx^2$ or $y^2 = px - kx^2$; he explicitly gives the general law for the progression of the coefficients, determines the circumference of the ellipse from the arcs corresponding to the two halves of the semi-major axis, and also expresses x as a function of s. Finally he indicates how one can easily, by changing the sign of k, pass to the hyperbola. To find the area of a hyperbola[8] he takes

$$xy = ab, \quad \int_a^x y \,.\, \mathrm{d}x = \int_0^t ab \,.\, \mathrm{d}t/(a + t)$$

[3] *NC* **2**: 22–4. The examples are taken from the *De Analysi* (1669); the numerical one is printed in *NP* **2**: 218–20, the algebraical one *ibid.* 222–4.

[4] *NC* **2**: 25. The series for the inverse sine appears first in the autumn of 1665 (*NP* **1**: 125), then again in the *De Analysi* (*NP* **2**: 232) where the sine-series also occurs.

[5] *NC* **2**: 25. As can be seen in *NP* **1**: 476–81, Newton in 1664–5 was stimulated by his reading of Viète's *Ad angulares sectiones* (1615) = *Opera* (1646): 294–5, 291 to express, where $x = 2 \cos \theta$, $x_n = 2 \cos n\theta$, by means of $x_{n+1} = x \,.\, x_n - x_{n-1}$, step-by-step x_n as a function of x, and later to explore $y = 2 \sin \theta$, $y_n = 2 \sin n\theta$: this may have led him – though it is not explicitly stated in his papers – to the recursion $y_{n+2} = (2 - y^2)y_n - y_{n-2}$ and hence to an expression of y_{2n+1} in terms of odd powers of y. The relation given by him on *NC* **2**: 25 arises from comparing the numerical coefficients, with an intuitively formed law for their formation to be used in Wallisian style for a fractional index n. Newton's formula appears in print first in the letter-extract in *WO* II: 384. It was first proved by De Moivre in *PT* **20**, No. 240 for May 1698 (1699): 190–3, then in a treatise by Jakob Bernoulli of 1702 (printed 1704; *BKC*: 926) where (on p. 928) a direct reference is given to the place in *WO* II.

[6] *NC* **2**: 25–6. No earlier occurrence is known.

[7] *NC* **2**: 26–7. The rectification of the ellipse is first dealt with in *De Analysi* (1669): *NP* **2**: 232; the present text follows the treatment in the *Methodus* (1671): *NP* **3**: 326.

[8] *NC* **2**: 27. The calculation appears in *De Analysi*, *NP* **2**: 234 and in the *Methodus*, *NP* **3**: 58.

and by inversion of the series thereby obtained finds the number as a function of its logarithm. For the quadratrix[9] $y = x \cot(x/a)$ he gives the series-expansion of y itself, and those of the area $\int_0^x y \, dx$, of the arc $\int_0^x \sqrt{[1 + (dy/dx)^2]} \, dx$ and the contrary subtangent

$$y - x \, dy/dx.$$

Newton ends with the series-expansion for the volume of the so-called second segment[10]

$$\int_0^d \int_0^c b\sqrt{[1 - (x/a)^2 - (y/b)^2]} \, dx \, . \, dy$$

of an ellipsoid of revolution

$$(x/a)^2 + (y^2 + z^2)/b^2 = 1.$$

When we look at each of the theorems or examples cited we find that Newton has deliberately communicated only those results that were previously known in one form or another. The note in the *Commercium epistolicum* contains merely a vague indication.[11] The series for the sine and inverse sine, together with certain remarks on the quadratrix, were (surely with Newton's approval?) sent by Collins to Gregory,[12] who for his part (following hints by Collins) soon after achieved the rectification of the ellipse and the hyperbola[13] and found the volume of the second segment of the ellipsoid of revolution;[14] Gregory's solution of the Kepler problem was achieved wholly independent of Newton. The latter had nowhere laid explicit claim to any of the results due to Gregory, but then again he had found it nowhere necessary to name the Scotsman, with the consequence that it was generally believed until recently that these were all Newton's own discoveries. We have indeed no reason to accuse him of any conscious deception in this matter, since on the one hand Gregory had expressly renounced any claim to priority in these results,[15] and on the other had recognized Newton as the original

[9] *NC* **2**: 27; *compare De Analysi*, *NP* **2**: 236–40, *Methodus*, *NP* **3**: 326–8.
[10] *NC* **2**: 28; *compare Methodus*, *NP* **3**: 288–90.
[11] *CE* (*1712*): 49, note *. We now know that Newton has employed binomial expansions as early as 1664–5; *compare* ch. **19**: note 10.
[12] Collins–Gregory, 24 Dec. 1670 (3 Jan 1671) (*GT*: 155).
[13] Gregory–Collins, 15 (25) Feb. 1670/1; 17 (27) May 1671 (*GT*: 171, 189–90).
[14] Gregory–Collins, 17 (27) May 1671 (*GT*: 188–9).
[15] Gregory–Collins, 15 (25) Feb. 1670/1 (*GT*: 171).

discoverer of the method of series, declaring his intention of with-
holding his own researches from publication till after Newton's
awaited account came out,[16] while lastly, Newton believed he had
no need to cite Gregory's independent discovery since, in his opinion,
only first discoverers (and none of their successors) possessed any
rights to an invention.[17] When, moreover, he wrote his letter to
Oldenburg, Newton had as yet no detailed knowledge of the contents
of the 'Abridgment'; all he was aware of was that Leibniz was to
receive excerpts from the letters exchanged between Gregory and
Collins. Conceivably the intention was to transmit in it all the
separate subjects treated by him and, the rule for the coefficients of
the series for the inverse sine apart, this is in fact what happened in
the case of a question treated simultaneously and independently by
Newton and Gregory, namely Kepler's problem: by a strange
coincidence this very piece was missing when the 'Abridgment' was
sent to Newton. It is not quite certain whether Newton even heard
anything from Collins about the series for $\sin nt/\sin t$; the problem of
angular section was at the time very much a subject of general
discussion.

Concerning the method employed, it is certain that in developing
the series Newton proceeded by successive approximations while
Gregory used repeated differentiation coupled with cleverly chosen
rationalizing parameters. He had developed this method following the
publication of his *Geometriæ pars universalis* in the summer or
autumn of 1668, and had combined it with his method of interpolation invented at about the same time: these results he had
arrived at quite independently of Newton and had reached a more
comprehensive viewpoint. Collins had indeed expressly confirmed[18]
that it was possible to treat the same topic in a differing way but
evidently with equal success by Newton's method of series and by
Gregory's method of interpolation. It will not do to see in Gregory a
'second' discoverer who had been influenced by Newton: he had
arrived at his decisive discoveries almost simultaneously and wholly
independently. Newton from the start works in an analytical manner,
but not so Gregory. In his *Geometriæ pars universalis* everything is
still presented verbally in purely geometrical form and it cost
Gregory much effort to free himself from this traditional way of

16 Collins–Strode, 26 July (5 Aug.) 1672 (*CE* (*1712*): 29).
17 Newton expressed this opinion most emphatically in the 'Account': *PT* **29**,
   No. 342 (for January–February 1714/15): 215, and subsequently.
18 Collins–Gregory, 24 Dec. 1670 (3 Jan. 1671) (*GT*: 156).

demonstration. Of all this – only now properly to be comprehended – Newton was wholly ignorant, otherwise he would scarcely have made his priority claim.

In the last section of his letter[19] Newton applies the method of series-expansion to the approximate construction of transcendental problems. In sequel to Huygens' results in the *De circuli magnitudine inventa* (propositions 12 and 16) he gives the relations

$$\theta = \frac{16 \sin \theta/4 - 2 \sin \theta/2}{3} + \frac{\theta^5}{16.5!} + \cdots$$

and

$$\frac{\theta \operatorname{versin} \theta}{\theta - \sin \theta} = 3 - \frac{\operatorname{versin} \theta}{5} - \frac{6 \operatorname{versin}^2 \theta}{175} - \cdots,$$

and from the equation $y^2 = 2px + kx^2$ develops approximations to a segment of a circle and an arc of a conic. Newton here continues a theme inaugurated in the *Methodus* but is everywhere careful to remain within the bounds of the binomial series stated at the beginning of the letter. Oldenburg's desire for a communication on his part had thus been satisfied and yet everything was done to prevent Leibniz from, as it were, improperly penetrating the world of Newton's thought. All that had been exposed were a few series already otherwise known, here augmented on occasion by a rule for their formation, but nothing was said of the central problems – nothing of his *methodus fluxionum* or the differential equations into whose solution by power series Newton already possessed considerable insight.

As if to prevent his autograph being transmitted to Leibniz, Newton added on the last page of his letter a few remarks on Lucas' objections to his theory of colour.[20] He expected Oldenburg to forward a careful copy of the letter reasonably quickly to Paris and repeated at the end for the benefit of the copyist a few passages that had become difficult to read because of the corrections he had made. Later, he asked Oldenburg to return the original. A transcript of the original made by Collins went to Baker,[21] another, made for Oldenburg, was given to Wallis[22] who took over some portions of it into his English *Algebra* in 1685. Newton's letter[23] reached Oldenburg on

[19] *NC* **2**: 29–31; compare *Methodus*, *NP* **3**: 288–90.
[20] *NC* **2**: 32.
[21] It is apparent from Collins–Baker, 19 (29) Aug. 1676 (*CR* II: 4), that already a fortnight earlier an excerpt from Newton's letter with results on series had been dispatched; on 21 Sept. (1 Oct.) 1676 there followed additionally the procedure for solving numerical equations (*CR* II: 10).
[22] *WO* II: 368.
[23] This follows from Oldenburg's note on Newton's letter, *NC* **2**: 41, note 1.

13 (23) June 1676, was read at the Royal Society on 15 (25), acknowledged and answered on 17 (27); and after some delay it was forwarded to Leibniz on 26 July (5 August) 1676. Through a printing error (not adequately corrected in the Errata) the date it was forwarded is given as 26 June (6 July) 1676 in the third volume of Wallis' *Opera*: this date is then repeated in the two editions of the *Commercium epistolicum*[24] and might be taken to demonstrate that the original was not again inspected. It was reasserted by Newton in his (anonymous) English review of the *Commercium epistolicum*[25] and was often to be repeated later on; it has only finally been refuted since the rediscovery of the original covering letter. It remains incomprehensible that in 1712 Newton did not think of looking again at the original which was in his hands, nor in 1722 when a new edition of the *Commercium epistolicum* was issued. The 'editor' of the *Commercium epistolicum* evidently wants the reader to believe that Newton's letter was sent to Leibniz along with the 'Abridgment' on 26 June (6 July) 1676, followed by the 'Historiola' on 26 July (5 August) 1676. Confirmation of this view seems to come from a note on the first sheet of the 'Historiola' asking that this document should be returned by Leibniz after he had perused it.[26] But this is all wrong. The 'Historiola' was never sent to Leibniz, though he inspected it during his second visit to London in October 1676, and the above-mentioned note refers to this occasion. We shall later revert to this whole affair, and then produce detailed proof to show that Leibniz cannot have seen the 'Historiola' any earlier than October.

[24] *CE (1712)*: 49–56 = *CE (1722)*: 131–44; the erroneous remark about the forwarding date is at the head of the letter.
[25] 'Account' (1715): 186, *compare* note 17.
[26] *CE (1712)*: 46.

17

LEIBNIZ' REPLY

To ensure its safe delivery Oldenburg was unwilling to entrust the letter-packet of 26 July (5 August) 1676 to the post but gave it to a young German mathematician from Breslau, one Samuel König,[1] who was just leaving for Paris. As happened frequently when correspondence was given to a private individual to deliver, this led to an awkward delay in its transmission. Not finding Leibniz at home, König deposited the packet at the German apothecary's shop where Leibniz found it when he chanced to call on 24 August. Leibniz remarked briefly on this in his acknowledgment, but this passage was deleted by Oldenburg as unimportant in transcribing it in the Royal Society's Letter Book and so it was omitted from the standard text of the letter published by Wallis in the third volume of his *Opera* in 1699: the full version has only recently been printed.[2] The fact of the late arrival of Oldenburg's letter-packet only came to light after the autograph of Leibniz' reply of 27 August was rediscovered and the original draft of 24 August deciphered. In this reply, as in its second draft, Leibniz writes that he had received the packet 'yesterday' (on the 26th); in the first draft he says more accurately that it had been handed to him 'on Monday' (the 24th) and that he wished to

---

[1] *LBG*: 192.

[2] For Leibniz's letter two (hitherto unpublished) drafts exist, of 24 Aug. 1676 (*C* 1505 A, 1505 B), as well as the final version, *C* 1505, which was copied by Collins with numerous misreadings. The transcript sent to Newton has been preserved, though not those sent to Wallis on 14 (24) Sept. and to Baker a week later. In September 1697 Newton lent his copy to Wallis as is evident from David Gregory's note of 12 (22) Sept. 1697: *compare* Scriba (1969): 74. Wallis in a letter to Leibniz, 21 (31) Oct. 1697 (*LMG* IV: 44) asked for his consent to publish it. While readily granting this, 24 Mar. (3 Apr.) 1697/8 (*LMG* IV: 44), Leibniz hoped it would be remembered that he had written this and other letters to Oldenburg with youthful fervour and haste before attaining the full height of the necessary mathematical knowledge. The reproduction of Leibniz' letter in *WO* III: 628–33 is based on the faulty transcript for Newton as are the later reprints of which *CE* (*1712*): 58–65 is important because of its footnotes and because it carries Leibniz' own autograph notes. The full original text appears only in *NC* 2: 57–71 and our references will be to this (without further specification) and to *C* 1505 A, 1505 B for the drafts.

reply at once so that he could dispatch his letter by the regular post 'on Wednesday' (26th). Both drafts are hastily written, as is the fair copy. The text is properly thought out only in parts, and carelessly structured.

As soon as he glanced over the contents of Newton's letter Leibniz grew aware that its communications were extraordinarily wide-reaching and important, and that any delay in his reply might – indeed, must – cast the darkest suspicion on his motives. This thought, together with Oldenburg's appeal for a quick answer, induced him to an extempore reply, so avoiding the suspicion of having beforehand thoroughly studied the letter. The first draft of 24 August was presumably meant to be turned into the fair copy to be sent the next day, but, not satisfied with it, he re-wrote and enlarged it. When he remarks that he had received Oldenburg's packet 'yesterday' (*heri*) Leibniz perhaps means to signify that to speed his reply he has dashed off its text in great haste, thus excusing his far from easily legible handwriting and a style of writing which he himself felt to be rather uneven. He had every reason to make an apology: mistakes in the writing of formulae are left uncorrected, and Collins, when he came to transcribe it for Newton, could not completely decipher it and the copy he forwarded contained a number of serious errors. Of this we shall have more to say below.

Newton accepting the erroneous date of 26 June (6 July) for its dispatch later maintained that it had taken Leibniz more than six weeks after receiving Oldenburg's letter to reply on 27 August, using the intervening time to appropriate the essential portion of what he had received and then impart it in an altered form as his own invention: this imputation is utterly false. Newton was misled into making such an assertion by his biased reading of the printed excerpts before him; had he cared to look at the originals (easily accessible to him) he would immediately have recognized that his assertion was quite untenable. Oldenburg's letter was answered by the next post, if not actually on the day after its receipt as Leibniz pretends. In one day Leibniz could not have managed to make the two drafts and lengthy fair copy and discuss the letter with Tschirnhaus[3] (who saw the reply and adapted his own to it). In reality there was an interval of exactly three days between Leibniz' receiving the letter and his reply. Not even a superhuman mathematical genius could in so short a time have invented the arithmetical quadrature of the circle if he

[3] Tschirnhaus–Oldenburg, 22 Aug. (1 Sept.) 1676: for this *see* ch. 18.

had not had it before. Furthermore, the crucial series for $\int_0^x dt/(1 \pm t^2)$ was not amongst the results received by Leibniz; Newton's general method for series-expansion had been represented by a single example in a form not easily intelligible, and the numerical co-efficients in the various series given were disfigured by many errors due to the copyist's inattention. Leibniz had, as he says himself, looked only fairly superficially at Oldenburg's letter before he began almost at once to work up an answer from his own notes.

In his final version Leibniz points out[4] that Oldenburg's letter contains more of the new analysis than all the fat tomes hitherto published on the subject. In first draft[5] he had added a remark that it would have been desirable to have everything explained more thoroughly, though this had evidently been impossible in so narrow a space: but this is now suppressed, as is a sentence later on in the second draft[6] where he expresses the hope that Collins, the only person competent to do so, might write a book on higher analysis: there was nobody in Paris capable of so doing now that Huygens had departed. In the reply as sent,[7] Newton's general theory of series is strongly praised as an achievement fully worthy of the author of a novel theory of colours[8] and the inventor of the reflecting telescope.[9] The method of solving equations or effecting quadratures by means of infinite series is acknowledged to be quite different from Leibniz' own,[10] and it seemed surprising to him that the same goal could be reached by two routes so very different. Mercator had expanded rational functions into series by long division and subsequent integration, and Newton had now established other series by root extraction. His own method, Leibniz goes on to say, has grown out of a general theory[11] by which he can always reduce a quadrature by rational transformations to the integration of a function of at most the third degree in y (!). In his first draft Leibniz had stressed more

[4] NC 2: 57. [5] C 1505A.
[6] C 1505B. [7] NC 2: 57.
[8] Leibniz knew of this long and lively discussion only the pieces printed in the PT up to the spring of 1676; for a reference to the relevant letters see ch. 8: note 66. Leibniz had access to the twenty-one letters published from 6 (16) Feb. 1671/2 over several years to 29 Feb. (10 Mar.) 1675/6 in the PT (6, No. 80 to 11, No. 123), now reprinted in NC 1: 92–425 and HO 7: 242, 303, involving apart from Newton, Oldenburg and Huygens also Pardies, Linus, Moray and Gascoines.
[9] Five letters from Newton to Oldenburg had appeared in the PT 7, Nos. 81–3, 6 (16) Jan. 1671/2 to 4 (14) May 1672, reprinted in NC 1: 74–155.
[10] NC 2: 57–8.
[11] Though he alludes to his transmutation, Leibniz has deliberately replaced it by a special ad hoc procedure to disguise the crucial step.

strongly how much he owed to Mercator, adding that, once Mercator had expanded a function given in fractional form as an infinite series, it became highly probable that we would be able to do the same for radicals and rational functions also. Newton had further developed the procedure for radicals, Leibniz the reduction to rational functions (as exemplified, for instance, in his rationalizing transformation of the quadrature of circle and ellipse). In revision this first version may well have appeared to Leibniz as too arrogant.

In sequel Leibniz stresses[12] that he had always endeavoured to select from the numerous transformations possible the one most suited to the problem in hand, such that the new ordinate became expressible as a rational function of the new abscissa, so enabling one to derive the series by a Mercator's division. He believed a general theory of transformations to be of the highest importance for analysis, since it by no means always produced only an approximating series-expansion but also on occasion exact geometrical solutions. He writes all this freely and openly in the hope that English mathematicians will in return not deny him the opportunity of sharing their own effective methods of analysis. The essential point in squaring an area, he says in summary of his principle of integration by substitution, is to dissect it into an aggregate of infinitesimal parts of any form that can be suitably transformed into other equivalent parts of an equal area of another more easily manageable figure – as, for example, when rectangular inscribed or circumscribed step-polygons are changed into triangular ones. In this way, he claims, it is possible to accomplish all quadratures hitherto discovered by a unified and comprehensive method, alluding thereby to his method of 'transmutation'.

In the *Commercium epistolicum*[13] all this is declared to be in no way new; Leibniz' method is of the same type as that of Gregory in his *Geometriæ pars universalis* and Barrow in his *Lectiones geometricæ* by whose means the quadrature of conics may be rationalized. But this was not a general method; in fact, apart from a few special cases, a transformation of this type leaves the degree of an equation unaltered. It is evident that the 'editors' of the *Commercium episto-licum* did not grasp the truth of the matter; 'they' presumably thought that it was here a matter of changing to a new set of rectangular coordinates – hence the insistence on the invariance in degree – or to polar coordinates, whereas in reality Leibniz is

---

[12] *NC* **2**: 58.
[13] *CE* (*1712*): 58, note.

thinking of transformation of a quite general nature; also, his own particular transmutation is essentially not metric but affine.

Somewhat changing his stance, Leibniz in sequel demonstrates a rational quadrature for the circle[14] defined by $y^2 = 2ax - x^2$; on setting $ay = xz$ he obtains $x = 2a^3/(a^2+z^2)$, $y = 2a^2z/(a^2+z^2)$ and $y\,dx = [-]8a^5z^2\,dz/(a^2+z^2)^3 = [-]xy^2\,dz/a^2$, whence he can derive a series-expansion, and so after integrating term by term, the quadrature. About the connection of z with the tangent to the circle he wisely says nothing so as not to have to divulge his general theorem of transmutation, but he really has succeeded in arriving at an equivalent rational transformation of the integral. The English did not grasp the deeper meaning underlying his procedure. In the *Commercium epistolicum* it is alleged – rather ineptly – that Leibniz had described his procedure at uncommon length when it could all have been done in a few words, so that (it was concluded) he could not then have yet found his new analysis.[15] Leibniz refutes this wholly untenable remark in a marginal note, indicating that he was here writing for others who could not yet know his method. We must point out however that Leibniz originally (in the first draft) took $a = 1$ and developed $8z^2/(1+z^2)^3$ in the form $8z^2/(1 + 3z^2 + 3z^4 + z^6)$ according to the Mercator rule. Later he saw that the binomial theorem allowed a much shorter exposition.[16]

This method,[17] Leibniz continued, is not in any way restricted to curves of the second degree but can also be employed for curves in whose equation y rises to the third degree – though probably he actually worked only through the example $y^3 = x^2(2a - x)$ where the same substitution $ay = xz$ suffices. In the case of the central conic[18]

$$(y/b)^2 = x(2a \pm x)/a^2$$

the general sector can (when $ab = 1$) be expressed in the form

$$t \pm \frac{t^3}{3} + \frac{t^5}{5} \pm \frac{t^7}{7} + \dots$$

where $z = bt$ is the segment on the y-axis made by the tangent at the end of the arc. In particular the ratio of the area of the circle to the circumscribed square (again taken to be unity) will be

$$(1 - \tfrac{1}{3} + \tfrac{1}{5} - \dots) : 1,$$

14 *NC* **2**: 58–9; similarly in *C* 1438, 1458.
15 *CE* (*1712*): 59, note. Leibniz' marginal remark says: *Sed scribebat aliis, quibus ea non erat nota.*
16 Leibniz' marginal notes, written late August 1676 (*LBG*: 180) on Newton's letter of 13 (23) June 1676. 17 *NC* **2**: 59. 18 *NC* **2**: 60.

a result which, Leibniz reports, he had communicated to his friends some three years earlier. The *Commercium epistolicum*[19] here refers to Gregory's series for the inverse tangent and its transmission to Leibniz, but it overlooks the essential importance here of describing the method by which it was derived. This general representation of the area of a conic sector is taken from the 'Quadratura arithmetica circuli, ellipseos et hyperbolæ' which Leibniz had meanwhile completed and whose main result was later published in the 'Quadratura arithmetica communis'.[20] In his letter Leibniz gives next[21] the summation of the series $\sum_{k=2}^{\infty} 1/(k^2-1) = \frac{3}{4}$ (*compare* p. 60 above) and of the partial series obtained by omitting every other term or every group of three terms out of four. For this the *Commercium epistolicum* refers to the *Vera circuli et hyperbolæ quadratura*.

On the other hand, Leibniz declares,[22] he has observed the following to hold for the hyperbola: on setting the hyperbolic logarithm of $1-y$ equal to x [that is, $x = \log 1/(1-y)$] we obtain

$$y = x - \frac{x^2}{2!} + \frac{x^3}{3!} - \dots,$$

but on setting the hyperbolic logarithm of $1+y$ equal to x [that is, $x = \log(1+y)$] then it will be

$$y = x + \frac{x^2}{2!} + \frac{x^3}{3!} + \dots,$$

as Newton had also noticed. The first series yields a better approximation than the second, and by changing $1+y$ to $1/(1-y)$ the second can be reduced to the first. The *Commercium epistolicum* here comments[23] that Leibniz really took everything from Newton; that he came from the second series to the first simply by changing the signs of x and y; in his own reply, Newton says he cannot see why the series for e^{-x} should converge better than that for e^x.

Unfortunately Leibniz has expressed himself in an over-obscure manner in this passage since he wished to preserve the secret of his way of deduction. His difference in method from Newton's, who had

[19] *CE* (*1712*): 60, notes.
[20] *C* 1233. The printed version appeared in the *AE* (April 1691): 178–82 (*LMG* v: 128–32).
[21] *NC* 2: 60. The reference in the footnote to *CE* (*1712*): 61 relates to the presentation in the *AE* (February 1682): 45–6 (*LMG* v: 121–2).
[22] *NC* 2: 61.
[23] *CE* (*1712*): 62. Leibniz has added the remark: *Sed cum hæc scriberet, jam aliam invenerat methodum.*

achieved his variant of the result by inverting the logarithmic series, cannot therefore be seen. Leibniz' procedure becomes clearer from his first draft.[24] He had found

$$y = x + \frac{x^2}{2!} + \frac{x^3}{3!} + \cdots$$

by the iteration

$$\int_0^x y \, dx \left[= \frac{x^2}{2!} + \frac{x^3}{3!} + \cdots \right] = y - x$$

and knew that the corresponding differential equation $y = dy/dx - 1$ could only be satisfied by the logarithmic function $x + k = \log(1 + y)$. His remark that the series for e^{-x} gives a better approximation than that for e^x is to be understood as follows: for a fixed positive x the error when the series is broken off at a certain term is smaller in the case of e^{-x} than e^x; for Leibniz the reciprocals $e^{\pm x}$, being numbers not logarithms, are completely equivalent.

In regard to the determination of the trigonometric functions from the arc, Leibniz continues,[25] he had hit first of all upon the cosine-series; afterwards he had noticed that from it by integration he could find the sine-series and so, from $x - \sin x$ the area of the double circle segment. By breaking off the cosine-series at its third term, we can derive the approximation $x \approx \sqrt{(6 - \sqrt{[12 + 24 \cos x]})}$ where the error in the original equation will not reach $x^6/720$. Again the *Commercium epistolicum* comments[26] that all this can be obtained by easy transformations from Newton's series. Numerous other results could here be added, says Leibniz,[27] but in the main he was interested only in the general method and was content to leave its application to others. If so desired he declares that he is ready to explain everything more fully – for instance, the solution of affected equations by series[28] (which, he states in his first draft, he believed he had been the first to develop from Viète's method); he does, however, ask Newton for a more accurate explanation of the origin of the binomial theorem[29] and

[24] The details, documented and interpreted in Hofmann (1957), make it evident that Leibniz had in fact already at that time invented for the instances of the exponential, cosine and sine functions the method of successive integration, as he indicates in the letter to Johann Bernoulli, 7 Jan. 1697 (*LMG* III: 352).

[25] *NC* 2: 61.   [26] *CE (1712)*: 62: note.

[27] *NC* 2: 61–2.

[28] This topic is dealt with in *C* 1467–8.

[29] How Leibniz took up and assimilated the binomial expansion may be understood from his marginal notes on Newton's letter of 13 (23) June 1676 (*LBG*: 180) and his addenda to the second draft, *C* 1505B; moreover, the expansion is used in a supplement to his arithmetical circle quadrature, *C* 1233, recognizable in the 'Compendium' from the autumn of 1676 (*LMG* V: 112–13).

the development of his series-solutions of affected equations,[30] and also for his technique of series inversion (in the case, for instance, of passing from the logarithmic to the exponential series). He confesses to having read Newton's letter only superficially so far and to having not yet examined it really thoroughly – indeed he was not quite sure whether he would be able to puzzle it all out even by careful study, and it would be right, he thought, for Newton himself to give the supplementary explanations he wanted.

Leibniz now turns to the items communicated by Collins. He would find it desirable if a proof of Gregory's approximations for the circle-arc were added;[31] Leibniz himself had a method avoiding the use of angle bisections which he believes to be simpler but where it is still necessary to proceed to infinity. He hopes that Gregory's papers are being well preserved, since, even if his proof of the impossibility of the analytical quadrature of circle and hyperbola were unsatisfactory he had nevertheless been an excellent pioneer of the analytical method. Leibniz recalls that he had already in the spring of 1675 given a method for treating generalized Cardan equations,[32] though it is in fact completely general only for cubic equations: the details of the investigation he had left for Tschirnhaus to pursue since his researches in this field had gone deeper than his own.[33] It was, to be sure, not all that difficult for him and Tschirnhaus to reconstruct the essence of Gregory's method from Collins' indications of it. Leibniz confirms that it will not always be possible to avoid the imaginary even where the roots are real; as an example he gives the illustration[34]

$$\sqrt{(1 + \sqrt{[-3]})} + \sqrt{(1 - \sqrt{[-3]})} = \sqrt{6},$$

where (in Leibniz' opinion) the root of neither of the binomials can be extracted, and similar instances might be cited from cubic equations. Thus Descartes[35] and certain other mathematicians were mistaken

[30] In C 1528 Leibniz deals critically with Newton's remarks in the letter of 13 (23) June 1676 (NC 2: 23–5).

[31] NC 2: 62. The addendum on approximations for the circle-arc is presumably a reference to the series for the inverse tangent.

[32] This is an allusion to the results mentioned on p. 146 which arose out of his research from autumn 1674 to spring 1675. The summary in C 911 compiled in February 1675 (ch. 10: note 33), was probably shown to Huygens in the spring of that year. It is at this place in the reproduction of his own letter on CE (1712): 63 that Leibniz wrote in the margin: *Hoc postea etiam invenit Moivræus* (*compare* ch. 11: note 17).

[33] Possibly this is a reference to Tschirnhaus' results mentioned in ch. 12: note 69.

[34] NC 2: 62–3; *compare* ch. 11: note 17, and the text p. 146.

[35] *Geometria* (1659): 95.

when they believed that Cardan's formulae were in this respect unique. The imaginary can no more be avoided here than in any attempt[36] to deduce the quadrature of the circle $\left[\int_0^x \sqrt{(1-t^2)} \, . \, dt \right]$ from that of the hyperbola $\left[\int_0^x \sqrt{(1+t^2)} \, . \, dt \right]$. Leibniz can, he claims,[37] reduce the general equations of eighth, ninth and tenth degree to ones of seventh degree, but the amount of calculation required is immense even if one is able to abridge some of it by an adroit arrangement of the computation. He is ready to communicate his general procedure if someone will take the trouble of performing the detailed manipulations. We need, essentially, a tabulated list of all analytical functions[38] – this will be no less important and useful than computing a sine-table. A well thought out symbolic notation and especially a careful analysis of the axioms are indispensable.[39] Leibniz hopes that Pell will not any longer keep back his general considerations, in particular what he had collected in his 'Cribrum Eratosthenis'; even if he could ultimately not keep all his promises it would still be regrettable if what he had achieved were to perish unheeded. The same applied to his theory for delimiting the roots of equations and his use of sine-tables in solving equations. The final sentence, in which Leibniz expresses surprise that Pell had in all these years not communicated more of his achievement in mathematics has been marked by Oldenburg to be omitted from the copy in the

---

[36] *C* 1134.

[37] The same remark appears also in the draft of a letter Leibniz–Tschirnhaus, mid-May 1678 (*LBG*: 522). The passage in *NC* **2**: 63 is referred to again in Wallis–Leibniz, 16 (26) Jan. 1698/9 (*LMG* IV: 57) where Wallis, while not denying the truth of Leibniz' assertion, yet considers that it could only be verified by an immense effort of computation. In his reply, 30 Mar. (9 Apr.) 1699 (*LMG* IV: 62) Leibniz does not take up this issue but he points out that Descartes had in his *Géométrie*: 403–11 [*Geometria* (1659): 97–100] talked of the treatment of the general equation of sixth degree and had tried in a letter to Dotzen, 15 Mar. 1642 (*DO* III: 553–5) – mentioned also in Leibniz–Wallis, 29 Dec. 1698 (8 Jan. 1699) (*LMG* IV: 53–4) though unpublished at the time – to reduce the equation of sixth to one of fifth degree, unsuccessfully in his opinion. We may presume that Leibniz was led to his fallacious assertion by some transformation like $x = y_1 + y_2 + \ldots + y_{n-1}$, where an error in his working led him astray; but no documentary evidence has yet come to light.

[38] *Compare* the passages of similar content cited in Couturat (1901): 478–80.

[39] Leibniz refers on this occasion clearly to the fundamental ideas developed in his *Ars combinatoria* (1666). *Compare* also Leibniz–Mariotte, August 1676 (*LSB* II. 1: 269–70) and the remarks in an undateable letter to Oldenburg (*LSB* II. 1: 239–42), probably never actually sent. Wallis, in a letter to Leibniz of 22 July (1 Aug.) 1698 (*LMG* IV: 51) fully agrees with his thoughts on axiomatics.

Letter Book[40] – Pell's touchiness was well known and nobody wanted to provoke him. Leibniz cannot agree with Newton's opinion[41] that the theory of series had overcome the main difficulties in analysis with the exception, perhaps, of those involved in resolving Diophantine problems. There are, he says, many remarkable, complicated questions which depend neither on equations nor on quadratures, such as, for instance, the inverse tangent problems which Descartes himself said he could not master.[42] The *Commercium epistolicum* concludes[43] at this point that Leibniz had evidently at this time not yet comprehended the nature of differential equations, otherwise he could not have stated that inverse tangent problems did not depend on equations. Leibniz later added in the margin that he meant the term 'equation' to be taken in its everyday sense; Newton, too, had used the word in this way.

That Leibniz had indeed advanced to differential equations proper follows at once from the example he gives: he mentions a letter by Descartes[44] in which, inter alia, he considers De Beaune's problem[45] of finding a curve whose subtangent is constant. The text reads *curva huius naturæ*, but Collins transcribed this as *ludus naturæ*,[46] and so it went into Wallis' *Opera* III, and from there into the *Commercium epistolicum*. It gave occasion for an attack on Leibniz by Newton[47] and was of course taken up in the *Commercium epistolicum*;[48] Leibniz pointed out the correct wording in vain[49] for in the second edition of the *Commercium epistolicum* in 1722 the erroneous reading

[40] *NC* 2: 63.
[41] Newton–Oldenburg, 13 (23) June 1676 (*NC* 2: 29).
[42] In a note of July 1676 (*LBG*: 201) Leibniz quotes the relevant passage from Descartes–De Beaune, 20 Feb. 1639 (*DC* III: 409) word for word.
[43] *CE* (*1712*): 65, footnote. Leibniz has written in the margin: *nempe tunc vulgo receptis, qualibus utebatur et Newtonus.*
[44] Descartes–(?), June 1645 (*DC* III: 449).
[45] The wording of this problem – the second of four contained in a now lost challenge pamphlet – is recognizable from a letter, De Beaune for Roberval, 16 Oct. 1638 (*MC* 8: 142–3); in rectangular Cartesian coordinates it is required to find the curve through the origin satisfying $dy/dx = (x-y)/a$. From Descartes–De Beaune, 20 Feb. 1639 (*DC* III: 409–16) follows that though Descartes had not been able to determine the curve as a whole he yet had achieved its construction by points, equivalent to the representation of the logarithm by a series of unit-fractions. For the details *see* Scriba (1961): 411–16.
[46] *CE* (*1712*): 65 (*NC* 2: 64); corrected by Leibniz in his own copy.
[47] Newton–Oldenburg for Leibniz, 24 Oct. (3 Nov.) 1676 (*LBG*: 224 = *CE* (*1712*): 86): *Sed hos casus vix numeraverim inter ludos naturæ.*
[48] This comes in a lengthy footnote to Leibniz–Oldenburg, 21 June (1 July) 1677 (*CE* (*1712*): 93 = *NC* 2: 216).
[49] Leibniz–Conti, 9 Apr. 1716 (*LBG*: 281–2).

and the criticism it had attracted were reproduced without alteration. Now in his present letter to Oldenburg, Leibniz does not
produce a solution: all he says is that he found one quickly but that
he is not yet quite satisfied with it.[50]

We know that Leibniz had been directed to study Descartes'
inverse tangent method by Huygens, initially in the belief that
Descartes had a procedure which he had not revealed and which not
even the clever Hudde had been able to discover; but later he changed
his opinion on this.[51] Leibniz had come on these investigations of
Descartes in connection with a study of his ovals,[52] rightly recognizing the infinitesimal character of this type of problem,[53] he had
later looked up the relevant letters on De Beaune's problem in
Descartes' correspondence.[54] Leibniz deals with the differential
equation in the following way (fig. 28): where $P(x, y)$ is a point on the
curve in a rectangular coordinate system, through $Q(x, x)$ draw
$QT = a$ parallel to the x-axis: then PT will touch the curve at P
while T is a point on the asymptote $y = x - a$ of the curve; QP
produced meets this asymptote in S so that $ST = a\sqrt{2} = \alpha$ and
$SP = a - (x - y) = \eta$. Let p be a point on the curve near to P and let
its ordinate meet the parallel to ST through P in s then the triangles
psP and PST will be very nearly similar and so $ps/Ps = d\eta/d\xi = \eta/\alpha$,
whence $\log (\eta/\alpha) = \xi/\alpha$; thus the problem has been reduced to a
logarithmic curve in Cartesian coordinates, and has indeed been
solved by *certa analysis*.[55]

In the next passage of the letter Leibniz asks for further news of
research in chemistry and physics in England; he hopes Boyle will
freely publish his discoveries and not as hitherto keep back the best
of them. Leibniz would never have allowed this passage to stand if

---

[50] He remarks specifically that he had reached his solution by *certa analysis*; the
phrase is repeated in Leibniz–Conti, 9 Apr. 1716 (*LBG*: 281) with the interpretation that it meant treatment by infinitesimals; but this is altogether denied –
though mistakenly – in Newton–Conti, 19 (29) May 1716 (*LBG*: 290, 294).

[51] *C* 844. [52] *C* 832–3.

[53] *C* 845. [54] *See* note 42.

[55] In his letters to Fabri, late 1676 (*LSB* II. 1: 300) and to Tschirnhaus, late May 1678
(*LBG*: 375), Leibniz alludes, though only briefly and without diagram or formulae,
to this treatment of De Beaune's problem. He is somewhat more precise in the *AE*
(October 1684): 467–73 (*LMG* V: 226) where he especially mentions Descartes'
attempt but characterizes it as incomplete. Again he draws no diagram, but we get
what amounts to the essential differential equation. (For the further history of the
problem *see* Hofmann (1972): 14 ff.) Leibniz therefore is fully aware of the significance of the problem and its solution but withholds the solution so as not to be
'found out' as he writes in a different context to Bodenhausen, 23 Mar. 1691
(*LMG* VII: 359).

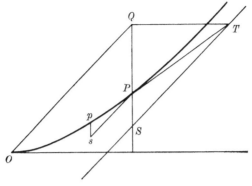

Fig. 28

he had not been in such a hurry. It is, as we might expect, missing in the copy in the Letter Book, but we there find Leibniz listing a number of topics about which he hopes to write at greater length as soon as he can find time:[56] these include the reduction of mechanics to geometry,[57] and investigations into elasticity,[58] the motion of liquids,[59] the pendulum,[60] projectile motion,[61] the resistance of solid bodies,[62] friction[63] and the laws of motion.[64] All originate from a single metaphysical axiom which plays no less important a rôle in physical thinking than the axiom of the whole and its parts in the doctrine of magnitudes. (A note here in the margin of his copy of the *Commercium epistolicum*[65] reveals that Leibniz had in mind the principle of conservation of energy.) Finally there is a reference to the latest work on centres of gravity[66] and his promise of a reply from Tschirnhaus by the next post.[67]

[56] *NC* **2**: 64. The passage is repeated in the review written jointly by Leibniz and Pfautz of *WO* III (1699) in the *AE* (May 1700): 196.

[57] *C* 1503–4.   [58] *C* 973.   [59] *C* 1133, 1141–2.
[60] *C* 975.   [61] *C* 976.   [62] *C* 976–8.
[63] *C* 945–6, 965, 1189.   [64] *C* 943–4, 969.

[65] *CE* (*1712*): 65: *Nempe effectum æqualem causæ.* A first version occurs in an incidental remark on *C* 1311: *proprietas omnis quæ continet efficientem rei causam seu generationem, sufficit ad omnia ejus attributa invenienda.* Leibniz only published his principle in the February 1687 issue of the *Nouvelles de la République des lettres* (*LD* III: 197): *il y a toujours une parfaite équation entre la cause pleine et l'effect entier.* This text is a slightly changed reproduction of a letter to the editor of the *Nouvelles,* Bayle, 19 Jan. 1687 (*LPG* III: 45–6). A Latin version appeared in the *AE* (May 1690): 228–39 (*LMG* VI: 201). Leibniz refers to this passage in the review of *WO* III in *AE* (May 1700): 196: *Axioma arbitramur esse id cujus mentionem postea in Actis nostris fecit, nempe causam plenam effectui integro æquipollere, uti totum omnibus partibus simul æquatur.* See also note 56.   [66] *C* 1419–21, 1474.

[67] Tschirnhaus–Oldenburg, 22 Aug. (1 Sept.) 1676 (MS: London, RS: LXXXI, 33).

The reader who today has at his disposal the extensive supplementary material of which Leibniz' contemporaries could have had no idea, will find in this great letter by Leibniz interesting testimony to the success with which he managed at the age of thirty, after creating a suitable symbolic notation, to attack the central problems of higher analysis. From his still unpublished manuscript drafts we may gather certain further details. His determination of the surface of an oblique circular cone suggested by a conversation with Roberval[68] we mention only in passing. Far more important was his preparatory work for the final version of the arithmetical circle quadrature, leading to often renewed but always ultimately unsuccessful efforts to prove its analytical impossibility, together with his simultaneous study of previous attempts to solve the circle-squaring problem.[69] This includes a critical discussion of Wallis' interpolation method in his *Arithmetica infinitorum*,[70] and his statement that the infinite product for $4/\pi$ occurs also in Mengoli's *Circolo* of 1672. Leibniz for quite a while laboured, in vain,[71] on Wallis' problem of interpolating equidistant ordinates of successive magnitudes 1, 6, 30, 140, 630 …, having found the question in Wallis' *Arithmetica infinitorum* (prop. 133)[72] and in the collection of letters[73] published by him regarding the debate with Fermat and others on number theory. On the whole he agrees with Fermat's rather negative judgment.[74] It appears interesting that he believes the problem to be insoluble because it is too indeterminate, although he is aware of the law by which the numbers $6 = \dfrac{1.6}{1}$, $30 = \dfrac{6.10}{2}$, $140 = \dfrac{30.14}{3}$, $630 = \dfrac{140.18}{4}$, … are found. The connection with the quadrature of the circle we can now easily establish by setting $f(n) = 1 \Big/ \displaystyle\int_0^1 y^{2n}.\,\mathrm{d}x$

---

[68] *C* 1095, 1426. *Compare* also Leibniz–Varignon, 18 Jan. 1713 (*LMG* iv: 191) and the posthumously edited treatise in the *Miscellanea Berolinensia* 3 (1727) (*LMG* vii: 345–7).   [69] *C* 1245–7.   [70] *C* 1164, 1202.

[71] *C* 1383–4; *compare* also Collins–Gregory, 28 May (7 June) 1672 (*GT*: 231–2). No-one had noticed at the time that Mengoli's proof is vitiated by a basic error.

[72] *C* 1200.

[73] The question had already occurred in correspondence Wallis–Schooten, autumn 1652, as is evident from Huygens–Schooten, 26 Dec. 1652 (*HO* 1: 208–10), though neither Schooten nor Huygens reached a solution.

[74] While in the letter Fermat–Digby, 15 Aug. 1657 (*WCE*: 25–6), the question is declared to be too indeterminate, a proper exposition is given in Wallis–Digby, 21 Nov. (1 Dec.) 1657 (*WCE*: 48), without clearly indicating the method employed. The whole matter is analysed – justifying Wallis, as it were – in Whiteside (1961): 237–40; *compare* also *NP* 1: 99–101 concerning Newton's related notes of 1664–5.

where $y^2 = x(1-x)$, whence we have the recurrence relation

$$nf(n) = 2(2n+1)f(n-1) = [6+4(n-1)]f(n-1),$$

so that $$f(\tfrac{1}{2}) = 1 \Big/ \int_0^1 \sqrt{[x(1-x)]} \, . \, \mathrm{d}x = 8/\pi$$

is the interpolated value, correctly stated by Wallis.[75] The 'Quadratura arithmetica circuli, ellipseos et hyperbolæ' was probably completed some time in July or August 1676; his preliminary research for the last piece to be included, the 'Trigonometria sine tabulis', was written down towards the end of June[76] while the scholium to prop. 29 in which Leibniz refers to the method of series in Newton's first letter to him was added subsequently.[77] This treatise[78] combines

[75] On working through WO I (1695) and WO II (1693), later fairly thoroughly reviewed by him in AE (June 1696): 249–59, Leibniz encountered the remark in Wallis–Digby, 21 Nov. (1 Dec.) 1657 (WO II: 786 = WCE: 51) where it is pointed out that the 'mean' to be interpolated between the first two terms in the sequence

$$1, \; 5/6, \; 31/30, \; 209/140, \; 1471/630, \; 10\,625/2772, \; \dots$$

depends on the quadrature of the hyperbola. He therefore evidently asked in a lost letter, 6 (16) Dec. 1695 for more details, but Wallis' reply, 21 Nov. (1 Dec.) 1696 (LMG IV: 5–6) fails to provide these. In fact, where $g(n) = \int_0^1 y^{2n} \, . \, \mathrm{d}x$ with $y^2 = x(1+x)$ the 'interpolated' value will be $g(\tfrac{1}{2}) = \tfrac{3}{4}\sqrt{2} - \tfrac{1}{4}\log(1+\sqrt{2})$. See NP 1: 116–17 for Newton's early encounter with this problem, and ibid. note 94 for explanation and reference to Leibniz' enquiry.

[76] C 1460–2.   [77] See Mahnke (1931): 596, note 17.

[78] The registers in LSB I. 2 and I. 3 give information on the fate of the manuscript C 1233. Leibniz left it before his own departure with his young friend Soudry who was to transcribe it and get it printed in Paris, but he died during the campaign in the Netherlands in 1678. Sauveur inspected the manuscript without Leibniz' knowledge in the house of Soudry's brother, himself an abbé. Later the paper came into the hands of another of Leibniz' Parisian friends, the Danish nobleman Hansen who asked for instructions about Leibniz' intentions regarding publication: it might be supervised by Comiers or Mariotte. When Hansen had to leave Paris for England he gave the manuscript to the Hanovarian resident in Paris, Brosseau, who intended to return it with some other of Leibniz' possessions still left in Paris in the autumn of 1679 through the good offices of I. Arontz, a Jew, about to travel to meet his cousin C. E. L. Berens in Hanover: but he was attacked by robbers – the manuscript seemed lost, yet sometime after 1683 it must have found its way back to Leibniz. He had indeed hoped to get it published in Paris, as is evident from letters to Gallois in September 1677 (LSB II. 1: 387), and to Placcius, early July 1679 (LSB II. 1: 421). He had also hoped by it to gain his entry into the Académie as the letters to Huygens, 18 Sept. 1679 (HO 8: 215–16, 219) and December 1679 (HO 8: 249) make clear. Since publication in Paris was no longer opportune, now that the manuscript was not there any more, Leibniz approached Elseviers in Amsterdam through La Loubère to try and get it printed by them, but the publishers first wished to see and approve the manuscript – and the plan came to nothing: La Loubère–Leibniz, 10 Aug. 1680 (LSB I. 3: 415). Ozanam published the 'Leibniz series' without naming the author in his Géométrie pratique (1684): 192–6, causing Leibniz to protest both in the review in the AE (October 1685): 481–2 and in a letter to Foucher, 1686 (LPG I: 381). At the time

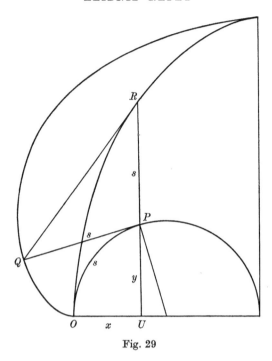

Fig. 29

all the earlier studies of Leibniz; it contains in particular a fully rigorous and unexceptionable proof of the general transmutation method.

On this occasion he again takes up his earlier notes, improving them considerably and at the same time shortening them; this is especially noticeable in the summary recapitulations about the

Leibniz believed that Ozanam had got the details from Tschirnhaus. Foucher writes on 31 Dec. 1691 (*LPG* I: 400), that Ozanam acknowledged to have been shown the idea of Leibniz' circle quadrature but without any proof; having worked it out afterwards all by himself he claimed as much right to the discovery as Leibniz. Perhaps the basis for this is the document *C* 779 which contains, together with some writing that may be in Ozanam's hand, the circle-series and a note on binary arithmetic, the binary representation of the first few natural numbers and a simple example of addition and multiplication. In his *Cours de mathematique* III (1693): 163–7, Ozanam also gives a deduction of the circle-series without naming Leibniz. A reference to a theorem from the circle quadrature tract in the letter Leibniz–Johann Bernoulli, 8 Aug. 1698 (*LMG* III: 522) elicited a request in Bernoulli's reply, 26 Aug. 1698 (*LMG* II: 528), for publication of the treatise. We must assume that the manuscript was meanwhile again in Leibniz' hands as he does not mention its loss; nevertheless he declared, 1 Sept. 1698 (*LMG* III: 537), that he would not get it printed now as it had been overtaken by newer developments and would be of interest to beginners only. To the present day it has not been published in full; all that exists is the much abbreviated edition we have mentioned in ch. 6: note 58.

quadratrix[79] and the cycloid.[80] One result of this is the elegant general theorem about the relation (see fig. 29) between a curve $P(x, y)$, a corresponding involute $Q(x - s\,dx/ds, \; y - s\,dy/ds)$ and a correlated cycloid $R(x, y + s)$, which proves to be an ingenious extension of prop. 8 in Gregory's *Geometriæ pars universalis*. Ozanam appropriated the result for the semicircle (to which our diagram refers) on page 96 of his *Dictionnaire mathématique* without naming the original author. We can here do no more than simply mention the existence of a number of other studies in differential geometry,[81] and also of certain pieces – excerpts from these are partly published already by Gerhardt in his *Entdeckung der Differentialrechnung* – applying his newly created Calculus,[82] together with researches on mortality tables[83] and on dice playing[84] undertaken for the Duke of Roannes.

Finally we must briefly mention the meeting with Römer, a member of the Académie des Sciences since 1672. Römer's studies of the epicycloids of a circle and their use in the construction of gear wheels are mentioned by Huygens[85] in 1674. Römer lectured at the Académie on mechanical topics[86] in the summer of 1675 and on gear wheels[87] in the spring of 1676, but could not bring himself to publish his notes (which were later probably lost in a fire at Copenhagen in 1728). Leibniz heard from Huygens about Römer's work[88] and when he was given the second part of the manuscript, he excerpted its main contents.[89] In the year 1694 La Hire, who had by his own account occupied himself with this same subject for twenty years, published an extensive paper on the topic without, however, naming Römer. Although Leibniz urged him to do so, Römer would not be moved either to assert his priority or to publish his earlier manuscript, even when Leibniz emphasized that La Hire's investigations nowhere

[79] *C* 1386, 1464.
[80] *C* 1103, 1144, 1465.
[81] *C* 1125, 1131, 1157, 1208, 1277, 1469–70, 1475.
[82] *C* 1176, 1242, 1244, 1280, 1482.
[83] *C* 1197; *compare* for this Biermann (1955b).
[84] *C* 1259: Biermann (1955a); 1281: Biermann (1954, 1956, 1965a). For the further development of these topics in Leibniz' hands and later *see* Biermann–Faak (1957) and Biermann (1965b), (1967).
[85] *HO* 18: 607.
[86] AS meeting, 6 July 1675: *compare HO* 19: 182.
[87] AS meetings, 11 Jan., 1 Feb., 8 Feb., 15 Feb., 22 Feb., 7 Mar. 1676; *compare HO* 19: 185.
[88] Leibniz' letters to Johann Bernoulli, 28 Jan. 1698 (*LMG* III: 477); to Römer 31 Jan. 1700 (*LD* IV. 2: 115); to Johann Bernoulli, 24 June 1707 (*LMG* III: 816).
[89] *C* 1187–8.

went as deep as Römer's.[90] The manner in which Leibniz stands up
for the discoverer of the velocity of light reveals the high opinion he
had gained in Paris of Römer's ability; this may be confirmed by an
interesting passage in the *Acta Eruditorum* of 1700.[91] Römer had
previously spoken also of his demonstration of the law of the lever,[92]
which differed from that given by Archimedes (whose deduction in
the *De æquiponderantibus* is known to be unsatisfactory),[93] and in
connection with combinations of gear wheels had given an account
of his method for the best approximation of fractions in large
numbers by ones in small numbers: in this he perhaps anticipated
some of the results on continued fractions which Huygens gathered
in the description of his 'automatic' planetarium.

Thus from the autumn of 1675 when he achieved his decisive
mathematical success by his invention of the Calculus, we see Leibniz
incessantly occupied with gathering, deepening and broadening his
own studies – not however living in quiet seclusion, but joining in
lively conversation with everybody interested in questions of
natural philosophy and ever alert to an outside stimulus with ready
curiosity. More justifiably than before he could now feel himself
superior to the other Paris mathematicians; he is able to recognize
and assess the type of problem and mode of treatment characteristic
of each one of them, aware that they all lack one crucial element
which he himself possesses: a total panoramic vision from a higher
viewpoint. He has reached maturity through conversing with them;
now he will no longer be able to learn much from them. From the
members of the older school (of whom Hardy, Carcavy, Roberval,
Boulliau, Frénicle and Billy were still living representatives) he did
not draw any significant stimulus through his personal contact; he
gained rather more from men in their prime, in particular from
Mariotte in physics and from Huygens in mathematics. In him, the
benevolent mentor of his earliest mathematical studies and unprejudiced critic of his first own pieces of research, Leibniz had come
to know, esteem, admire and honour a man of great charm and a
truly creative genius. Their relationship had, it is true, grown somewhat cool by the autumn of 1675 but though their ties were now

[90] Leibniz' letters to Johann Bernoulli, 28 Jan. 1698 (*LMG* III: 477); to Römer,
3 July 1703 and 1 Oct. 1705 (*LD* IV. 2: 123, 125); to Johann Bernoulli, 1 Feb.,
24 June 1707 (*LMG* III: 811, 816); to Römer, 23 July 1710 (*LD* IV. 2: 127); to
Kortholt, 1 Feb. 1715 (*LD* V: 322).
[91] *AE* (May 1700): 198–208 (*LMG* V: 342).
[92] *C* 1555; *compare* Leibniz–Huygens, 6 Feb. 1691 (*HO* 10: 15).
[93] For this *see* Krafft (1971).

slightly looser they were by no means severed. Then Huygens was once more stricken by a severe illness which forced him for several months to leave off all forms of mental exertion and finally to seek rest and to convalesce at home. Thus Leibniz had no chance to discourse about his latest discoveries with one of the few men capable of properly understanding and appreciating them. Whether he might at that moment have succeeded in obtaining from Huygens more than his polite agreement – this was later impossible – it would be difficult to say.

For the first time Leibniz becomes aware that he will be quite alone on his future journeys in science; that of the circle of his contemporaries – Ozanam, La Hire, Mohr, Foucher, Römer – none would be ready or able to follow him, not even Tschirnhaus. With Huygens' departure from Paris a new note creeps into his correspondence with Oldenburg; might it be possible to accomplish in England what it had become hopeless for him to do in France?

18

TSCHIRNHAUS' REACTION

As soon as possible Oldenburg's letter was shown to Tschirnhaus as well. He glanced through it in Leibniz' rooms, read Leibniz' reply[1] and gave evidence on his behalf in the first part of his own answer of 1 September 1676. Probably Tschirnhaus saw no reason to give this to Leibniz to read through. We surmise this from some infelicities of expression of which Leibniz would never have approved and which were immediately taken up by the other side and later brought up against Leibniz. In the *Commercium epistolicum* Tschirnhaus' letter is reproduced from a transcript made, not without a number of errors, by Collins[2] who was, to be sure, particularly interested in the algebraical portion but found it hard to decipher Tschirnhaus' handwriting very accurately, even the very formulae. For this reason the copy was passed on to Wallis[3] who corrected[4] some of Collins' erroneous readings.

In his introduction[5] Tschirnhaus says that he had for months been most impatiently awaiting an answer to his last letter[6] without ever receiving any indication at all that it had actually arrived. He acknowledges that he feels highly flattered to be allowed to participate in the debate between Leibniz and the English; he had quickly glanced over the letter now received and observed as a main point that it did not contain the series which Leibniz had given for the quadrature of a circle-sector, adding that he had not hitherto met with a simpler procedure than this of Leibniz'. To this the *Commercium epistolicum*[7] ripostes that it seems unbelievable that Tschirnhaus had not seen Gregory's related series for the inverse tangent.[8] This

---

[1] Leibniz–Tschirnhaus, September 1684 (*LBG*: 457).
[2] *CE* (*1712*): 66; slightly fuller version *NC* 2: 84–5.
[3] Collins–Wallis, 9 (19) Sept. 1676, traceable from *CR* II: 591.
[4] Wallis–Collins, 11 (21) Sept. 1676 (*CR* II: 591–7).   [5] *NC* 2: 84.
[6] Tschirnhaus–Oldenburg, mid-June 1676 (MS: London, RS: LXXXI, 25).
[7] *CE* (*1712*): 66, footnote.
[8] Referring to remarks in Gregory–Collins, 15 (25) Feb. 1670/1 (*CE* (*1712*): 25 = *GT*: 170).

comment is quite out of place; only the series for the tangent is given in the 'Abridgment', that for the inverse tangent not at all, and in the 'Historiola' (which the English then believed also to have been sent) the whole passage is missing. This instance shows only too clearly the carelessness of the attempts to frame an accusation of plagiarism against Leibniz.[9]

For Newton's theory of series Tschirnhaus has nothing but praise, even though he believes that there exists a universal method for representing all types of quantity without the use of series, which appear to him merely as a contingent feature[10] (*per accidens*) – a similar reservation is set on Gregory's studies. Probably Tschirnhaus thought that his own notion of nested radicals could be made universal. This opinion would certainly not have met with Leibniz' approval, and the form in which it was presented even less. The predictable consequence was a sharp rejoinder from Newton,[11] not addressed to Tschirnhaus, whom Newton saw only as a convenient man of straw, but directed against Leibniz himself.

As for the elimination of intermediate terms in an equation, Tschirnhaus remarks that this is quite easy. Thus for instance, in the equation $x^3 - px^2 + qx - r = 0$, the second and third term can be simultaneously removed if $p^2 = 3q$; in the equation

$$x^4 - px^3 + qx^2 - rx + s = 0$$

the second and third term again if $3p^2 = 8q$, and the second and fourth term if $9p^2 + 8r = 36q$. Where no given relation between the coefficients exists, the problem will be exactly as difficult as finding the solutions of the equation themselves, but, he wrongly believes, it will not necessitate a rise in its degree. Gregory's remark in his letter of 11 (21) September 1675 about reduction of the degree when some roots depend rationally from the rest[12] is correct and easy to demonstrate as soon as the relevant relations between the roots are given. But Tschirnhaus asserts that one can go further than Gregory and ask what conditions the coefficients will have to satisfy so that a rational relation between the roots holds. If in an equation of third degree, for instance, the condition

$$2p^3 + 27r = 9pq$$

[9] *Compare* Mahnke (1932).
[10] *NC* **2**: 84.
[11] Newton–Oldenburg for Leibniz, 24 Oct. (3 Nov.) 1676 (*LBG*: 212).
[12] *GT*: 328.

is satisfied, then the solutions will be successive terms in an arithmetical progression. The general determination of the roots of higher equations is acknowledged to be laborious: Tschirnhaus has to confess that he has so far mastered only some special cases of quintics, though he knows rules for the general case also: he has abandoned this research for the time being but hopes to advance more rapidly later on.

He now refers back to a method he had previously sent to Collins concerning the solution of higher equations whose roots are successive terms in an arithmetical progression;[13] this is followed by an enumeration of the simplest cases of generalized Cardan equations together with their solution in the form of radicals. Tschirnhaus here mentions Leibniz' related studies in which the conditions for the coefficients formed by himself had been confirmed. He has, he says, exactly like Leibniz long before, used imaginary quantities also and had found such transformations as

$$\sqrt{[p/4 + \sqrt{(-q/4)}]} + \sqrt{[p/4 - \sqrt{(-q/4)}]} = \sqrt{[p/2 + \sqrt{(p^2/4 + q)}]}.$$

He has a step-by-step procedure for establishing the general angle-section equation and solving it by radicals, and is very interested in Wallis' work on this topic. In pursuit of these questions he has attained some interesting theorems on the circle that went much further[14] than what was to be found in prop. 69 of Gregory's *Geometriæ pars universalis*. Let the circle be divided, for instance, into n equal parts, and let the lines joining one corner with all the others be $s_1, s_2, s_3, \ldots s_{n-1}$; Tschirnhaus then asserts that

$$\Sigma s_{2k}{}^2 = \Sigma s_{2k+1}{}^2 = nr^2$$

(where r is the radius). He also announces that he possesses a very simple instrument for dividing angles.[15]

Tschirnhaus writes only very briefly about Pell's results; he does not know the 'Idea of Mathematicks'. It would interest him to hear whether by Pell's method a final decision can be made on the possibility or impossibility of squaring the circle or of expressing the roots of the general equation of nth degree by radicals; the scope of the

---

[13] *Compare* Collins for Gregory, 3 (13) Aug. 1675 (*GT*: 317–19).

[14] *See* Hofmann (1971): 102–4, 108–9.

[15] From Tschirnhaus–Gent, 6 Nov. 1675 (MS: Amsterdam, Wisk. G. 49f) we see that Tschirnhaus employs the same technique as Newton had done in generalization of a scheme by Viète (*NP* 1: 484); *see* Hofmann (1971) and *compare* ch. 19: note 56.

method will become immediately apparent from the way it deals
with such problems. Clerselier, so Tschirnhaus reports, knows nothing
of any remark in Descartes' letters about Pell's *Idea matheseos*. His
efforts to find Desargues' 'Leçons de tenebres' have unfortunately
proved fruitless, the book is unobtainable. He concludes by an-
nouncing that he is going to Italy in about five weeks' time[16] and he
offers to carry out any commission for Oldenburg en route.

For Collins this letter from Tschirnhaus was far more important
than the one from Leibniz. He had a copy made and forwarded it to
Newton;[17] he sent another to Wallis[18] with the request that he look
through the text, which he could not himself quite make out, and
enlighten him where he could. This was not an especially exacting
demand since Collins knew that Wallis was currently at work on his
Algebra and would certainly be interested in Tschirnhaus' letter. This
assumption proved to be correct; and in his answer Wallis makes the
required emendations.[19] He also by way of introduction to the letter
refers to a visit Tschirnhaus made him in 1675 and proposes a
facetious etymology of the name Tschirnhaus as deriving from 'churn-
house'. The theorem on chords of a regular polygon might well be
deduced, Wallis continues, from a proposition in his own *Sectiones
angulares* of 1648, which he had deposited with Collins several years
ago and of which his own memory was now somewhat unsure. The
proposition went like this: if lines are drawn from an arbitrary point
on the circumference of a circle to the corners of a regular polygon
of n sides, then these joins will be the roots (to be taken partly
positive, partly negative) of an equation on degree n and their sum
will be zero.[20]

In regard to circle quadrature Wallis contends that he has long ago
shown in the scholium to proposition 190 of his *Arithmetica infinitorum*
that the circle cannot be squared analytically (though the proof is in
fact fallacious); it is just as impossible as halving an odd number in

16 Tschirnhaus left only on 21 Nov. 1676 as is clear from Tschirnhaus–Leibniz,
17 Apr. 1677 (*LBG*: 328).
17 Enclosed in a letter of mid-September 1676, mentioned in Collins–Newton,
9 (19) Sept. 1676 (*NC* 2: 99).
18 Enclosure in Wallis–Collins, 9 (19) Sept. 1676 as is evident from Wallis' reply,
11 (21) Sept. 1676 (*CR* II: 591).
19 Wallis–Collins, 11 (21) Sept. 1676 (*CR* II: 591–7). This letter, however, was dispatched together with Wallis–Collins, 16 (26) Sept. 1676 (*CR* II: 598).
20 Wallis–Collins, 11 (21) Sept. 1676 (*CR* II: 595). Collins, however, in his draft reply
to Tschirnhaus, 30 Sept. (10 Oct.) 1676 (*CR* I: 217) rightly points out that Wallis
in his *Sectiones angulares* deals only with special instances; *see* Scriba (1966) 24, 33
for a confirmation of this.

integers or determining equations of fractional degree, which should then have more than n but less than $n+1$ solutions, and the like. It was, of course, right and proper to advance towards new concepts, for instance to the definition of such 'hypergeometric' progressions as

$$1, \; \tfrac{6}{1}, \; \tfrac{6}{1} \cdot \tfrac{10}{2}, \; \tfrac{6}{1} \cdot \tfrac{10}{2} \cdot \tfrac{14}{3}, \; \ldots, \quad \text{or} \quad 1, \; \tfrac{3}{2}, \; \tfrac{3}{2} \cdot \tfrac{5}{3}, \; \tfrac{3}{2} \cdot \tfrac{5}{3} \cdot \tfrac{7}{4}, \; \ldots,$$

where the mean between the first two terms[21] in the first progression will be $\pi/8$, in the second $4/\pi$; the infinite product and the infinite continued fraction for $4/\pi$ can be expanded as an infinite series in a variety of ways.[22] Wallis believes that the new theory of series has received a not insignificant impulse from such considerations though perhaps only indirectly. After he had received the letter from Oxford, Collins was in a better position to understand Tschirnhaus' communications and to appreciate their import; so he now made haste to gather the material required by Oldenburg for his answer. The unexpected turn of events may have appeared highly flattering to him as he now found himself placed as the only effective intermediary between the foremost mathematicians of all Europe, and he made every effort to keep the correspondence going. We know at present only the English original[23] which was translated into Latin by Oldenburg before being forwarded to Tschirnhaus;[24] the final version has unfortunately not yet come to light. By some mistake Collins' draft is dated 30 September (10 October) 1675 but the contents prove that the year must be 1676.

Right at the start Collins praises the excellence of Leibniz' method of transformation and his series for a conic sector, which indeed is not amongst the results communicated by Newton;[25] but notes that the series converges rather slowly for the whole circle. In fact Collins repeats this remark from a letter by Wallis[26] to which we shall return. Newton's method is also applicable to arithmetical questions, as for instance Davenant's problem, to which Tschirnhaus' attention had been drawn at the time of their conversation on (9 August 1675);

[21] *Compare* pp. 244–5.
[22] A similar remark occurs in Wallis–Leibniz, 6 (16) Apr. 1697 (*LMG* IV: 17–18). On the equivalence of Brouncker's continued fraction for $4/\pi$ with Leibniz' series, *see* ch. 5: note 33. We might add that obviously Wallis failed to see the difference between the result of an iterative process and that of a series expansion.
[23] Collins–Oldenburg for Tschirnhaus, 30 Sept. (10 Oct.) 1676 (*CR* I: 211–20).
[24] Oldenburg–Tschirnhaus, 10 (20) (?) Oct. 1676, excerpt made by Leibniz (MS: Hanover LBr 695, 66r).
[25] *CR* I: 211.
[26] Wallis–Collins, 16 (26) Sept. 1676 (*CR* II: 598–9); *compare* p. 259.

Baker in particular had occupied himself extensively with this topic and successfully solved it:[27] he should be the right man, suggests Collins, to convert the general method indicated by Tschirnhaus into a computational procedure. It would suffice to send him the general directions – which must of course first be entered in the Register Book of the Royal Society.

Collins now gives a lengthy survey of the methods hitherto tried for Davenant's problem.[28] This asks for four successive terms a, b, c, d in geometrical progression where the sum Q of their squares and sum C of their cubes are given. The question had already been sent to Leibniz[29] in the spring of 1675 and had elicited from him a note[30] from which we gather that the problem reduces to an equation of ninth degree and will be considerably simpler when it is restricted to three terms. We know that Strode made an attempt in which at first the sum Q only was to be considered,[31] and Collins had proceeded from there.[32] Gregory had believed that the problem required an irreducible equation of the thirtieth degree[33] but Dary had used[34] the substitution

$$b = at, \quad c = at^2, \quad d = at^3,$$

and so

$$Q = a^2(1+t^2+t^4+t^6), \quad C = a^3(1+t^3+t^6+t^9),$$

and had thus arrived at

$$a^6 = \frac{Q^3}{(1+t^2+t^4+t^6)^3} = \frac{C^2}{(1+t^3+t^6+t^9)^2};$$

Collins hoped for further simplification of this. Newton solved the last equation by an expansion in series[35] and spoke appreciatively of Baker's method.[36] Baker alone noticed that a factor $(1+t^2)^2$ can be

[27] *CR* I: 212.
[28] *CR* I: 212–13. The problem was first communicated by Oldenburg to Sluse in the summer of 1673 but he in his reply, 22 Nov. 1673 (*SL*: 683), pointed out that one would need to specify what means for a solution were to be admissible. Thus by eliminating b and c two equations for a and d could be gained and by drawing the curves to these equations, a and d were determined by the coordinates of their intersections whereupon b and c followed from $b^3 = a^2d$ and $c^3 = ad^2$.
[29] Oldenburg–Leibniz, 12 (22) Apr. 1675 (*LBG*: 120); *compare* p. 138.
[30] *C* 928–9.
[31] Strode–Collins, 28 July (7 Aug.) 1675 (*CR* II: 452–3).
[32] Oldenburg–Leibniz, 26 July (5 Aug.) 1676 (*LBG*: 175–6).
[33] Gregory–Collins, 26 May (5 June) 1675 (*GT*: 303); *compare* p. 222.
[34] Collins–Gregory, 19 (29) Oct. 1675 (*GT*: 341–2).
[35] Collins–Baker, 19 (29) Aug. 1676 (*CR* II: 4–10); extract in *NC* 2: 82; according to *NC* 2: 83, note 6, no more details are now known.
[36] Newton–Collins, 5 (15) Sept. 1676 (*NC* 2: 95).

cancelled so that the result is a symmetrical equation of fourteenth degree which he reduces to one of seventh degree.[37]

The elimination of intermediate terms in an equation, Collins continues, had been undertaken already by Harriot,[38] and Kinckhuysen had employed the technique in various ways.[39] Whether Gregory was right when he asserted that the intermediate terms could be removed only by raising the degree of the equation, or whether, on the contrary, Tschirnhaus was correct in maintaining that there was no justification for such a rise in the degree, Collins is unable to decide; he will, he says, like Viète,[40] be satisfied in this matter with approximate solutions. He reports that Pell had repeatedly affirmed his ability to find the logarithms of the solutions from the limits for the roots, but for this one needed very accurate tables of logarithms and antilogarithms, in the calculation of which Pell had achieved a great deal. Suppose it were required to solve $y = f(x)$, given that two pairs (x_1, y_1) and (x_2, y_2) nearly satisfying the equation are known, then the *regula falsi* could be applied: this leads to

$$\log (y/y_1):\log (y_2/y_1) = \log (x/x_1):\log (x_2/x_1).$$

When the given equation happens to be of the form

$$(y/b)^n = (x/a)^m,$$

the result will be exact; otherwise it is necessary to restrict oneself to branches of the curve that can be approximated by parabolic arcs of this form and then work with their chords and tangents.

The theorem on dividing an angle (constructing chords in a regular polygon) is acknowledged as new by Collins, and not contained in Wallis' *Sectiones angulares* (whose publication its author had not very urgently attempted). This in any case is not in Collins' opinion of much use in calculating tables, yet the topic must not be neglected. In England the tool for angle-section was the well known 'sector' with movable scales on whose divisions the result is to be

---

[37] An allusion to this occurs in Baker–Collins, 20 (30) July 1676 (*CR* II: 3); the full exposition is given in Collins–Newton, 31 Aug. (10 Sept.) 1676 (*NC* 2: 89–90) while Baker's own original paper has not survived.

[38] In Wallis–Collins, 11 (21) Sept. 1676 (*CR* II: 591) we are told that the systematic elimination of the second term in a cubic equation is due to Harriot (1631): 29–34, 89–96, whence Descartes is said to have taken it over, *Géométrie* (1637): 376–7 = *Geometria* (1659): 73–4. Wallis overlooked however that Cardan had taught in the *Ars magna* (1545): chs. XIV–XXII how to remove the second term.

[39] The passage in his *Algebra* (1661): 52–8 = *NP* 2: 325–8 in Mercator's translation is taken over from Descartes.

[40] *Numerosa potestatum resolutio* (1600) = *Opera* (1646): 162–228.

read; he suspects that Tschirnhaus' instrument has a purely geometrical basis and will be glad to know some details. Collins adds
that questions on compound interest and on angle-section can be
solved by means of the logarithmic spiral – he here names it 'ser-
pentine' – which can be projected from the equatorial plane into a
loxodrome on the sphere when the centre of projection is at a pole.
The length of this spiral can be found once its tangent is known – a
problem which itself according to Barrow in his *Lectiones geometricæ*[41]
depends on the quadrature of the hyperbola.

Finally, Collins gratefully accepts Tschirnhaus' offer of help in
acquiring books for him, and singles out La Hire's Conics, Viviani's
Solid Loci and Borelli's edition of Archimedes – but he is of course
interested in other new mathematical books. He apologizes for the
delay in his reply and for the absence of Newton's expected answer.

This letter ends the correspondence between Tschirnhaus and
Oldenburg. The letter reached Paris punctually[42] but was still unanswered in April 1677;[43] and so for ever it remained. For in Rome
Tschirnhaus did not find the time to write out a printable text of his
planned treatise on transforming the equation $f(x) = 0$ of degree n
by the substitution $y = x^{n-1} + p_2 x^{n-2} + \ldots + p_n$ which was to be
published in the *Philosophical Transactions*, and Oldenburg died in
early September 1677. We do not at present know whether Tschirnhaus ever again, as he had evidently intended, came into close
personal contact with mathematicians in the Royal Society.[44]

Seen as an episode in the general history of mathematics the
discussion with Oldenburg appears as largely fruitless; it nowhere
goes beyond ideas and concepts that were in general currency at the
time and leads nowhere. For the purposes of the Leibniz scholar it is,
however, of the greatest importance because it gives us a characteristic picture of his milieu and also because it provides evidence
that Leibniz learnt through Tschirnhaus nothing of any significance
regarding the results reached by the English mathematicians. The
very last of Oldenburg's letters gives us Collins' first, still unprejudiced opinion on Leibniz' quadrature of the circle and his rationalizing transformation of integrals: it is wholly positive and contains
the notable admission that the series in question had not been among
those communicated by Newton. Of this passage, in its Latin version,

[41] Appendix to Lectio XI: 110–13.
[42] This is to be concluded from Tschirnhaus–Leibniz, 17 Apr. 1677 (*LBG*: 332).
[43] *Ibid.* 333.
[44] Tschirnhaus–Leibniz, 27 Jan. 1678 (*LBG*: 339–40).

Leibniz made a special note – during, we may reasonably assume, his second visit in London when Oldenburg may have shown him the Latin original of the letter he had since dispatched to Tschirnhaus.[45] About most of the topics dealt with in that letter Leibniz was already well informed; he was interested in the approaches to the solution of Davenant's problem it sketched and added some further calculations of his own in his notes; but nothing new came of this.[46] When he later looked through his papers again Leibniz wrote the date 'November 1676' on this excerpt; but this is clearly an insignificant lapse of memory.

[45] *Compare* note 24 above.
[46] *Compare* notes 29–37 above, and the text pp. 255–6.

19

NEWTON'S SECOND LETTER
FOR LEIBNIZ

We can easily understand that Collins, in transcribing the letters he had received from Leibniz and Tschirnhaus, became aware of the importance of the communications with which he was dealing and of his own inability to judge their full value. Hence he found it necessary to send a copy not only to Newton but to Wallis and Baker as well.[1] In particular, he had doubts about the correctness of the Leibniz-series, though not about its independent invention.[2] He failed to notice that it was in fact a special case of Gregory's series for the inverse tangent. Wallis acknowledged that the series, though converging rather slowly, was probably correct.[3] Neither Baker[4] nor Newton reacted immediately.

Unfortunately in the autumn of 1676 Collins suffered from a troublesome attack of acute blood poisoning which caused inflammation and swelling in his right arm, so that he was not merely less alert in his reactions but also was severely hindered in writing letters for a long time.[5] During this time he encountered Leibniz, a meeting to which we shall refer more closely below. Collins gained a very good impression of this young German who had an intelligent answer to all his questions and knew how to turn the conversation to suit his own wishes – while Newton still remained silent.

What had happened? Newton had gone carefully over the letter-excerpts he had received and immediately discovered that he need not take Tschirnhaus seriously as a partner in the debate. Regarding Leibniz, he had convinced himself that he spoke extremely cleverly

---

[1] *Compare* ch. **17**: note 2, and ch. **18**: note 18. Tschirnhaus' letter was sent to Baker together with one from Leibniz, on 21 Sept. (1 Oct.) 1676; *see CR* ii: 10.

[2] *Compare* ch. **18**: note 25.

[3] *Compare* ch. **18**: note 26.

[4] Baker's reply, 27 Dec. 1676 (6 Jan. 1677) (*CR* ii: 10–13) relates to Tschirnhaus' letter, to the Davenant problem and questions on algebra, but does not touch on Leibniz' letter at all.

[5] Collins–Strode, 24 Oct. (3 Nov.) 1676 and 8 (18) Feb. 1676/7 (*CR* ii: 454, 455); Collins–Wallis, 16 (26) Feb. 1676/7 omitted from the portion printed in *CR* ii: 604–6.

and quite generally on infinite series, that in his rationalizing integral transformation he had contributed an interesting idea, but that he had not produced anything essentially new, either from a general viewpoint, or as an individual result. The thing smelled of plagiarism. After a thorough study over several weeks – this is how it appeared to Newton – Leibniz was well able to reconstruct the analytical content of his first letter from the details given to him. That there was here no independent invention, nor even a rediscovery, but simply an attempt at plagiarism seemed certain to Newton when Leibniz asked him to explain again more explicitly the decisive points in his letter, namely his methods of series-expansion and series-inversion. Had Leibniz – so we may conjecture ran Newton's argument – already been in possession of these techniques, then he had no cause to ask for further details: hence he cannot have had them himself. There was also the passage with the phrase *ludus naturæ*,[6] which had for Newton an unpleasant taste of metaphysics, and Tschirnhaus' remark on the contingent significance of his doctrine of series:[7] perhaps the rash German nobleman had revealed more by this remark than could have pleased Leibniz? If this were so, then it was obvious that his partner had not penetrated to the central ideas of the new analysis. Newton had no intention of allowing him to continue this present game and meant to end the discussion with an unworthy opponent by an answer that was – this went without saying – politely phrased but unmistakable in its tone. So much for the general motive underlying Newton's second letter of 24 October (3 November) 1676.[8]

---

[6] *Compare* ch. **17**: notes 46–7, 49, and the text p. 241.

[7] *Compare* ch. **18**: notes 10–11, and the text p. 251.

[8] This letter, Newton's famous 'epistola posterior', is the outcome of painstaking thought. Newton could not refuse Oldenburg's request to give Leibniz the information he desired. The letters of 26 Oct. (5 Nov.) and 14 (24) Nov. 1676 (*NC* **2**: 162–3, 181–2) contain additions and corrections. Some passages made illegible in the original can be restored with the help of an autograph copy. An incompetent copyist's transcript (*LBG*: 203–25) was dispatched together with a letter from Oldenburg only on 2 (12) May 1677 (*LBG*: 238–9), entrusted to Schröter for delivery in Hanover, where it arrived on 21 June (1 July) 1677 – and Leibniz sent his reply on that same day. Collins promised to Baker in his letter 10 (20) Feb. 1676/7 to send him a copy he had himself made of the 'epistola posterior': neither this nor the copy sent to Wallis, of which Collins' letter of (?) 1677/8 (*NC* **2**: 242) shows evidence, have survived. This copy provided the text for the excerpts in Wallis' *Algebra* (1685): 318–20, 338–47 and in *WO* II: 357–9, 368–71, 381–90, 390–6. During the summer 1684 Newton started to compile two papers on fluxions and series with the title 'Matheseos universalis specimina' and 'De computo serierum' but they remained unprinted until the recent edition in *NP* **4**: 526–617. The letters exchanged between Newton and Leibniz in 1676–7 were meant to be incorporated

260

In his introduction[9] Newton speaks in somewhat smooth and vague phrases of Leibniz' method of series. He himself knows three ways of expanding in series and scarcely expects there to be any more. Originally he here made the significant admission that he had previously known nothing of Leibniz' series; in the copy sent to Oldenburg this phrase is heavily crossed out, but it is still recognizable in the copy retained by Newton – unfortunately, we have no information on the circumstances surrounding this crucial alteration. Newton then enumerates his three methods: the first, by step-by-step substitution, had been already described in his first letter to Leibniz; the second by numerical induction, he had used initially but subsequently abandoned; the third by series-inversion, finding the coefficients step-by-step, he now preferred.

Newton describes in some detail[10] how he came upon his second method as he went through Wallis' *Arithmetica infinitorum* (propositions 113, 121 and 169), and how from the expansion of

$$\int_0^x (1-t^2)^n . dt$$

for positive integral n he obtained by interpolation the expansion of $\int_0^x \sqrt{(1-t^2)} . dt$ (that is, for $n = \frac{1}{2}$) and correspondingly that of $\int_0^x \sqrt{(1+t^2)} . dt$; from there he went on to deduce by differentiation the expansion of $\sqrt{(1-x^2)}$ and checked its correctness by squaring.

by way of introduction, and so Newton had a copy made by Wickins of his own transcript of the 'epistola posterior', himself revising the script (*NP 4*: 618–33). This text became the basis of the first full publication in *WO* III: 634–45 and the reprint, with notes, in *CE (1712)*: 67–86 and subsequent reprints. The original was first published, with ample notes, in *NC 2*: 110–29. Here we base ourselves on the version received by Leibniz, as printed in *LBG*.

[9] *LBG*: 203.
[10] *LBG*: 204–7. Newton's notes of 1664–5 contain quadratures of curves $y = (1-x^2)^k$, the binomial expansion for $\sqrt{(1-x^2)}$ (*NP 1*: 104–11), followed by the quadrature of the hyperbola from $y = 1/(1+x)$ (*ibid.* 112–15). In the autumn of 1665 appears the interpolation array of the 'arithmetical triangle' of binomial coefficients with the insertion of single intermediate terms (122–6), also the continuation of the arithmetical triangle for negative indices (128–31), the interpolation of pairs of intermediate terms (132–4), lastly the computation of the coefficients of the logarithmic series accompanied by related numerical calculations to a high number of decimal places (134–42). During the summer 1665 Newton found the inversion of the logarithmic series (320), the general binomial expansion for fractional positive index, the expansion of $1/(a+b)$ by long division and of $\sqrt{(a^2+b)}$ by the algorithm of arithmetical root-extraction (321). Later, *c.* 1667, Newton resumes the computation of an hyperbola area to even more decimal places (*NP 2*: 184–9).

From then on he had abandoned the interpolation method (which perhaps did not seem quite reliable to him) in favour of the straightforward process of root extraction, extended to literal expressions. Subsequently the far simpler derivation of series by division and the extension of the technique to higher equations had also been found by him. These details of how he had originally derived the binomial theorem, he had, he writes, scarcely been able to recall; only when he some weeks ago (in connection with his first letter for Leibniz) looked again through his earliest notes had they once more sprung vividly to mind. All this, Newton says, was effected before Mercator's *Logarithmotechnia* appeared in print; when that happened,[11] Newton now reports, he had, through Barrow, sent his *De Analysi* to Collins, who ever since had been urging him to publish his researches. In fact Newton had written an enlarged version (the *Methodus*) some five years before and this he planned to publish in appendix to his optical lectures. He had, however, unwisely allowed himself to be induced to communicate his theory of colour to the Royal Society in conjunction with his reflecting telescope, and to allow its publication. The controversy which arose in sequel had dissuaded him from his original intention:[12] he now blamed himself for his lack of forethought in that in pursuit of a vain phantom he had endangered his tranquillity, a far more important possession.

Newton goes on to relate how Gregory had at that time rediscovered the same method after long study, taking his lead, as he himself had confessed, from a single series which Collins had sent him; he had left a manuscript tract on all this and it was hoped to have his friends publish it, as he had contributed a great many very valuable discoveries of his own, the loss of which would be regrettable.[13] This description is not quite correct; in reality Gregory had already, before 1668, wholly independently discovered the binomial theorem[14] and the method of series using Taylor's theorem[15] as the basis for his expansions. The long study referred, as has been explained above,[16] merely to the structure of the series for the circle-zone; this Gregory

---

[11] *LBG*: 207–8.
[12] Concerning public knowledge of the discussion about Newton's theory of colour *compare* ch. **17**: note 8, about the reflecting telescope, ch. **17**: note 9.
[13] Newton knew of Gregory's successes in power series already in 1670 through the accounts given by Collins to Barrow; more information came in Collins' letter 5 (15) July 1671 (*NC* **1**: 65–6).
[14] *See* pp. 134–5.
[15] *See* ch. **10**: note 58.
[16] *See* p. 220.

at first sought to construct by combining his various interpolation series for the circle, until he noticed its link with the binomial expansion. Of a compendium of the method of series by Gregory we know nothing; presumably Newton here alludes to the 'Historiola' which Collins intended to have printed.

The *Methodus*, Newton continues, he left unfinished after abandoning his intention of publishing it. It still lacked a chapter in which he had meant to deal with problems that cannot be reduced to quadratures (differential equations in immediately integrable form); on these he had made only a few basic observations. Other topics there discussed also needed fairly extensive supplementation, for instance the tangent-method,[17] which Sluse had communicated two or three years ago to Oldenburg only to be told that Newton possessed the same rule,[18] each having reached it in a different way; Newton's procedure did not lack proper proof.[19] If one set out with his directions, it would be impossible to determine tangents in any other manner without deliberately deviating from the right way. His rule comprehended irrational expressions also and it was accordingly quite unnecessary to eliminate these beforehand; similarly it could deal with questions on extreme values and other topics, which Newton does not now wish to go into. The basic principle, hidden by Newton in the form of an anagram which he later resolved in a letter to Wallis, states:[20] from an equation containing fluents to find the fluxions, and vice versa.

Without going into too much detail we may here briefly remark that Newton had, to his discredit, provoked a dispute about the tangent-method, voicing his conjecture that Sluse's method might have originated from the report which Oldenburg had made about his own:[21] this had been politely but very firmly refuted by Sluse.[22] The matter seems to have been discussed at a council meeting of the

17 *LBG*: 208; *compare* ch. 3: note 13, also the reference in ch. 6: note 43, and the text p. 72.
18 Oldenburg–Sluse, 29 Jan. (8 Feb.) 1672/3 (*CE* (*1712*): 31 = *LBG*: 232–3) where a Latin extract from Newton's 'tangent' letter of 10 (20) Dec. 1672 (*NC* **1**: 247) is reproduced. The Latin text in *CE* (*1712*): 29–30 is somewhat fuller.
19 This remark alludes to the lack of a proof in Sluse–Oldenburg 23 Apr. (3 May) 1673 (*SL*: 678–9) (partially printed in *PT* 8, No. 95 of 23 June (3 July) 1673: 6059).
20 Newton–Wallis, 27 Aug. (6 Sept.) 1692 (*compare WO* II: 391). The anagram, with its solution, is found in Newton's 'Waste Book', entered in October 1676 (*NP* **2**: 191, note 25).
21 Oldenburg–Sluse, 16 (26) Dec. 1672 (not printed); its contents can be surmised from the reply, 7 (17) Jan. 1672/3 (*SL*: 677).
22 Sluse–Oldenburg 23 Apr. (3 May) 1673 (*SL*: 878).

Royal Society[23] and culminated in an official apology by Newton[24] in which he says that, like Oldenburg, he now believed that Sluse had already discovered his method several years before the second edition of his *Mesolabum* was published in 1668; having then communicated it to his friends, he was entitled to claim full priority for it. The rule agreed completely with Newton's but probably originated in a different conceptual approach. Newton cannot tell if Sluse's method reaches as far as his own, which permits immediate application of the procedure to irrationals within an equation without any need first to remove them. Sluse was generous enough to accept this none too satisfactory explanation;[25] Newton himself never relinquished the reservation contained in it, as is proved by a remark he added in the second edition of the *Commercium epistolicum* (to an excerpt from a letter by Gregory)[26] to the effect that Sluse's tangent-method was dependent from his own.

Newton next proceeds to represent the binomial integral[27]

$$J_\theta = \int z^\theta (e + fz^\eta)^\lambda \, dz$$

by the series

$$\frac{z^{\theta+1-\eta}}{(\theta+1)f}(e+fz^\eta)^{\lambda+1} \cdot \left(\underbrace{\frac{r}{s}}_{A} - \underbrace{\frac{r-1}{s-1}\cdot\frac{eA}{fz^\eta}}_{B} + \underbrace{\frac{r-2}{s-2}\cdot\frac{eB}{fz^\eta}}_{C} + \cdots \right)$$

where $r = (\theta+1)/\eta$ and $s = (\theta+1)/\eta + \lambda$. No explanation is given. Leibniz indicates in a marginal note[28] how he thinks it was worked out: setting $e+fz^\eta = w$, he transforms by an integration by parts; his procedure (in which there is a verbal lapse of no importance) then leads in modern notation to the reduction formula

$$J_\theta = \frac{1}{\eta f s} \cdot z^{\theta+1-\eta} \cdot w^{\lambda+1} - \frac{r-1}{s}\cdot\frac{e}{f}\cdot J_{\theta-\eta},$$

from which the rest is evident. For positive integral values of r and s, Newton continues, the series will break off after a finite number of terms; for others it does indeed go on to infinity. All this is an exten-

[23] RS Council meeting, 18 (28) June 1673 (*BH* III: 92). Perhaps the outcome of the discussion is incorporated in Collins–Newton, 18 (28) June 1673 (*NC* 1: 288–9).
[24] Newton–Oldenburg, 23 June (3 July) 1673 (*NC* 1: 294), forwarded with Oldenburg–Sluse, 10 (20) July 1673 (*CE* (*1712*): 31–2).
[25] Sluse–Oldenburg, 26 July (5 Aug.) 1673 (*SL*: 679).
[26] The reference is to the footnote *CE* (*1722*): 95–6 to Gregory–Collins, 5 (15) Sept. 1670, asserting that Sluse's and Gregory's tangent methods are dependent on Barrow's in the *Lectiones geometricæ* (1670): 80.
[27] *LBG*: 208.   [28] *LBG*: 208–9.

sion of a section in his *Methodus*.[29] That similar formulae are already indicated in the *De Analysi*, as a note in the *Commercium epistolicum* asserts,[30] is not evident from the general remarks made there;[31] nor is it clear from the similarly worded text of a letter to Strode[32] to which Newton refers in the second edition of the *Commercium epistolicum*. All that is said in each case is that by his method Newton can square any area, either exactly (that is by a closed expression) or at least in such a way that the error can be made arbitrarily small. Newton here gives as examples the integrals

$$\int \sqrt{(az)}\,.\,\mathrm{d}z, \quad \int az^{m/n}\,.\,\mathrm{d}z, \quad \int \frac{a^4 z\,.\,\mathrm{d}z}{(c^2-z^2)^2}, \quad \int a^5 z^{-5}\,\sqrt{(b^2+z^2)}\,.\,\mathrm{d}z,$$

and

$$\int \frac{b\sqrt[3]{z}}{\sqrt[5]{(c-a\sqrt[3]{z^2})}}\,.\,\mathrm{d}z.$$

In all these cases the quadrature can be found in a closed expression, since r is a positive integer; but when s proves to be a positive integer, an expansion in the contrary sense has to be used. When the given area cannot be squared geometrically, because neither r nor s are – nor can be made – positive integers, then at least a reduction to the quadrature of a conic or some other reasonably simple curve will be possible. Trinomial integrals can be similarly dealt with. These remarks relate to a corresponding section in the *Methodus* [33] and the last is elaborated in a subsequent letter on 8 (18) November 1676 of Newton's to Collins.[34] As an instance of this type of reduction Newton cites his rectification of the cissoid[35] $xy^2 = (a-x)^3$: construct (fig. 30) the hyperbola $y^2 = a^2 + 3x^2$ and on it determine the points C $(a, 2a)$ and $Q\,(\sqrt{(ax)}, \sqrt{(a^2+3ax)})$, and draw the tangents CD and QT. Then $a.\widehat{BP} = 6$ area $TQCD$. In the *Methodus*, whence this example has been taken, Newton determines

$$2x^2\sqrt{[1+(\mathrm{d}y/\mathrm{d}x)^2]} = a\sqrt{[ax+3x^2]}$$

[29] The original, now reproduced in *NP* 3: 236–65, contains a large catalogue of formulae with numerous examples; the first formula is missing in Wallis (1685), but appears on *WO* II: 390–1.
[30] *CE (1712)*: 72.
[31] *NP* 2: 240–2.
[32] Collins–Strode, 26 July (5 Aug.) 1672 (*CE (1712)*: 28; enlarged in *CE (1722)*: 102–3).
[33] *See* note 29 above.
[34] *NC* 2: 179.
[35] *LBG*: 211; *compare Methodus* (*NP* 3: 320–2).

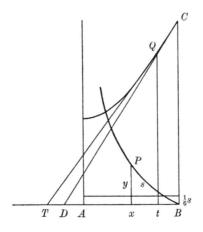

Fig. 30

and then substitutes $x = t^2/a$ so obtaining

$$\int_t^a \mathrm{d}s = \int_t^a (a/t^2)\, \sqrt{(a^2 + 3t^2)}\,.\,\mathrm{d}t,$$

and the rest follows easily.

Newton can scarcely believe that Tschirnhaus has good expectation of a simpler or more general representation of quantities than by series.[36] They are not a contingent innovation in the solution (*per accidens*) but exactly render the general character of the solution; the transition from decimal numbers to the doctrine of series is comparable to that from arithmetic to algebra. It will moreover also be possible to expand a quantity as a series in several variables, or, equally, to follow a procedure analogous to the functional iteration employed by Gregory for squaring the circle and hyperbola; but all this involves laborious calculation which Newton feels not inclined to undertake.

He now warns that there are indeed other problems for which an expression in series will not be the simplest method of solution, as for instance to determine an algebraic curve through a certain number of points – a conic through five points, say, or a cubic through eight.[37] Here Newton originally added that he possessed a construction method for the latter problem also,[38] but later desired this phrase to be cancelled; in the meantime, however, Newton's letter had already

[36] *LBG*: 212–13; *compare* p. 251.
[37] *Compare* for this Newton's treatment of the cubic with a double point (*NP* 2: 134–51, especially 150).
[38] Newton–Oldenburg, 14 (24) Nov. 1676 (*NC* 2: 182).

been copied (on 4 (14) September 1676) and the correction was not there made, though it was made subsequently in the original, probably after its return, and by Newton himself. The passage appears in the text reproduced in the third volume of Wallis' *Opera*[39] and in the *Commercium epistolicum*,[40] where the 'eight' is replaced by 'seven'.

Newton now presents certain theorems cognate to the Leibniz series for the conic sector, asserting that by their aid the following quadratures can be reduced to conic areas[41]

$$\int \frac{x^{kn-1}.\,\mathrm{d}x}{e+fx^n+gx^{2n}}\,,\quad \int \frac{x^{kn-1}\sqrt{x}.\,\mathrm{d}x}{e+fx^n+gx^{2n}}\,,\quad \int x^{kn-1}\sqrt{(e+fx^n+gx^{2n})}.\,\mathrm{d}x\,,$$

$$\int \frac{x^{kn-1}.\,\mathrm{d}x}{\sqrt{(e+fx^n+gx^{2n})}}\,,\quad \int \frac{x^{kn-1}\sqrt{(e+fx^n)}}{g+hx^n}.\,\mathrm{d}x\,,$$

$$\int \frac{x^{kn-1}.\,\mathrm{d}x}{(g+hx^n)\sqrt{(e+fx^n)}}\,,\quad \int x^{kn-1}\sqrt{\frac{(e+fx^n)}{(g+hx^n)}}.\,\mathrm{d}x.$$

Some of these integrals reduce to a single conic quadrature, others are composed of several. The matter is, says Newton, quite complicated and he cannot really think that the transformations given by Gregory and others (meaning, of course, Leibniz) will by themselves be successful. He had, he remarks, gathered these results long ago (*dudum*); the subject is extensively dealt with in a tabular array in the 1671 *Methodus*.[42] There is again a reference to it in the *Commercium epistolicum*[43] pointing out that it follows from this that the theorems in question had been found a long while (*diu*) before 1676.

In particular, writes Newton, from $\int_0^1 \mathrm{d}x/(1+x^2)$ there follows Leibniz' series and, again,[44] from $\int c.\,\mathrm{d}x/(x^2+x\sqrt{2}+1)$ the less slowly convergent series

$$\pi/2\sqrt{2} = 1+\tfrac{1}{3}-\tfrac{1}{5}-\tfrac{1}{7}+\tfrac{1}{9}+\tfrac{1}{11} - \ldots.$$

Intended is the series-expansion of

$$\int_0^x \frac{(1+t^2).\,\mathrm{d}t}{(1+t^4)} = \int_0^x \frac{\mathrm{d}t}{1+(t\sqrt{2}+1)^2} + \int_0^x \frac{\mathrm{d}t}{1+(t\sqrt{2}-1)^2}$$

$$= \frac{1}{\sqrt{2}}\tan^{-1}\!\left(\frac{x\sqrt{2}}{1-x^2}\right).$$

[39] *WO* III: 639.   [40] *CE (1712)*: 76.
[41] *LBG*: 213–14; he alludes here to further results from his *Methodus* (*NP* 3: 236–65); compare also note 29.   [42] *NP* 3: 244–55.   [43] *CE (1712)*: 76, footnote.
[44] *LBG*: 214–15; the original, written in the autumn of 1676, is reproduced in *NP* 4: 204–13.

Oldenburg was to be doubtful about the sequence of signs, but Newton confirmed their correctness.[45] In the *Commercium epistolicum* it is commented that Brouncker was the first to square the hyperbola[46] by the series

$$\frac{1}{1.2}+\frac{1}{3.4}+\dots \ = \ 1-\tfrac{1}{2}+\tfrac{1}{3}-\tfrac{1}{4}+\dots;$$

Mercator had added a new demonstration, Gregory the circle series

$$\pi/4 \ = \ 1-\tfrac{1}{3}+\tfrac{1}{5}-\tfrac{1}{7}+\dots$$

and Newton the series

$$\pi/2\sqrt{2} \ = \ 1+\tfrac{1}{3}-\tfrac{1}{5}-\tfrac{1}{7}+\tfrac{1}{9}+\tfrac{1}{11}-\dots.$$

Concerning Brouncker and Mercator we have already said all that is necessary; there is really no connection between their two methods. Gregory had, it is true, given the general series for the inverse tangent,[47] but not the particular Leibniz series.

Methods, Newton continues,[48] are to be judged by their computational usefulness, and on this criterion his sine-expansion is decidedly preferable. In order to calculate 20 places of $\pi/2\sqrt{2}$ it is necessary to add about $5\,000\,000\,000$ terms of the series $1+\tfrac{1}{3}-\tfrac{1}{5}-\tfrac{1}{7}+\dots$, while the like computation from $\sqrt{2}\int_0^{1/\sqrt{2}}\mathrm{d}x/\sqrt{(1-x^2)}$ requires no more than 55 to 60 terms of $1+\tfrac{1}{12}+\tfrac{3}{160}+\dots$, and for $\pi/6$ a computation from the series-expansion $\tfrac{1}{2}+\tfrac{1}{48}+\tfrac{3}{1280}+\dots$ of $\int_0^{1/2}\mathrm{d}x/\sqrt{(1-x^2)}$ will be very simple.[49] For small values of x the calculation of the half segment of a conic is best effected from $\int_0^x \sqrt{(t\pm t^2)}.\,\mathrm{d}t$; thus with $x=\tfrac{1}{4}$ it suffices to take 20 terms to gain an accuracy to 16 decimal places.[50] The Leibniz series, too, can be adapted to practical computation,[51] for instance by combining it with Newton's series in the form

$$\pi(1+\sqrt{2})/8 \ = \ 1-\tfrac{1}{7}+\tfrac{1}{9}-\tfrac{1}{15}+\tfrac{1}{17}-\tfrac{1}{23}+\tfrac{1}{25}-\dots$$

[45] Newton–Oldenburg, 14 (24) Nov. 1676 (*NC* **2**: 181).
[46] *CE* (*1712*): 77, footnote. *Compare* p. 96.
[47] Gregory–Collins, 15 (25) Feb. 1671 (*GT*: 170 = *CE* (*1712*): 25).
[48] *LBG*: 214–15.
[49] *Compare Methodus* (*NP* **3**: 224–6), and *see* also *NP* **4**: 24–6.
[50] This expansion for the circle-arc occurs in *De Analysi* (*NP* **2**: 238) and, independently, in Gregory–Collins, 17 (27) May 1671 (*GT*: 190). It had been communicated to Leibniz on 12 (22) Apr. 1675 (*LBG*: 114–15); *compare* ch. **10**: notes 51–2, and the text p. 133. The ensuing computation (*LBG*: 215–16) is repeated from the *Methodus* (*NP* **3**: 224).   [51] *LBG*: 216–17.

or computing from

$$\pi/6 = \int_0^{1/\sqrt{3}} dt/(1+t^2).$$

Even more advantageous is the representation of

$$\pi/4 = \tan^{-1}\tfrac{1}{2} + \tfrac{1}{2}\tan^{-1}\tfrac{4}{7} + \tfrac{1}{2}\tan^{-1}\tfrac{1}{8}$$

by combining the previous series: namely, as the value for $x = \tfrac{1}{2}$ of

$$\left(\frac{x}{1} - \frac{x^3}{3} + \frac{x^5}{5} - \frac{x^7}{7} + \cdots\right) + \left(\frac{x^2}{1} + \frac{x^5}{3} - \frac{x^8}{5} - \frac{x^{11}}{7} + \cdots\right)$$

$$+ \left(\frac{x^4}{1} - \frac{x^{10}}{3} + \frac{x^{16}}{5} - \frac{x^{22}}{7} + \cdots\right);$$

that is, in exact equivalent,

$$\tan^{-1} x + (\sqrt{x}/\sqrt{2}) \tan^{-1} [\sqrt{(2x^3)}/(1-x^3)] + x \tan^{-1}(x^3):$$

this is the first historical occurrence of a relation of this type and it is no less excellently contrived than similarly constructed modern ones.

Newton turns next to Leibniz' two exponential series,[52] one of which ensues from the other by a change of sign, while both are determinable by addition or subtraction of the series

$$x + \frac{x^3}{3!} + \frac{x^5}{5!} + \cdots \quad \text{and} \quad \frac{x^2}{2!} + \frac{x^4}{4!} + \frac{x^6}{6!} + \cdots;$$

Newton does not, he says, understand the remark about the better convergence of the series for e^{-x} compared with that for e^x. In order to compute the logarithms of the first few primes[53] it is convenient to set $2 = \sqrt[10]{(9984.1020/9945)}$, $3 = \sqrt[4]{(8.9963/984)}$, and similarly for others. Again, the approximation $x \approx \sqrt{(6 - \sqrt{[12+24\cos x]})}$ does not meet with Newton's approval:[54] if we set $1 - \cos x = y$ then its difference from the true value comes to nearly[55] $y^3/90 + y^4/194$; a closer approximation will be $x \approx \sqrt{(2y)}(120-17y)/(120-27y)$ with an error of the order $61y^3\sqrt{(2y)}/44\,800$. To find circle-segments Newton

[52] LBG: 217–18 referring to Leibniz–Oldenburg 17 (27) Aug. 1676: NC 2: 60–1; compare pp. 237–8.
[53] LBG: 219; the original is the passage on approximations in NP 4: 22–35, especially 26.
[54] LBG: 218, referring to Leibniz–Oldenburg 17 (27) Aug. 1676 (NC 2: 61); compare p. 238.
[55] Newton had originally written $y^3/90 + y^4/140$, later correcting the error in Newton–Oldenburg, 26 Oct. (5 Nov.) 1676 (NC 2: 162).

believes it is easier to work with tables, since the corresponding series-expansion (of $x - \sin x$) is not very convenient. To construct a sine-table Newton's advice is very simply to find the sine and cosine of a sufficiently small angle by the series; the rest is then best achieved by applying the formulae for sines and cosines of multiple angles.[56]

From these particular computations Newton passes on to treat a topic of far more general interest, namely the famous parallelogram,[57] named after him, by which one finds the first approximations to the series solution of an algebraic equation in two unknowns, one in terms of the other. He elaborates the basic idea and a first example using the manner of description with which he had introduced it into his earlier *Methodus*, confining his attention in each case to a single series. He warns that the calculation will not always be straight-forward but Leibniz will, he is sure, be able to cope with the diffi-culties arising. With this we have come full circle back to Leibniz' original question concerning the formal methods used by Newton in his expansions. Now at last the third procedure, the step-by-step inversion of a series,[58] is illustrated by an example. When it is given that

$$y = x + \frac{x^2}{2} + \frac{x^3}{3} + \frac{x^4}{4} + \cdots,$$

we then must compute in turn

$$y^2 = x^2 + x^3 + \frac{11x^4}{12} + \cdots,$$

$$y^3 = x^3 + \frac{3x^4}{2} + \cdots,$$

$$y^4 = x^4 + \cdots.$$

By continually eliminating, stage by stage, successively higher powers of x we obtain

$$x = y - \frac{y^2}{2} + \frac{y^3}{6} - \frac{y^4}{24} + \cdots.$$

In the same manner we can pass from one unknown to another; we may proceed, for instance, from the series for the inverse sine to that for the inverse tangent, by combining

$$s = r \sin^{-1} \frac{x}{r} = x + \frac{x^3}{6r^2} + \frac{3x^5}{40r^4} + \cdots$$

[56] *LBG*: 219–20, evidently in reaction to the instrument for angle-section mentioned by Tschirnhaus (*compare* ch. **18**: note 15) Newton next describes the arrangement he has been using for this purpose, for which *see NP* **4**: 22–4.
[57] *LBG*: 221–2, from *Methodus* (*NP* **3**: 48–56).
[58] *LBG*: 222–3, from *Methodus* (*NP* **3**: 58).

with

$$t = r \tan \frac{s}{r} = \frac{rx}{\sqrt{(r^2 - x^2)}} = x + \frac{x^3}{2r^2} + \frac{3x^5}{8r^4} + \dots$$

The direct deduction is of course to be preferred and this application is mentioned only to illustrate the scope of the method. To end this paragraph[59] Newton gives the general schemes for inverting the series

$$y = ax + bx^2 + cx^3 + dx^4 + \dots \quad \text{and} \quad y = ax + bx^3 + cx^5 + \dots$$

exemplyifying these by inverting the series for the logarithm $\log(1 + x)$ and for the inverse sine $r \sin^{-1}(x/r)$. He hints that he has yet another method of passing from areas to their corresponding ordinates (namely by means of the calculus of fluxions), but he does not wish to reveal it here.

Returning to his earlier remark about the scope of problems amenable to treatment by infinite series, Newton now explains that he had had especially in view those problems with which mathematicians usually occupy themselves, and in which typical mathematical methods of deduction are employed; he admits that it will doubtless be possible to devise such complicated initial circumstances that the tools used hitherto will no longer suffice. Still he believes it is not too much to assert that he is master of the inverse tangent problem and other still more difficult questions. For this he uses two methods, one particular, and one more general; these however he merely hints at in a second anagram.[60] The solution was first printed in the third volume of Wallis' *Opera*:[61] the first method relates to finding the fluent from an equation involving the fluxion (that is to the general integration of differential equations), the other to deriving infinite series with indeterminate coefficients. Both are developed in his earlier *Methodus*.[62]

But, for instance, to find the curves of constant tangent-length, says Newton, neither will be required; the problem is solved by a mechanical curve (the tractrix) which depends on the quadrature of the hyperbola.[63] Of a similar nature is the question to find the curve

[59] *LBG*: 223.
[60] *LBG*: 224.
[61] *WO* III: 645, note e. Presumably it was sent to Wallis together with the text of the two letters in May 1695; *see* also note 20.
[62] *See Methodus* (*NP* 3: 70–120).
[63] *LBG*: 224. Starting from a differential equation $dy/ds = y/c$, equivalent to $dy/dx = y/\sqrt{(c^2 - y^2)}$, the substitution $c^2 - y^2 = t^2$ will lead to $x = \int t^2 . dt/(c^2 - t^2)$. There is no evidence amongst Newton's papers to tell us his procedure: *NP* 3: 26, note 31. The problem was shown to Leibniz by Perrault in Paris during 1676 and at once solved by him according to the account in *AE* (Sept. 1693): 385–92 (*LMG*

whose subtangent is a given function of x; yet Newton would not describe such a problem as a *ludus naturæ*.[64] When a relationship between any two sides of the triangle formed by tangent, subtangent and ordinate is expressed by an equation, the problem will be resolvable without applying the general method (that is to say, it depends only on quadratures, and not on a differential equation in not immediately integrable form); the matter is altogether different when the subtangent is involved with the tangent and ordinate in more complicated expressions.[65] Newton evinces strong interest in Leibniz' general method for solving equations and in particular he wishes to hear how he deals with equations containing terms with fractional or even irrational indices.[66]

Originally, Newton ended the letter with a word of greeting to Leibniz and Tschirnhaus, but this was cancelled in the final version – evidently because he had meanwhile been told that Oldenburg had already written to Tschirnhaus, and because Newton had no wish to continue his dialogue with Leibniz.[67] This is suggested by a short accompanying note to Oldenburg in which he explains that to avoid becoming burdensome he had only glanced at some topics, and had omitted others altogether; he believes that Leibniz will be satisfied, adding that he himself has now come to be more interested in other subjects.

Very obviously when Newton began his letter he had formed a by no means poor impression of Leibniz' competence. His judgment of the arithmetical circle quadrature was reserved but friendly; we have the feeling that, in reporting on his earliest discoveries in mathematics he seems to recapture the happy, cheerful mood of more than ten years before when one beautiful insight after another had come to him. Only subsequently, where he begins to comment on details in Leibniz' letter, does another sentiment gain the upper hand. A first hint of annoyance that Leibniz' series converges so badly is momentarily suppressed when the practical aspect of the problem comes

v: 296–7) though no trace of his original work exists now in his papers. Priority in publication belongs to Huygens, in a letter to Basnage of 1693 printed in the *Histoire des ouvrages des sçavans* for December 1692/February 1693 (*HO* **10**: 408–9, with addenda 418–20).

[64] *Compare* p. 241.

[65] For this *see* Scriba (1964a): 125–30.

[66] *LBG*: 225. This passage led Leibniz in his reply, 21 June (1 July) 1677 (*NC* **2**: 217–18) to put forward the example of finding x and y from the pair of equations $x^y + y^x = xy$, $x^x + y^y = x + y$, and being himself unable to solve the problem he wondered if Newton could cope with it.

[67] Newton–Oldenburg, 24 Oct. (3 Nov.) 1676 (*NC* **2**: 110).

to the fore holding his attention for a while, but he soon tires of it and irritation sets in; the matter of the trigonometrical and logarithmic functions he takes in ill part and does not bother to conceal his disapproval. We might suppose that at this point he re-read the opening paragraphs of his letter and now deleted his original approving comment on the arithmetical circle quadrature: that it has become clear to him that Leibniz' work is unoriginal and that he passes off the thoughts of others as his own. It is with some degree of reluctance that Newton produces his tabular parallelogram technique for expanding an algebraic function in series and explains his third method of series inversion. The unfortunate misreading *ludus naturæ* introduces for Newton an unpleasant turn in their exchange; he coldly dismisses it as an irrelevant consideration. His colleague should know that he has been acquainted only with some of the simpler peripheral problems, not with the central ideas which are concealed in his anagrams. This letter is thus a curious blend of a singular conflict of emotions and compactly described scientific results – it allows us an unusually penetrating glimpse into the heart of this enigmatic man, who could be both intimately confiding on one occasion and yet shy and forbidding on another.

A new misunderstanding soon arose to drive them further apart. As Newton knew only afterwards, Leibniz stayed for a few days in London in October 1676, during which time he had a conversation with Collins and gave him some papers on algebraical topics:[68] news of this he gained from a subsequent letter of Leibniz' to Oldenburg,[69] which also contained details of his method for finding the derivative y' as a function of x alone if $f(x, y) = 0$ is the implicit equation of a curve of nth order; at the same time he learned that Hudde had known Sluse's tangent-rule in an improved form for a long time,[70] and further Mercator's quadrature of the hyperbola[71] since 1662. He did perhaps not know the exact dates of Leibniz' visit, but he would

[68] These are the essays *C* 1032 on solving cubics by Cardan's formula, *C* 1042 (with later addenda in *C* 1138, 1344) on equations of higher degree for whose solution an algorithm can be devised – anticipating, though badly expounded, De Moivre's formula – and *C* 1044 containing a synopsis of his work on certain special higher equations. All these papers were known to Newton through transcripts.
[69] Leibniz–Oldenburg, 18 (28) Nov. 1676, enclosed in Collins–Newton, 5 (15) Mar. 1676/7 (*NC* 2: 198–201).
[70] *Compare* Leibniz' minute of November 1676 on his conversation with Hudde in *LBG*: 228–9. Hudde's rule is contained in his letter to Schooten, 21 Nov. 1659, printed in a French version in the *Journal literaire* for July/August 1713 (*LBG*: 234–7).
[71] Hudde–Van Duyck, 1662; *compare* Leibniz' note *LBG*: 228.

NEWTON'S SECOND LETTER FOR LEIBNIZ

assume that his second letter for Leibniz had been conveyed to Hanover by a reliable intermediary[72] towards the end of March 1677; he would not guess that the letter had in fact been dispatched only about the middle of May and did not arrive in Hanover before the end of June 1677. In Leibniz' reply[73] he found the general determination of a tangent to a curve $f(x, y) = 0$ given in series-form, avoiding Sluse's rule by the use of the differential calculus, and also the extension of the procedure to irrational quantities by the rule for differentiating the function of a function. He could also see from it that Leibniz was by now certainly in full possession of the doctrine of series – this emerges even more clearly from a supplementary letter some three weeks later[74] – and that Leibniz' first letter had been written immediately after he had received Oldenburg's large letter-packet. Neither Newton nor Collins planned an immediate reply,[75] and Oldenburg's death soon afterwards terminated the correspondence altogether, scarcely contrary to Newton's wishes.

From the communications he had received Newton could hardly form a clear picture of Leibniz' mathematical ability or achievement. On the really crucial point, his invention of the calculus, he was only very superficially informed. Moreover, in contrast to Leibniz, he attached little value to the symbols to be used and consistently retained.[76] Accordingly all that remained for him was a collection of individual results, and by that scale of comparison Leibniz necessarily made a poor show. Moreover, their differences in the methodological treatment of problems under discussion were not very remarkable. So we may understand – if not support – Newton's judgment only by surface appearances with no attempt to penetrate to the root of the matter: doubly so when we recall that he was mistrustful from the first, and that Leibniz' latest communications served only to confirm him in his opinion that Leibniz had contributed nothing essentially new to the whole group of problems.

There is a footnote in the *Commercium epistolicum* in connection with Leibniz' tangent-method[77] which considers whether Leibniz

72 Collins promises in the letter to Newton, 5 (15) Mar. 1676/7 (*NC* 2: 201) to forward the letter which had not yet been dispatched within the next week.
73 Leibniz–Oldenburg, 21 June (1 July) 1677 (*NC* 2: 213–15). The letter is erroneously dated as of 11 June O.S. in *NC* 2: 212 where the editor had overlooked that in Hanover the Julian calendar was still being used until 1700.
74 Leibniz–Oldenburg, 12 (22) July 1677 (*NC* 2: 231).
75 Oldenburg–Leibniz, 9 (19) Aug. 1677 (*LBG*: 253).
76 *CE* (*1722*), Annotatio: 247: *Methodus fluxionum utique non consistit in forma symbolorum.*
77 *CE* (*1712*): 89, footnote to Leibniz–Oldenburg, 21 June (1 July) 1677.

might possibly (*forte an*) have profited from a perusal of other papers of Newton's during the London visit; here we have indeed an instance of the emotion poisoned by dark suspicion which animated the 'editor' of the *Commercium epistolicum*. In his review of the book in the *Philosophical Transactions*[78] Newton still writes as though his second letter came into Leibniz' hands towards the end of the winter of 1676–7 or early the next spring, but in the 'Recensio libri' (in the second edition in 1722)[79] it is said still more strongly that Leibniz had seen Newton's letter already at the beginning of November 1676 during his second visit to London. For such a hardening in attitude – after Leibniz' death – there was a good reason, and Leibniz himself had given cause for it, having pointed out in a letter to Conti[80] that while in London he had, through Collins' assistance, seen some of Newton's correspondence and on that occasion had come upon a passage where Newton admitted that in the field of rectification of curves he himself had found little of essential importance apart from his result for the arc-length of the cissoid. This passage, Newton now stated,[81] occurs in his second letter for Leibniz:[82] therefore this must have been amongst the papers that Leibniz saw at the time in London – a serious reproach if it were justified.

We know however that Leibniz had already left London[83] on 29 October 1676 – before, that is, Newton's letter was ready to be communicated to him; we have indeed a letter written by Collins to Strode in which he reports that Leibniz had been in London the week before and paid him a visit.[84] This letter is dated Tuesday, 3 November 1676 – the same day that Newton dispatched his second letter to Oldenburg for onward transmission to Leibniz; the meeting with Collins can therefore only have taken place at some time during 'the week before', that is, after 25 but before 29 October. The draft of Collins' letter to Strode which came into William Jones' possession about 1708 (and still exists in the Macclesfield collection) was possibly accessible to Newton; if so, it would have shown him that Leibniz

---

[78] 'Account': *PT* **29**, No. 342 (January–February 1714/15): 194.
[79] *CE (1722)*: 25.
[80] Leibniz–Conti, 6 Dec. 1715 (*LBG*: 264).
[81] Newton–Conti, 26 Feb. (7 Mar.) 1715/16 (*LBG*: 272).
[82] Newton–Oldenburg, 24 Oct. (3 Nov.) 1676 (*LBG*: 211: *Sed in simplicioribus vulgoque celebratis figuris vix aliquid relatu dignum reperi, quod evasit aliorum conatus.*
[83] Leibniz–Kahm, 14 (?24) Nov. 1676 (*LSB* I. 2: 3, l. 16).
[84] Collins–Strode, 24 Oct. (3 Nov.) 1676 (*CR* II: 453–4).

could not have seen his second letter in Collins' hands. Leibniz' remark on rectification presumably refers to the passage which Newton mentions himself and was doubtless due to a slip in memory[85] not really to be wondered at after a lapse of almost forty years. It does not appear to be based on any excerpt made by him in London – at any event the remark in question is not amongst the many notes that Leibniz made when he then saw Collins. A final opinion may possibly only be passed when all of Newton's papers have been edited; it is not likely that any excerpts made by Leibniz during his London visit have been lost without trace.

[85] Leibniz–Conti, 9 Apr. 1716 (*LBG*: 276): *je me suis trompé dans l'exemple que j'ay cité, n'ayant pas pris garde, ou ayant oublié qu'il s'y trouvoit.*

20
THE SECOND VISIT TO LONDON

Leibniz left Paris on 4 October 1676 and arrived in London two weeks later.[1] First of all, naturally, he called on Oldenburg. The Royal Society had not yet reassembled after the summer recess and we know nothing, apart from a conversation with Collins, of any other meeting between Leibniz and Fellows of the Society. Leibniz' frequently mentioned draft about his universal language[2] perhaps belongs to the period of this second stay in London, though it was not passed around, certainly not in the form in which we now know it; it is nowhere mentioned in the rest of Leibniz' correspondence with his English friends even though the topic would have found an interested echo, in Pell at any rate. Indeed Oldenburg's last remarks on Pell's 'Idea of mathematicks' may have provided the incentive for Leibniz to fix his thoughts on his own *Characteristica universalis*. He expresses his pleasure that the idea of the universal language is now coming to meet with more interest: it is a subject of great importance, but will have to be treated in a different way from any hitherto proposed; not even Wilkins, who had achieved such considerable progress therein, had, according to Leibniz, found the right way to approach it. The appropriate choice of symbols will be crucial; the possibility of international communication is, for Leibniz, a secondary problem – far more important is the construction of an ideographic script which will make all ambiguity impossible, which will signpost a safe path to accurate thinking and hence towards truth and whose very syntax will render all error impossible. Here, as it were in a nutshell, is expressed what Leibniz intended and actually achieved in one field by his invention of the formalism of the Calculus.

On some occasion or other Oldenburg managed to see Leibniz' calculating machine and received a few mathematical papers from

[1] The dates are evident from Leibniz–Kahm, 14 (?24) Nov. 1676 (*LSB* I. 2: 3).
[2] Leibniz–Oldenburg, 1676 (?) (*LSB* II. 1: 239–42).

him which had been promised to him for quite a while.[3] On the other hand through Collins Leibniz gained sight of certain papers by Newton and Gregory. An accurate understanding of his excerpts from these is of crucial significance in coming to a correct appreciation of the whole situation. Let us begin with his excerpts from Newton's *De Analysi*.[4] Here are entered the 'areas'

$$\int ax^{m/n}.\,\mathrm{d}x \;=\; anx^{(m+n)/n}/(m+n)$$

and
$$\int_0^x \mathrm{d}t/(1+t^2) \;=\; x-\frac{x^3}{3}+\frac{x^5}{5}-\frac{x^7}{7}+\dots$$

('for sufficiently small values of x') or alternatively

$$\int_\infty^x \mathrm{d}t/(1+t^2) \;=\; -\frac{1}{x}+\frac{1}{3x^3}-\frac{1}{5x^5}+\frac{1}{7x^7}-\dots$$

('for sufficiently large values of x') – and Newton's technique for expanding
$$\sqrt{(a^2+x^2)} \;=\; a+\frac{x^2}{2a}-\frac{x^4}{8a^3}+\dots$$

and
$$\sqrt{[(1+ax^2)/(1-bx^2)]} \;=\; 1+\tfrac{1}{2}(a+b)x^2+\dots$$

by division of the two component square root series are given, together with the advice that it is better to multiply top and bottom by $\sqrt{(1-bx^2)}$ beforehand. Leibniz adds that surely the result will be the same. Newton's description of his method of series-expansion for solving equations is transcribed in full for the numerical example $x^3-2x-5 = 0$ with all supplementary remarks carefully recorded, while the schematic algorithm is omitted for the literal example presumably since Leibniz remembers that he has already received this in Newton's first letter. Also copied is the representation of the hyperbola-area in the form

$$y \;=\; \int_0^x \mathrm{d}t/(1+t) \;=\; x-x^2/2+x^3/3-\dots$$

and inversion of this series is fully given; furthermore the inversion of the series for the arcsine, the quadrature of the quadratrix and lastly Newton's (rather summary and incomplete) proof of the approximation procedure. Leibniz' excerpts are confined exclusively to series-expansions and the general remarks accompanying them:

---

[3] *Compare* ch. 19: note 68.   [4] *NP* 2: 248–9.

the sections relating to infinitesimals in the *De Analysi* remain completely disregarded – for the obvious reason that they offered nothing new to Leibniz.

Immediately following his notes on the *De Analysi* Leibniz has excerpted a 1672 letter of Newton's to Collins,[5] in which he describes the application of logarithmic scales in calculations preliminary to the solving of equations – one of the subjects subsequently brought up for discussion by Leibniz during the summer of 1675.[6] At this point he also transcribes Newton's remark that he had checked Gregory's series for the second segments[7] of an ellipsoid of revolution and could not devise a better one.[8] The way in which this passage from the 1672 letter is joined on to his excerpts from the *De Analysi* makes it all but certain that the two pieces were written out in immediate sequel.

Far more significant are the excerpts which Leibniz made from Collins' 'Historiola',[9] for they prove conclusively that it had not previously been sent to Leibniz. The text, written in great haste in a very tight script, reproduces portions of the original (which is mainly in English) partly in English, partly in hurried Latin translation and contains many lapses of the pen and a number of errors in transcription, translation and comprehension. Leibniz begins his excerpts at the Introduction and opening pieces; there he encountered questions relating to interpolation theory not now detailed in the portion of the 'Historiola' preserved in the Royal Society, which derive from Gregory's letter of 23 November (3 December) 1670;[10] so Leibniz inserted excerpts from that letter, then went on to copy later pieces in the 'Historiola', but subsequently, when he looked again through his excerpt and compared it with the original, he added further single items from the 'Historiola' and from other letters by Gregory (some no longer extant). If Leibniz had already known the 'Historiola' it would have been enough for him to

[5] Newton–Collins, 20 (30) Aug. 1672 (*NC* 1: 229–30). Leibniz' excerpt is at LH 35. VIII. 19, fol. 2.

[6] Oldenburg–Leibniz, 24 June (4 July) and Leibniz–Oldenburg, 2 (12) July 1675 (*LBG*: 130, 131). Leibniz had devised an 'analogue' instrument, mentioned in ch. 10: note 27, for solving numerical equations which works by moving triangles of variable scale but is still rather imperfect in its original form and despite Leibniz' optimism about its possibilities not generally usable. Newton's method employed movable parallel 'Gunter's scales', that is logarithmically divided rulers, and is indeed a practicable tool.

[7] Gregory–Collins, 17 (27) May 1671 (*GT*: 188–9).

[8] Newton–Collins, 20 (30) Aug. 1672 (*NC* 1: 229).

[9] LH 35. VIII. 23, fols. 1–2; further LBr 695, fol. 63.

[10] Gregory–Collins, 23 Nov. (3 Dec.) 1670 (*GT*: 119–21).

transcribe only those passages from Gregory's letters that are not contained in the 'Historiola'. The originals of these remained in Collins' hands, and hence Leibniz' excerpts of them can only have been made during his second visit to London. The manner in which the individual items are interlaced with the other excerpts from the 'Historiola' wholly excludes the possibility that they originated elsewhere. The note on the cover of the 'Historiola',[11] asking Leibniz to return the manuscript after he had inspected it, refers, therefore, not in the least to its dispatch to Leibniz, but to his seeing it while in London. In the *Commercium epistolicum* the beginning of the dedicatory letter to Oldenburg is reproduced in its original English and in Latin translation. There singled out for comment are Gregory's 'arithmetical circle quadrature' as communicated in his letter to Collins[12] on 15 (25) February 1670/1 and Newton's letter to Collins[13] of 10 (20) December 1672, where, it is said, Newton explains his tangent-method by an example and points out that he possesses a general procedure for constructing tangents, quadratures of curves, and the like: the *Commercium epistolicum* goes on to assert that this method is that which Leibniz later called the differential one. In his own copy Leibniz has underlined this phrase and added in the margin,[14] that this is silly (*ineptum*); the method indicated here is not at all the differential one, but something differing *toto cælo* – a world away. In the new edition of the *Commercium epistolicum*[15] in 1722 Newton also drew attention to Gregory's letter of 5 (15) September 1670[16] which is included in the 'Historiola', adding the further assertion that the 'Historiola' had been dispatched[17] to Leibniz on 26 June (5 July) 1676.

We now turn to a brief survey of the content of the excerpts. First, from Collins' Introduction to the 'Historiola' Leibniz notes that

[11] This request is quoted in Latin and English in *CE* (*1712*): 46–7, followed by Collins' introductory remarks and a survey of what appears as especially important to the editor of the *CE* (that is Newton).
[12] Gregory–Collins, 15 (25) Feb. 1670/1 (*GT*: 170).
[13] Newton–Collins, 10 (20) Dec. 1672 (*NC* **1**: 247–8).
[14] *CE* (*1712*): 47.
[15] *CE* (*1722*): 128.
[16] Gregory–Collins, 5 (15) Sept. 1670 (*GT*: 103). Gregory expresses his admiration for Barrow's work on tangents, similar to his own more general method. This was the basis for Newton's astonishing note in the *CE*, mentioned in ch. **19**: note 26 above.
[17] The incorrect date taken from the reproduction of the letter by Wallis is due to a misprint in *WO* III: 622; the date, 26 June, is corrected in the *Errata* at the front of the volume to 23 July (but another misprint there refers it to p. 662 instead of 622). Though the original with the correct date was in Newton's hands at the time, he evidently never consulted it.

during his stay in Padua, Gregory lived at the house of the (Scottish) professor of philosophy, Caddenhead, and had tried in vain to get some rare books for which Collins had asked, amongst them Maurolico's edition of Apollonius. In this context mention is made of Barrow's recent edition of the first four books of the *Conica* and various Arabic manuscripts of the other three surviving books, which we have already referred to above. Mengoli, Leibniz writes, had (in the *Novæ quadraturæ arithmeticæ*) indicated the problem of summing the terminating harmonic series, which he acknowledged to be beyond his powers; he had proved that the sum of the whole series was infinity by means of the inequality

$$\frac{1}{n-1}+\frac{1}{n}+\frac{1}{n+1} > \frac{3}{n}$$

in this manner: the first three terms of the harmonic series $\frac{1}{2}+\frac{1}{3}+\frac{1}{4}$ are together more than 1, so similarly are the 9 following, the 27 next following, and so on. Leibniz records his admiration by adding the adjective 'ingenious'. Mengoli, the extract goes on, had further, in his *Geometria speciosa*, revealed the connection between logarithms and partial sums of the harmonic series, mirrored in the inequalities

$$\frac{1}{n+1}+ \ldots +\frac{1}{kn} < \log k < \frac{1}{n}+ \ldots +\frac{1}{kn-1},$$

which derive from considering inscribed and circumscribed steppolygons in and around a hyperbola. This is further linked to the question of finding the present value of an annuity restricted to simple interest;[18] for instance, an annuity of £100 to be reckoned for an interest rate of 6%. Guldin remarks in his *Centrobaryca*, says Collins, that Soverus had already promised a quadrature of the hyperbola in the year 1630 (in his *Curvi ac recti proportio*), one presumably involving inscribed triangles.[19] Strode, in a manuscript on conics then ready to be printed, had used a similar procedure.[20] This passage is marked with a + in the margin by Leibniz.

He goes on to note Gregory's acknowledgment after the *Logarithmotechnia* appeared that Mercator's quadrature of the hyperbola was better than his, but also that he had objected to the latter's obscure

---

[18] Collins had himself published a 'paper of interest', his *Decimal arithmetick* in 1665 (and reprinted later), alluded to in Collins–Gregory, *c.* March 1668 (*GT*: 46–7).

[19] Gregory–Collins, 26 Mar. (5 Apr.) 1668 (*GT*: 49–50).

[20] The manuscript is referred to in Collins–Beale, 20 (30) Aug. 1672 (*CR* I: 195); Beale–Collins, 17 (27) Oct. 1672 (*CR* II: 445) and Strode–Collins, 31 Mar. (10 Apr.) 1673 (*CR* II: 447–8).

expression of 'composition of proportions'. Collins had thereupon induced Gregory to perform the determination of

$$\int_{-x}^{x} \mathrm{d}t/(1+t) = \log\left[(1+x)/(1-x)\right] = 2(x+x^3/3+\dots),$$

$$\int_{0}^{x} \mathrm{d}t/\cos t = \log \tan (\pi/4 + x/2)$$

and $\qquad \int_{0}^{x} \tan t \,.\, \mathrm{d}t = \log \sec x$

in the *Exercitationes geometricæ*. Gregory later wrote, however, that he had gone wrong[21] at the end of this book in his approximate computation of these functions: to interpolate a function by ordinates set at equal distances c apart and to find the area beneath it one must rather proceed by the formulae

$$f(x) = f(0) + \frac{x}{c}\Delta f(0) + \frac{x}{c}\cdot\frac{x-c}{2c}\Delta^2 f(0) + \dots$$

and $\qquad \int_{0}^{c} f(x)\,.\,\mathrm{d}x = c[f(0) + \tfrac{1}{2}\Delta f(0) - \tfrac{1}{12}\Delta^2 f(0) + \dots]$

from which Mercator's quadrature of the hyperbola is an immediate consequence on taking $f(x) = a^2/(a+x)$. Gregory's interpolation generalizes the special case applied by Henry Briggs in his *Arithmetica logarithmica*. Again Leibniz has made a mark in the margin alongside these general formulas.

From this, Gregory continues,[22] the antilogarithm corresponding to a given logarithm can be found; then follows the computation of 23^3 from the tabulated differences of 10^3, 15^3, 20^3, 25^3, 30^3. In this context also belongs a passage – missing in the portion of the 'Historiola' extant in the Royal Society – where Gregory asserts that the problem of interpolation can easily be dealt with by means of the approximations (excellent for a function which is slow to vary)

$$f(n) \approx 2f(n-1) - f(n-2) \approx 3f(n-1) - 3f(n-2) + f(n-3) \approx \dots$$

and so forth; the computation of a numerical table will thereby be much shortened and the degree of accuracy considerably improved.[23]

Leibniz now makes some notes on Gregory's relationship with Barrow. First, he records a general comment of praise for the latter's

[21] Gregory–Collins, 23 Nov. (3 Dec.) 1670 (*GT*: 119); at this place follows a reference to Mercator and to Gregory's work on interpolation (*GT*: 119–20).

[22] Collins for Gregory, 23 Nov. (3 Dec.) 1670 (*GT*: 131–2); *compare* p. 134.

[23] Gregory–Collins, 23 July (2 Aug.) 1675 (*GT*: 313).

Lectiones opticæ,[24] and then the passage where Barrow's tangent-method is praised[25] and where Gregory reports how he has, in conjunction with his own tangent-method elaborated in proposition 7 of his *Geometriæ pars universalis,* developed from it a comprehensive geometrical procedure, requiring no calculation and consisting of twelve theorems only. The question posed by Barrow in his *Lectiones geometricæ,* how to find a curve when the length of its normal is given as a function of the abscissa (Lectio XII)[26] – that is, how to solve the differential equation $y \, \mathrm{d}s/\mathrm{d}x = f(x)$ – is acknowledged as extremely difficult; its solution will represent a valuable advance in geometry, for it will then be possible to determine a surface of revolution $2\pi \int y . \mathrm{d}s$ of given area $2\pi \int f(x) . \mathrm{d}x$. Gregory had later on enquired[27] what further Barrow had now to say on his problem and had received the following answer.[28] Suppose three curves $y_k = f_k(x)$, $k = 1, 2, 3$ to be given such that $y_1{}^2 + y_2{}^2 = y_3{}^2$, then for the corresponding sub-normals $n_k = y_k y_k{}'$ the relation $n_1 + n_2 = n_3$ will hold; conversely from $n_1 + n_2 = n_3$ the relation $y_1{}^2 + y_2{}^2 = y_3{}^2 + c$ will follow. If for any curve there is given $z = n + y$ (so that $(z - y)\mathrm{d}x = y \, \mathrm{d}y$), then for z constant a hyperbola (t, y), $t = z \, \mathrm{d}x/\mathrm{d}y$ defined by $(z - y)t = yz$ can be found and by its aid $xz = \int_0^y t . \mathrm{d}y$ determined. But when more generally $z = f(x)$, then it seems necessary to work with a whole system of hyperbolas. In his hurry Leibniz incorporated Gregory's critical comment on this last remark with Barrow's text itself; the whole passage is badly mutilated by some loss of writing at the edge of the manuscript.[29]

In the 'Historiola' there now follows a transcription of Newton's tangent-method from his 1672 letter. Leibniz extracts only the example[30]

$$\begin{array}{cccccc} 0 & 1 & 0 & 0 & 2 & 3 \\ \multicolumn{6}{c}{x^3 - 2x^2y + bx^2 - b^2x + by^2 - y^3 = 0} \\ 3 & 2 & 2 & 1 & 0 & 0 \end{array}$$

[24] Gregory–Collins, 29 Jan. (8 Feb.) 1670 (*GT*: 79).
[25] Gregory–Collins, 5 (15) Sept. 1670 (*GT*: 103); the passage was also copied in Oldenburg–Leibniz, 26 July (5 Aug.) 1676 (*LBG*: 174–5).
[26] Gregory–Collins, 5 (15) Sept. 1670 (*GT*: 103) continuing the preceding text, missing in *LBG*. Leibniz was obviously interested in this passage because it is connected with one of his own first major discoveries; *compare* p. 48.
[27] Gregory–Collins, 23 Nov. (3 Dec.) 1670 (*GT*: 121).
[28] Barrow–Collins, 10 (?20) Dec. 1670 (*GT*: 161–3).
[29] Gregory–Collins, 15 (25) Feb. 1670/1 (*GT*: 171–2).
[30] Newton–Collins, 10 (20) Dec. 1672 (*NC* 1: 247).

where the subtangent is

$$[t = y\,\mathrm{d}x/\mathrm{d}y = -] \quad (-2x^2y + 2by^2 - 3y^3)/(3x^2 - 4xy + 2bx - b^2),$$

and then adds the general rule which Oldenburg had previously sent him.[31] This shows just how little significance he then saw in a piece which Newton held to be one of the most important in deciding calculus priority. Gregory's subtle exploration of the logarithmic spiral also rates merely a casual mention; Leibniz cites only its definition:[32] when the angles progress arithmetically, then the corresponding radii are in geometrical progression, though in addition its relationship with the loxodrome on a sphere by stereo-graphic projection from a pole of a spiral in the equatorial plane is noted down.

There follow a group of items (nly) oosely connected; for instance a deduction[33] of Euclid's theorem III, 36 from III, 35, a note on the completed deduction of Newton's series for the circle-zone,[34] and again the approximation[35] of $\log \sqrt{[(1-x)/(1+x)]}$ by a comparison of its remaining terms when the series is broken off with those (summable to infinity) of a geometric progression [the example cited is approximated as $x + x^3/3 + x^5/5 + 9x^7/(63 - 49x^2)$]; next come series-expansions[36] – some of which he had been previously given, and which are not included in the portion of the 'Historiola' now in the Royal Society – of the inverse tangent, tangent, secant, logarithmic tangent and secant and their inverses, and finally Gregory's series for the second segments of the ellipsoid of revolution with remarks on its possible extension to the hyperboloid of revolution[37] and the arc of an ellipse.[38] There is also a reference to the rectification $\int_a^x \sqrt{[1 + (b/t)^2]}\,.\,\mathrm{d}t$ – again missing in the 'Historiola' – of the logarithmic curve[39]

---

[31] Oldenburg–Leibniz, 26 July (5 Aug.) 1676 (*LBG*: 172).
[32] Gregory for Collins, 20 (30) Apr. 1670 (*GT*: 95).
[33] Gregory–Collins, 23 July (2 Aug.) 1675 (*GT*: 314).
[34] Gregory–Collins, 19 (29) Dec. 1670 (*GT*: 148).
[35] Gregory for Dary, 9 (19) Apr. 1672 (*GT*: 230).
[36] Gregory–Collins, 15 (25) Feb. 1670/1 (*GT*: 170). Leibniz had heard of most of these results in Oldenburg's letters of 12 (22) Apr. 1675 and 26 July (5 Aug.) 1676 (*LBG*: 115 and 171–2). [37] Gregory–Collins, 17 (27) May 1671 (*GT*: 188–9).
[38] Gregory–Collins, 15 (25) Feb. 1670/1 (*GT*: 171).
[39] Gregory–Collins, 19 (29) Dec. 1670 and 17 (27) May 1671 (*GT*: 148–9 and 189–90). Much later this rectification was also performed by L'Hospital, *see* his letters to Huygens, 26 July and 23 Nov. 1692 (*HO* **10**: 305 and 343–4); to Leibniz, 14 Dec. 1692 (*LMG* II: 216–17); but the latter did not at that time remember having seen Gregory's solution. The most elegant exposition is that contained in Huygens–Basnage, February 1693 (*HO* **10**: 407–8).

$y = b \log (x/a)$ by reducing it to the quadrature of a hyperbola (by substituting $t = 2b^2u/(b^2-u^2)$); and the series-expansions of this rectification expression where $x > b$, and of its modified form

$$\int_a^x \sqrt{[1+(b/t)^2]}.\,dt = b \int_a^x dt/t + \int_a^x [\sqrt{(b^2+t^2)}-b].\,dt/t$$

when $x < b$ are also written down. In addition Leibniz notices Gregory's remark that from Barrow's *Lectiones geometricæ* we are to conclude that no 'analytical' (that is, algebraical) relation[40] can ever exist between corresponding zones of a circle $\left[\int_0^x \sqrt{(1-t^2)}.\,dt\right]$ and a hyperbola $\left[\int_0^x \sqrt{(1+t^2)}.\,dt\right]$; there follows a reference to the possibility of extending the method of series to 'mechanical' problems, like Kepler's.[41] Further notes relate to the dispute that had taken place in 1669–70 between Cassini and Mercator about the construction of mean motion in planetary orbits[42] and to the most convenient way of investigating the parallax of fixed stars.[43]

Turning to the section on theory of equations Leibniz now copies some remarks showing Gregory's high esteem of Hudde's methods,[44] others on the impracticality of all known attempts[45] to find solutions by tables of sines or logarithms, others still regarding Gregory's own new method and the difficulties arising in its practice,[46] and also his negative comment on Tschirnhaus,[47] he furthermore transcribes the remark (already known to him from the 'Abridgment') on the reduction of equations when some of the roots depend rationally on the others.[48] Leibniz also copies Gregory's observation[49] that the rational part of the roots of an equation always equals $-a_1/n$, that of

[40] Gregory–Collins, 5 (15) Sept. 1670 (*GT*: 102).
[41] Gregory–Collins, 23 Nov. (3 Dec.) 1670 (*GT*: 120).
[42] Gregory–Collins, 5 (15) Sept. 1670 (*GT*: 103). *See* further Cassini's paper in *JS* No. 3 of 25 Sept. 1669 and Mercator's related remarks at RS meetings, 20 (30) Jan., 27 Jan. (6 Feb.) 1669/70 (*BH* II: 415–17) and in *PT* 5 No. 57 of 25 Mar. (4 Apr.) 1670: 1168–75.
[43] Gregory–Oldenburg, 8 (18) June 1675 (*GT*: 306–7).
[44] Gregory–Collins, 17 (27) Jan. 1671/2 (*GT*: 210).
[45] Gregory–Collins, 15 (25) Feb. 1670/1 (*GT*: 169).
[46] Gregory–Collins, 26 May (5 June), 20 (30) Aug., 2 (12) Oct. 1675 (*GT*: 303, 326, 335).
[47] Gregory–Collins, 20 (30) Aug. 1675 (*GT*: 324).
[48] Gregory–Collins, 11 (21) Sept. 1675 (*GT*: 328); *compare* Oldenburg–Leibniz, 26 July (5 Aug.) 1676 (*LBG*: 173).
[49] Gregory–Collins, 17 (27) Jan. 1671/2 (*GT*: 211).

their squares equals $(a_1^2 - 2a_2)/n$ and so forth, and connects this with finding the sums of powers of the roots.[50] Finally he shows interest in Gregory's iterative solution of the problem of annuities.[51]

On surveying the scope of the material covered in these excerpts we have to grant that Leibniz' sharp eye has caught every essential facet of Gregory's individual results; and in particular, everything relating to expansions into series, interpolation and methods of approximation. What Leibniz had been looking for most of all – a general description of the methods used or their proofs – he had not found: so there was nothing else to do but make as complete an enumeration of the technical details as possible. There is no question here of a narrowly literal copy or even one faithful to the exact sense; for that there was probably not time enough. Naturally everything that Leibniz either had previously received (as far as he now remembered) or now considered inessential, is left out[52] – such general remarks as that about Gregory's efforts to discover the source of Newton's series for the circle-zone, or that on eliminating the penultimate term in an equation, or that on locating limits to the roots of a quartic. A characteristic omission is that of Gregory's profound, yet not immediately evident, investigations in infinitesimals of necessary and sufficient conditions for extreme values, of the relationship between the logarithmic curve and the logarithmic spiral and of the tangent to the *spiralis arcuum rectificatrix*. Leibniz believed he could already master these topics completely by his own general method, and that he was entitled to skip them since obviously no immediately tangible new results derived from them. This applies also to Gregory's manuscript from the winter of 1668, which was intended as a supplement to the *Geometriæ pars universalis* and aimed at evaluating $\int \sec t . dt$; the transformations contained in it had, by means of his own characteristic triangle, become trivial to Leibniz. For him to note the approximation for a hyperbolic segment following the method of proposition 25 of Gregory's *Vera quadratura* was also probably superfluous; for on the one hand Gregory himself had admitted that his method had been superseded by that of Mercator, and on the other Leibniz had very possibly then acquired a copy of the book itself. The two manuscripts on finding the intersections of conics and relating to the equation of the hyperbola were

[50] Gregory–Collins, 26 May (5 June) 1675 (*GT*: 303).
[51] Gregory–Collins, 2 (12) Apr. 1674 (*GT*: 279).
[52] That is, the communications in Oldenburg–Leibniz, 12 (22) Apr. 1675, 26 July (5 Aug.) 1676 (*LBG*: 114–16, 118–19, 170–6).

uninteresting to him because they nowhere went beyond what was at the time either generally known or routine developments. Thus there remain a few remarks on the type and degree of difficulty of the series-expansions to be used in solving equations,[53] a topic about which Leibniz talked at length with Collins.[54] On that occasion Leibniz promised more details of his general method for solving equations by series (using indeterminate coefficients) and Collins recommended Baker as a conscientious, well-informed algebraist who would carry out the detailed computations successfully. Leibniz' communication[55] about finding the equation $F(x, y') = 0$ that arises by differentiating the equation $f(x, y) = 0$ of an algebraic curve and eliminating y between this and the derived equation was intended for Baker. Leibniz first considers curves of second order, then those of third indicating that calculation of the continuation and also finding the sequence of the series will be very useful. It would seem that he aims at an expansion in the manner of a Taylor series. Leibniz' letter went on to Baker[56] who worked through it critically,[57] but Collins did not subsequently continue his discussion of mathematical topics with Leibniz (perhaps obeying Newton's wish), and after Oldenburg's death there was no other intermediary who would have been sufficiently interested in seriously continuing the correspondence between Newton and Leibniz.

For a short time, it is true, Leibniz was able to overcome the distrust with which Collins had viewed him; he even succeeded in gaining a sight of his well guarded scientific correspondence. This was surely an immediate consequence of the excellent impression which Collins gained during the course of their conversation. We learn certain details of their talk from a jointly written memorandum which Leibniz took away with him: it is a large folio-sheet, folded once, on which Leibniz has written Colliniana. On its four pages it contains calculations and diagrams with short interspersed passages of text.[58] Since Leibniz could read English but probably not yet speak it very well, while Collins' knowledge of Latin was not exten-

[53] Gregory–Collins, 15 (25) Feb. 1670/1, 17 (27) May 1671, 9 (19) Apr. 1672 (*GT*: 171, 188, 227–8).
[54] Collins–Wallis, 16 (26) Feb. 1676/7 (*see CR* II: 604–6); Collins–Newton, 5 (15) Mar. 1676/7 (*NC* 2: 198).
[55] Leibniz–Oldenburg, 18 (28) Nov. 1676 (*NC* 2: 199–200).
[56] It is promised in Collins–Baker, 10 (20) Feb. 1676/7 (*CR* II: 15), and judging by *CR* II: 16, was dispatched on 3 (13) Mar. 1676/7.
[57] Baker–Collins, 15 (25) Mar. 1676/7 (*CR* II: 19).
[58] LH 35. xv. 4, fols. 1–2.

sive, the conversation was in part conducted by continued reference to the writing on this sheet. Despite the sparse quality of this document the course of the discussion can be fairly clearly recognized.

To begin, Collins simply reports, with examples on compound interest computations similar to those contained in his recently printed introduction to bookkeeping for merchants. In what time, he asks, will £100 grow to £108 at 6 % ? Written down is the calculation $1·3208 \times 20 = 26·4160$ and the remark that the sum of £100 will at 6 % increase in 26·416 years to the same amount as it would at 8 % in 20 years. This question leads to the equation $1·06^t = 1·08$ and is solved by computing $t = \log 1·08/\log 1·06 = 1·3208$, a value again used without further explanation in the allied question of finding x such that $1·06^x = 1·08^{20}$. Next comes the calculation

$$\begin{array}{cc} £17 & 12 \\ \dfrac{1·0121 \times 17}{0·06} & \dfrac{1·0121 \times 8\frac{1}{2}}{0·029\,563} \end{array},$$

dealing presumably with the terminal value b of an annuity which runs at 6 % for 12 years; the interest factor is therefore $q = 1·06$, and $q^{12} = 2·0121$. In the first case an annual payment of $r = £17$ is assumed, whence $b = r(q^{12}-1)/(q-1)$; in the second case, a half-yearly payment of $\frac{1}{2}r = £8\frac{1}{2}$ is understood, whence

$$b = \tfrac{1}{2}r(q^{12}-1)/(\sqrt{q}-1)$$

where $q-1 = 0·06$, $\sqrt{q}-1 = 0·029\,563$. Next Collins asks at what rate of interest will £1 increase to £10 in 10 years? The question remains unanswered, but below it is written

$$1:1·06 = 1·06:1·1236 = 1·1236: ...,$$

that is to say, the increase is compounded in the form

$$1:q = q:q^2 = q^2: ....$$

Thereafter Collins turns to one of his favourite topics, the graphical solution of equations. He sets out values of $y = x^3 - 27x$ tabulated for all integral x, $1 \leqslant x \leqslant 12$, draws a corresponding diagram and explains his procedure for approximating x when $y = 700$. To do so he uses the base points P (9, 486) and Q (10, 730) and determines the intersection of the line $y = 700$ with the chord PQ at U and also with the tangent in Q at V to achieve a nearer solution: this latter procedure he has learnt from Newton's *De Analysi*, now giving it a geometrical interpretation.[59] This, incidentally, is the first place where Newton's method of approximation appears in explicit form. As a supplementary example Collins writes down

$$y = x^4 - 4x^3 - 19x^2 + 120x,$$

but in illustration draws an incorrect diagram of a curve with three extreme values; only one, a minimum, is in fact real.

Next the conversation passes to the substitution $z = -x$ by which negative roots can be turned into positive ones. Collins adds that the rule that every root is a factor of the absolute term of the equation is, according to Glorioso in his *Exercitationes mathematicæ*, already to be found in Nunez' *Algebra*. When the absolute term is indeterminate, an arbitrary solution can be prescribed and the absolute term then calculated. When, for instance, the given equation $x^3 + 7x^2 + 5x = y$ is to have the root $x = 4$, then we will have to set $[(4+7)4+5]4 = 196 = y$. When a root of an equation is known then reversing this procedure will at once give the long division by the relevant linear factor. This is indicated in the following example, where the known root is $x = 8$:

$$7x^4 - 2x^3 + 5x^2 - 11x = 27\,880$$

$$\underbrace{7 \times 8 - 2}$$

$$\underbrace{54 \times 8 + 5}$$

$$\underbrace{437 \times 8 - 11}$$

$$3485 \times 8$$

The quotient is $7x^3 + 54x^2 + 437x + 3485$; the method is hardly different from our present one of contracted division.

Coming on to circle computations, Collins refers to 'Davenant's' method of successive divisions for obtaining a continued fraction[60] providing an approximation to π usable in practice, setting the computation out in the accompanying scheme

$$100\,000)3{\cdot}141\,59\ (3$$
$$14\,159)\ \ 100\,000(7$$
$$887)\ \ 14\,159\ (15^-$$
$$(16^+$$

[59] *NP* 2: 218.
[60] This type of expansion for π from its decima lfraction is mentioned also both in Collins–Baker, 19 (29) Aug. 1676 (*CR* II: 8–9) and in Wallis–Collins, 1 (11) Sept. 1676 (*CR* II: 589–90) as Davenant's method; it is surprising to see it thus singled out when we remember from *WCE* that the technique of working with continued fractions seemed well established in England so that Brouncker could use it for solving the Fermat equation $x^2 - py^2 = 1$.

This is followed by a check to test that the better of the two expansions thus arising is in fact

$$\pi \approx 3 + \cfrac{1}{7 + \frac{1}{16}} = \frac{355}{113} \; .$$

Collins also mentions the approximate circle quadratures of Wallis[61] (making use of the subdivision of a quadrant by equidistant ordinates) and Dary[62] (employing a subdivision by ordinates at the corners of a regular chord-polygon), and talks of the very accurate new tables of sines and circle-segments which John Smith was then in course of compiling.[63]

And then at last some discussion of topics in higher geometry occurs, though unfortunately, since there is no accompanying text at all, we must draw our present conclusions only on the basis of the diagrams in the manuscript.

Collins seems first to refer to the *spiralis arcuum rectificatrix*[64] whose tangent Gregory could construct only with a great deal of labour. Then Leibniz demonstrates as an example of his inverse tangent method the determination of the curve which has a constant subtangent,[65] further specifying it by the condition that it shall cut the ordinate through the origin at an angle of 45°: thus his solution is the curve $x/a = \log (y/a)$; by means of $x = a\theta$, $y = r$ he transforms this to polar coordinates and so finds the particular logarithmic spiral which cuts the radii under an angle of 45°. Presumably at this point Collins mentioned from the 'Historiola' Gregory's[66] somewhat more complicated *evolutio* into Cartesian coordinates by a transformation which leaves the arc-length invariant. Next appears 'Bertet's curve'[67] – surely provoking from Leibniz a statement about its tangent and its area? – and finally the involute of a circle.[68] In two further diagrams Leibniz indicates how he is accustomed to embed a curve in a system of Cartesian coordinates, and then find its area through a division by equidistant ordinates.

[61] Wallis–Collins, 26 Sept. (6 Oct.) 1668, 19 (29) January and January–February 1668–9 (*CR* II: 505, 509–10, 512). He attempts to adapt Mercator's procedure for squaring the hyperbola in the *Logarithmotechnia* to the circle.

[62] Dary–Collins, (?) 1675 (*CR* I: 220–1).

[63] *Compare* ch. **10**: note 70.

[64] *Compare* ch. **15**: note 54.

[65] Leibniz had investigated the logarithmic curve and its tangent already since 1674 (*C* 827).

[66] Gregory for Collins, 20 (30) Apr. 1670 (*GT*: 93–6).

[67] *Compare* ch. **13**: note 39, 40, and the text pp. 195–6.

[68] References to this are contained in *C* 1144.

This probably ends the memoranda about their conversation; there is another note by Leibniz on finding $\Sigma k^2 = \Sigma[k(k+1)-k]$ by means of $\Sigma k = \frac{1}{2}n(n+1)$ and $\Sigma\frac{1}{2}k(k+1) = \frac{1}{6}n(n+1)(n+2)$, but this is squeezed in a smaller hand into a space between the diagrams and the remarks on Smith's tables, standing in marked contrast to the rest of the text (which is written in large and clear characters). In the text corrections are made to a proof of the formula for the sum of square numbers which Leibniz had, when still in Paris, given to Arnauld.[69]

Leibniz had yet another important request to make to Collins. When, in the spring of 1675, he had hoped to find an early opportunity of paying a second visit to England, he had made careful excerpts from Oldenburg's letter of 6 (16) April 1673 and its enclosures and was now eager to find out more about their detail, especially in regard to the work of the English mathematicians named in the letter. He now learned that Thomas Merry's studies were still – as they were to remain – unpublished, that Collins himself knew nothing further about Newton's method (later printed in his *Arithmetica universalis*) for reducing problems to equations, that Strode's 'Conics' had not yet appeared, and that Collins had in his own possession a manuscript by Barrow and another, larger one, by Newton on the subject of solving equations by means of conics and mechanical curves. He was further told about Dary's latest book on compound interest and solving equations logarithmically, which was just then at the printer's[70] and which came out[71] eventually in the spring of 1677. In return Leibniz gave Collins a copy of his letter to Périer on the content of Pascal's *Conica*[72] – intended perhaps as an introduction to their projected publication, and probably also presented a more widely ranging verbal report. In this connection Leibniz now made a note in his extracts from the Oldenburg letter of 6 (16) April 1673 that Wallis too had written on conics; as yet, apparently, he did not know of this little book.[73]

Collins was evidently well satisfied with his conversation with Leibniz. The personal impression he gained of the young German was

[69] *C* 1343, 1406.
[70] Collins–Strode, 24 Oct. (3 Nov.) 1676 (*CR* II: 454).
[71] It is clear from Oldenburg–Leibniz, 12 (22) July 1677 (*LBG*: 250) that the book was sent, at Collins' behest, on 2 (12) May 1677, that is together with the 'epistola posterior'.
[72] We conclude this from Collins–Bernard, 17 (27) Nov. 1676 (MS: Copenhagen, Boll. Ud 1), while the transcript of the letter Leibniz–É. Périer, 30 Aug. 1676, has not yet come to light.   [73] Wallis (1655).

vastly different from that afforded by the letters, especially those of the year 1673 in which Leibniz did not show up at all well. The opinion Collins had formed at that time was obviously false and in want of revision. He felt the need to oblige Leibniz and, with the best of intentions, readily acceded to his wish to be allowed to see Gregory's and Newton's papers, perhaps influenced in this decision by Oldenburg's intervention. Oldenburg may well have thought fit to do a favour for Leibniz now that the latter's calculating machine was standing complete before him – a weighty argument against the offensive attacks by Hooke which had just lately again caused him a good deal of trouble.[74] For his own part, Leibniz left London in a mood of great satisfaction; he had received a lively impression of the latest mathematical research in England and could hope in future always to be kept up to date with further news by way of Oldenburg and Collins. The doctrine of series seemed to be the most important field of English advance but they apparently possessed nothing comparable with his own invention, the Calculus. To his regret it had not been possible for him to meet Wallis or, above all, Newton; how fascinating it would have been to match oneself in a personal discussion against the inventor of the doctrine of series, a man of whose ability both Oldenburg and Collins had marvellous tales to tell. What answer would he give when he received Leibniz' letter?

Newton's reply was written during the very days when Leibniz set foot on Dutch soil to call on Hudde and Spinoza; and a short but significant supplement to this 'epistola posterior' is included in a note of his written soon afterwards to Collins,[75] expressing his thanks for the latter's extracts from Tschirnhaus' and Leibniz' letters. When he thinks of the endless discussions of his various letters on colour published in the *Philosophical Transactions*, he wishes only that he had never allowed anything of his theory to get into print, and so he now intends to keep his mathematical discoveries to himself. True, says Newton, the representation of roots by series of fractions of which Leibniz speaks, is good and so certainly is the procedure for finding solutions of affected equations: but these series are no better than his own, and the method of transmutation could really be dispensed with. In his reply, he now says, he had given only indications of some of his results but not of everything: he can, for instance, accomplish in a

[74] On these controversies *compare* ch. **9**: notes 42–3, and the text p. 122. His preoccupation with these prevented Oldenburg from collating the transcripts of Newton's two great letters for Leibniz thoroughly with the originals.
[75] Newton–Collins, 8 (18) Nov. 1676 (*NC* **2**: 179–80).

few minutes the quadrature of a curve given by $ax^p + bx^q + cy^s = 0$ or reduce it to the simplest possible curves of its type, and a similar observation holds for the corresponding equation in four terms.

Collins' reply came – because of his illness – only after an interval of more than three months;[76] it contains a report about Leibniz' visit to London and an extract from Leibniz' latest letter to him from Amsterdam. Nothing is said to reveal that Leibniz had, when he visited Collins, been given a chance to look at a selection of Gregory's and Newton's manuscripts. Collins was probably somewhat vexed that he had rather thoughtlessly permitted something that would never have met with Newton's approval – just as he had when he passed on Newton's series for the circle-zone to Gregory,[77] who had immediately on the mere basis – or so Collins imagined – of this one result discovered for himself the method of infinite series. Gregory's attitude had been entirely fair and he had readily conceded the right of priority to Newton; but from all one knew so far of Leibniz a similar disavowal was not to be expected from him. No wonder then if from now on he adopted a policy of the utmost restraint and altogether ignored Leibniz' attempt to continue their correspondence.

Leibniz' hope of remaining in contact with expert English mathematicians was still-born; no less well-informed than he, they purposely withdrew from him. Hanover for Leibniz meant narrowness and loneliness, and was to become the graveyard of his hopes and aspirations. There Leibniz became in future years a recluse, not from inner inclination but in consequence of the lack of concern, understanding and interest surrounding him.

[76] Collins–Newton, 5 (15) Mar. 1676/7 (*NC* 2: 198–201).
[77] Collins–Gregory, 24 Mar. (3 Apr.) 1669/70 (*GT*: 89).

21
CONCLUSION

There now remains the task of surveying in retrospect the finely woven, intricately intertwined fabric of relationships which we have been able to delineate only in broad strokes and nowhere describe in detail. If we have referred only infrequently to the older secondary literature devoted to our subject, this has not been occasioned by ignorance of all that has previously been gathered on our theme but from an awareness that much of the rich new material on which we have been able to base the foregoing exposition had not been at the disposal of earlier researchers. In these circumstances it would have been wrong and pointless once more to drag into the open the mistaken opinions of past historians based, as they were, on the fragmentary nature of the documents available to them; it is more appropriate to acknowledge the merits of these scholars with gratitude, while sparing the reader all the detours necessary to correct their errors and indeed the sidetracks down which the present author had himself to go before he was able to bring some measure of order into the whole complicated pattern.

What has induced us to confine the story of the development of Leibniz' mathematics to his time in Paris can be stated in a few words. In those few years he conceived his decisive ideas in mathematics, and brought them to a degree of completeness which allows us to view all his later researches as their offshoots, while from the chronological aspect his stay in Paris forms a distinct chapter in Leibniz' career which closes with his departure for Hanover. The multiplicity of his new tasks makes it quite impossible for him to continue his mathematical researches with the same depth or intensity as before; mathematics was not, indeed, altogether abandoned by him, but more and more it becomes a means of mental diversion and recreation and comes to acquire a golden sheen as the most intense experience of his youthful maturity. Time and again we read in his later papers that one idea or another had its origin in Paris, and this is fully confirmed by his dated notes or those that can be dated by the paper, watermarks, remnants of seals and other signs, or by references in letters. It was in this period of his life that Leibniz' philosophy, founded on mathematics and the natural sciences, was formed; then that his most significant and fundamental insights in psychology dawned upon him. It was the most intensively creative period of his life: a time when he was able to absorb with incredible ease all the sources of intellectual inspiration that flowed into his reach from all directions and when he knew to impose order upon them and fit them into a unified, organic system where each detail found a meaningful place and was to have its proper significance.

Still but a novice, Leibniz late in 1672 takes his mathematical departure from arithmetic and through his critique of the axiom of the whole and its parts gains a first general insight which persuades him to look more closely at the problem of infinity. An initial belief that he can master mathematics by superficial study and create it from his own untutored resources was shaken and then thoroughly destroyed by Huygens. Leibniz is faced with the choice either of giving up or of going on to work through it fully. He is captivated by a subject with which even the greatest minds had found it hard to wrestle. Driven by the ambition to do better, and conscious of his mental receptiveness and formative power, he dares to attempt the great enterprise of gaining a key height from which all previous results in infinitesimal mathematics can be coherently comprehended, viewed as a unity, and so be communicated as a systematic whole. In Blaise Pascal he finds an author whose facile pen can present even the most severe or brittle topic with grace and elegance. Under Leibniz' hand Pascal's corpus of thought acquires a new life; where Pascal – not yet fully aware of the underlying structure – sees the germ of a possibility, Leibniz creates its reality: his characteristic triangle originates by his abstracting a general idea from an excellently executed particular consideration; his transmutation method is the first truly independent achievement which grows out of an ingenious notion of his own; his arithmetical quadrature of the circle is at first nothing more than a surprising new detail, arising from the reworking of old results by a new method. These first successes provide an ever stronger stimulus to him to look for a comprehensive unification of the whole theory of infinitesimals. This demands surely that cognizance be taken of all approaches hitherto attempted, weighing one against another, thinking methodically about them and transforming them until at last their common *filum meditandi* has been discovered.

The persistent study of the available literature not only furnishes numerous factual details, but beyond these, shapes that instinctively right attitude towards infinitesimal problems without which a unifying synthesis would have been unthinkable. Leibniz' preoccupation with questions of algebra increases his algorithmic skill and opens up the way to the Calculus: the decisive feature is his introduction of suitable symbols to be formed on the model of algebraic ones. He now rethinks the basis of his *ars inveniendi* from a metaphysical, logical and mathematical viewpoint. In the treatment of one particular inverse tangent question the notation '*omn.*' for (infinite) summation which had hitherto been used is replaced by the $\int$ sign, and after a few further groping attempts the symbols $\int y \,.\, dx$ and dy/dx make their appearance; with their introduction the inverse character of the two operations they symbolize is recognized and the related formal apparatus is gradually constructed. Simple differential equations are changed into relations between integrals and then by a step-by-step integration their solutions are expanded into power series. By means of the newly introduced symbols the whole domain of infinitesimal problems so far explored is dealt with simply and elegantly.

Following a suggestion from Huygens, Leibniz embarks on the attempt to unravel the true nature of circle quadrature. With Descartes in his *Géométrie* he distinguishes between 'geometrical' (algebraic) and 'mechanical' (transcendental) problems and functions, without however excluding the transcendental ones – on the contrary he tries to classify them and to find their types. Like Wallis and Gregory, he is convinced that the trigonometrical and logarithmic functions (which he defines by quadratures) belong together with their inverses to the realm of transcendental functions; with a still imperfect technique, though with clear insight into essentials, he attempts to determine when a given function will or will not be algebraic. His activity in this direction leads him to the method of indeterminate coefficients. He is conscious of the full gravity but also of the very real difficulty of this line of approach, and recognizes the fallacy in Gregory's earlier reasoning on the matter without, however, being able to correct it. Already it occurs to him to use the limit-sum of a converging infinite series 'by definition' to characterize a quantity, and so, basing himself on the axiom of the whole and its parts, he develops a criterion for the convergence of the alternating series. In analogy with rational approximation of irrational and transcendental numbers he uses algebraic functions to approximate transcendental functions without thinking exclusively of expansion into power series. By rational integral transforms he is enabled to classify problems that 'depend on squaring the circle and the hyperbola', and while extending his transformation he discovers a method for characterizing the area of a sector by a power series expansion which is valid jointly for all central conics. His comparable attempt to reduce the arc-length of a conic to the quadrature of a conic area proves, however, unsuccessful.

In algebra the general validity of Cardan's formula for solving the cubic equation is established: for the reducible case the detachable linear factor is observed, and for the irreducible case the inevitability of the recourse to imaginary component parts which will cancel each other out in the end result is asserted, even when this cannot be made evident by actually extracting the roots. The example

$$\sqrt{(1+\sqrt{[-3]})} + \sqrt{(1-\sqrt{[-3]})} = \sqrt{6}$$

is used to make plausible the generalization that

$$f(x+iy) + f(x-iy)$$

is a real function of x and y. In the case of higher equations use is made of such substitutions as $x = u_1 + u_2 + \ldots$ and $x = \sqrt{v_1} + \sqrt{v_2} + \ldots$, but in the course of Leibniz' more detailed investigation it becomes clear that only the generalized Cardan equations (a class to which the general equations for angle-section belong) are solvable by substituting $x = u_1 + u_2$, while examples like the general quartic equation demonstrate the equivalence of the simple additive substitution and that involving radicals. Because of certain mistakes in his calculations Leibniz at first believes that higher equations can be reduced to those of seventh degree and that the quintic is solvable by means of an auxiliary equation of fourth degree, using for this task appropriate abbreviations for certain types of calculations (symmetrical functions, for instance, and simple invariant forms), but he is not sure of his ground and so omits to communicate any details. He determines solutions in integers of simple and simultaneous linear Diophantine equations and deals by algebraic substitutions with transformations of certain given expressions in square numbers. But though he does not arrive at any decisive results or fruitful methods in number theory, Leibniz owes to his concern with the solution of quadratic Diophantine problems the idea of a rationalizing integral transform.

On the whole we observe here a conscious progress from the older, purely geometrical point of view to the more modern, analytical (functional) attitude. The original naïve, verbally presented concept of indivisibles (that an area 'consists' of the totality of its ordinates) is superseded and replaced by the improved concept of his characteristic triangle; Leibniz' symbolism exactly mirrors this state of reinterpretation, and the ground is thereby prepared for successful future research in differential geometry. The class of problems capable of being so treated is greatly enlarged by including the consideration of moments and centroids. The question of the conditions which the functions to be considered will have to satisfy remains unclarified, but the importance of a methodical fully rigorous investigation is now recognized and emphasized; thus, for example, his general proof of the transmutation theorem is in its final version altogether unexceptionable.

True, Leibniz is concerned with practical questions also; he is certainly not uninterested in the accuracy of a construction or the convergence of a series-expansion, but the methodical, theoretical insight is still more important for him than any of these. Everywhere, he strives to attain the highest generality within his reach; he is fully aware that mathematical insights lead on not only to new gains in particular knowledge or computational improvement, but – what really matters – to deeper conceptual penetration. To this end the mechanization of all details pertaining merely to deductive technique and the classifying of basic problems will serve to permit progress with minimum effort from immaculate foundations to the most general theorems possible. Even where an individual performance of Leibniz miscarries, the whole tenor of his thought is nonetheless noticeable and worth our attention.

So much in broad outline for the internal mathematical content of his Paris studies. When it comes to the communication of his new-found results to others, his relationship with Huygens and correspondence with Oldenburg are to the fore for us. Huygens helps the beginner by advice and encouragement and through helpful criticism but later somewhat withdraws from the maturing young German. The reason for his so doing is probably to be found in the fact that the tendency of Leibniz' new ideas is not really to his taste. Leibniz is aiming at a technique of representation that should be simplified to the last degree by suitable symbolism and become virtually a part of logic; that cannot, however, be immediately comprehended but has to be gradually learnt. Whoever is able to acquire this technique will be superior to a degree undreamt of by the uninitiated, even when he has no particularly deep insight into the structure itself: the formalism will think for him. It was this very possibility that Huygens deemed undesirable. He was the last and most important representative of the old school of mathematicians who came to their results through truly ingenious but simple isolated considerations: his bent was not towards solely abstract, deductive reasoning but to the pure, geometrical chain of argument which starts from a single finite entity and proceeds, by application of the indirect Archimedean method, to a result on infinitesimals – a way of reasoning in which he was so well schooled that he had himself no need of the new, still rather clumsy, Leibnizian calculus. Along the difficult, though highly interesting and ingenious, path which he was wont to tread there was no longer any novel methodological possibility to be found: the future was to belong to the conceptual-algorithmic method. Huygens, now getting on in years, was unable to adapt his mathematical style, instinct and vision to the new trend; Leibniz' procedure appeared to him as a more or less capricious game which threatened to restrict his own freedom of thinking, and so he rejected it.

His relationship with Oldenburg brings Leibniz into personal contact with the English mathematicians, regarding whose researches he had at first been informed almost entirely through Collins' reports. His ill-considered promise concerning his calculating machine and his still rasher assertion that he was the first who could interpolate and sum any numerical series – a boast considered by Pell to be mere empty arrogance – leads him off to a bad start during his first visit to London in 1673, one not improved by his *faux pas* after he is told of his election into the Royal Society, whereby he causes Oldenburg acute embarrassment. A collection sent to him by Collins of recent, unpublished achievements by English mathematicians proves to be of no use to a novice who is as yet familiar with little of the relevant literature. Leibniz sees himself obliged to leave off his correspondence for more than a year afterwards; during this time he completes his calculating machine, discovers his general transmutation theorem and the arithmetical circle quadrature, works through the standard mathematical works of the period and familiarizes himself with the fundamental problems of infinitesimal mathematics. He lets it be known abroad that he is now well up in the whole field of mathematics and has achieved a number of new results whose value has been acknowledged by experts in Paris. The discussion leads to a renewal, at a much deeper level, of communication by Collins and Oldenburg of information on researches by British mathematicians, which prove to be comparable with Leibniz' own results. He now believes that he has already full news of the most recent discoveries by English mathematicians, whereas in fact he has been told only about problems that have been settled years before. He tries in vain to gain clearer insight into the analytical methods of Newton and Gregory; it remains particularly uncertain to him what exactly the English are able to represent by infinite series, and what in the form of closed expressions.

During these exchanges algebraical topics come more to the fore; the question is asked if and how an equation of general degree may be solved, and in particular the debate revolves round the general validity of Cardan's formulae, the avoidance of the imaginary, the application of logarithmic and trigonometrical tables in the solution of general equations, and the elimination of intermediate terms. Leibniz who had painstakingly consulted the sources and worked through the relevant sixteenth century Italian books brings a new slant, namely the historical one, to bear on the investigation. He reports in stages on his notions of the individual problems touched upon, and expects similar accounts of the views of Gregory, Pell and Wallis. At this point Tschirnhaus appears on the London scene; with his brilliant facility in computation and exuberant temperament he silences all objections against his methods, adumbrated by him only in particular examples, and promises the general representation of all numerical 'mathematical' quantities – including the roots of equations – by simple or compound radicals. As an ardent admirer of Descartes, whose wise restriction of mathematics to the sphere of the purely algebraic and thus exactly tractable he cannot praise highly enough, he is not particularly interested in the infinitesimal method of the English, which he values only as a procedure for approximation, and he has no desire to learn over much about it lest he jeopardize his own originality.

Leibniz comes to know Tschirnhaus well immediately after he has invented the Calculus, making friends with this amiable compatriot of wide-ranging interests. He is told of his algebrical methods and, seeing him successfully pursuing ways which he himself had been treading a few weeks earlier, he withdraws his attention a little from this topic to avoid the possibility of a conflict. Leibniz for his part wants to survey for Tschirnhaus his own central ideas – just then maturing into coherent form – but he finds his listener inattentive, devoid of the inclination to deal with any algorithms but the straight-forward algebraical ones he can manipulate.

A happy chance provides the opportunity for Leibniz to inspect the papers left to posterity by Pascal; he does so in company with Tschirnhaus and becomes absorbed in the *Conica*, where he is particularly fascinated by Pascal's habit of replacing parallel co-ordinates by a simple pencil of rays. Tschirnhaus' exaggerated Cartesianism rather repels him and induces him to point more strongly than before to the obvious defects of Descartes' system; on the other hand he is perfectly glad and willing to start a personal relationship with Clerselier, the affectionate editor and executor of Descartes' literary remains, in order to gain thereby sight of the still unpublished Cartesiana. By now only in loose contact with his home country, which anyway had no attractive offer to make for his services, he will take the risk of staying for good in Paris and of there buying himself an agreeable post; he thinks of seeking paid work at the Académie des Sciences or of taking up the Ramus professorship in mathematics vacated by Roberval's recent death. His connection, by Gallois' good offices, with the Duke of Chevreuse and Tschirnhaus' position as mathematics tutor to the young Colbert are to be used to that end; also Leibniz plans a series of publications on a variety of such individual mathematical discoveries, as the arithmetical quadra-ture of the circle, determination of a rational cycloid segment, rectification of a general conic arc, solution of generalized Cardan equations, axiomatic investigations of the whole and its parts, and the related treatment in this context of the harmonic triangle, or again the demonstration of the principles of his calculating machine and the construction of an 'algebraical instrument' – all intended to attract the eyes of the scientific world towards the new aspirant.

Through an indiscretion which cost Leibniz the good opinion and protection of the touchy Abbé the project not in itself at all hopeless comes to nothing; the hope of still achieving his aim by writing to Huygens, just then on the point of returning home to Holland after his recent serious illness also proves vain. Leibniz can find no em-ployment in Paris, either in a direct capacity or indirectly as the political delegate of some German prince. Though it has little to attract him he is forced to accept the offer by the Duke Johann Friedrich to run his Library in Hanover; the Duke, however, wants him to take up his new position immediately and only reluctantly allows him in the end a few weeks' postponement.

Meanwhile, through his conversations with Leibniz, Tschirnhaus has recognized that the transformation method held by him to be general can accomplish far less than he had initially expected and that he cannot therefore keep the promises he made in London. To repair the bad impression caused by his nine-months-old silence he writes a long letter on the value and significance of Descartes' methods in mathematics, insisting that to him is essentially due the re-awakening of modern mathematics – so much so that the achievements of present-day mathematicians amount to no more than extensions and refinements of Descartes' lines of thought. This induces Collins to prepare a careful reply in which he extols in general terms the methods of Newton and Gregory in particular and asserts that they go far beyond Descartes. Leibniz has also in the meantime written again; he offers his arithmetical circle quadrature in exchange for the English infinitesimal methods, expecting however not mere hints or single results but a factual description of the methods with proof, and will make a similar return. He is especially concerned that care should be taken to preserve the papers of the recently deceased Gregory, a man who in algebra, number theory and analysis obviously stood in the first rank.

Outside Scotland, Collins alone had been in Gregory's confidence in matters of science and had copies of nearly all his papers in his possession; he sets to work at once to abstract for Leibniz the most important results of his Scottish friend and so secure him the honour of first invention while at the same time enabling others to continue the train of his researches. Collins has now to satisfy Leibniz' request, and also to contradict Tschirnhaus' mistaken belief, to which he has still clung in his most recent letter, that the latest achievements of the English in mathematics are of no significance. Newton's general tangent-method was to be added as an item of special importance; furthermore, on Oldenburg's insistence an original contribution by Newton on the doctrine of series was to be sought. Collins' ensuing composition (the 'Historiola') proves in the event to be too thorough and too extensive for communication and has to be shortened. Newton's letter (the 'epistola prior') contained his binomial theorem and an insufficiently described procedure for solving numerical and literal equations, but otherwise only individual series that had already been treated in discussion with Gregory many years ago, and also a digression on the practical technique of approximately constructing roots: nowhere is there discussion of other than single examples. To Collins' letter communicating the final English draft of the 'Historiola' for his approval or criticism Newton answered only after some delay with a request for certain changes to be made. Thus several weeks went by before an answer to Leibniz and Tschirnhaus could be made ready. In view of the importance of the matter, Oldenburg thought it proper to retain the autograph papers in London and to forward to Leibniz only copies made by a secretary; these were disfigured by numerous errors in transcription and checked with insufficient care against their originals. Leibniz replied by return, now sending his arithmetical circle quadrature, not, however, in the form where it is based on his general transmutation theorem (which he wanted to keep private) but that using a rationalizing integral transformation. Before composing his answer Leibniz had looked rather too briefly through the detail of what he had received; he now effectively reproduced the series for the natural logarithm and sine and cosine functions sent to him by Newton in only slightly different form, emphasizing, it is true, the difference in their deduction yet not explaining his own method with sufficient clarity, and thus laying himself open to the suspicion of plagiarism. At the end of his letter he furnishes a summary of discoveries made by him in physics in consequence of his recently discovered principle of conservation of *vis viva*, and mentions by way of example for his treatment of inverse tangent questions the characterization of curves whose subtangent is constant in length. A few days later Tschirnhaus also wrote a letter, specially confirming that Leibniz' own circle-series was not contained among the results he had received from London; he there asserts that he takes the whole doctrine of series to be merely an 'accidental' method, since in fact every quantity is expressible by radicals or combinations of radicals. He follows this up with algebraical details of lesser importance.

Extracts from these two letters were sent on by Collins to Wallis and Newton, with a request for comments. Wallis corrected the misreadings in his copy of the Tschirnhaus letter and in the case of Leibniz' letter submitted that the circle-series converges very badly and that the identity $\sqrt{[1 + \sqrt{(-3)}]} + \sqrt{[1 - \sqrt{(-3)}]} = \sqrt{6}$ was not really satisfactory; because of the importance of the topics dealt with in them he hopes these letters will soon be published. Newton, in his own long delayed answer, begins by describing the genesis of his first discoveries in the field of series, cites a number of the more general propositions from his methods of quadratures and sharply criticizes the results sent by Leibniz: the series for the circle quadrature is all but useless because of its poor convergence, and can be made practicable for computation only by a fundamental transformation. For the rest, writes Newton, the results he indicates (including his expansion into power series of an algebraic function with the aid of his parallelogram) are no more than particular instances of a universal method (of fluxions and fluents in conjunction with expansions into power series) which he chooses to conceal in an insoluble anagram. He can, he states, classify all quadratures, and where they cannot be given by a closed expression he can expand them in a series; the situation is similar for the inverse tangent questions, for example where the tractrix is defined by the constancy of its normal.

Since Tschirnhaus had written of his impending departure for Italy, offering to undertake commissions there, Collins could not delay his reply till Newton's answer was received. He confirmed that the Leibniz series was not amongst the results the English had sent him, discussed certain algebraical details he had received from Tschirnhaus and added something about the approximate solution of equations. Leibniz later saw the copy of this letter to Tschirnhaus that Oldenburg retained and made a note of its acknowledgement of his own originality, as of prime importance. Towards the end of October 1676, on his return journey to Germany, he spent a few days in London where he handed over some of his promised manuscripts on algebra and demonstrated his calculating machine in Oldenburg's presence. With Collins, then laid low with an attack of blood poisoning, he conversed on questions of compound interest, methods for lowering the degree of equations and for solving them, the various attempts made in England at circle computation, and problems in infinitesimals. Collins also received the copy of a letter sent by Leibniz to Périer after he had studied Pascal's *Conica* and judged it ready for printing. Leibniz found means to disperse the doubt and mistrust with which Collins at first greeted him so skilfully that he was permitted to inspect not only the 'Historiola' and Newton's *De Analysi*, but in addition other manuscripts by Gregory and Newton. He made copious notes on Gregory's methods for solving equations, and also on everything relating to the doctrine of series and of interpolation, while from the *De Analysi* he copied the series there given together with the general rules of method, and particularly its proofs. The sections on infinitesimals in both Gregory's and Newton's papers he completely disregarded.

What, we ask finally, induced Leibniz to make these excerpts? He, too, had a method of series (thought by him to be general), proceeding by way of rationalizing transformations, with which he could derive step-by-step the power series expansion of an integral. Why did he expose himself in this way to the suspicion of having snatched the fruits of another man's mind, intending to appropriate them for himself? Here involved were not published accounts but manuscripts confided to Collins and carefully guarded by him. That Collins had consented to Leibniz' inspecting them could in itself already be considered a breach of confidence, and the existence of these excerpts of his threatens to confirm Newton's accusation of plagiarism to the full: they would seem to represent a documented proof of it weightier than the sum of all those cited in the *Commercium epistolicum*.

This line of argument would be valid if Leibniz ever subsequently made the least improper use of the excerpts he had taken. But this never happened; on the contrary we find Leibniz – in full accord with the facts – time and again admitting that he has learnt a great deal about series from the English. He wanted to make an accurate study of expansion methods employed by the English in order to compare them with his own and to evaluate their respective areas of application. In Gregory's papers he found everywhere mere statements of detail, without any rules of method to guide him in his investigation. The method – Taylor series expansion – which the ingenious Scotsman in fact employed has only this century come to light with the rediscovery of a significant portion of his papers. As for Newton, by carefully working through his first letter Leibniz could gather everything of significance it had to communicate, while from the other papers he later excerpted in London he learnt nothing of essential importance that he had not known before; hence the charge of plagiarism is insupportable.

There remains, however, the suspicion that even though no plagiarism in fact occurred, it had been his intention illicitly to appropriate Newton's and Gregory's findings. But the same suspicion might have been entertained against his excerpts in Paris from the papers of Pascal, Descartes, Roberval and many others: here there has never been any suggestion of intended plagiarization on his part, but they have always been accepted as interested private inspections of hitherto unprinted manuscripts devoid of ulterior motive. And so it is with both the 'Historiola' and the *De Analysi*. The action Collins here took is akin to his habit of dispatching a work still in press in sheets to his friends well before the eventual date of publication – as, for instance, with Kersey's *Algebra*. With regard to Newton's 1672 letter to Collins, expounding his logarithmic method for mechanically solving equations, which Leibniz was given to read, this is a perfectly innocuous item which Collins had inadvertently reported to Leibniz in such a way that its central idea had been misunderstood. Mindful of Collins' more usual caution in these matters, we think it highly unlikely that Leibniz could ever have had any other Newtoniana in his hands. We need to qualify this attitude only with regard to those letters by Gregory that are not contained in the 'Historiola': these, however, were in Collins' opinion not fit for publication and so he felt he could dispose freely of them by reason of the authority that Gregory had expressly given him to do so. There can therefore here be no question of a breach of confidence by Collins in allowing Leibniz to read them. The incident assumes an awkward aspect only because Collins' intended publication of his papers was indefinitely delayed – indeed the 'Historiola' remains unpublished to this day. But Leibniz was not to blame for this.

The accusation of plagiarism against Leibniz relates not at all to the doctrine of series (where the priority of the English discovery has never been questioned), but solely and entirely to the infinitesimal methods which paved the way to the calculus. That Leibniz here, wholly uninfluenced by others, gained his crucial insights unaided, is beyond all doubt. That he made no excerpts of any of the infinitesimal portions of the papers he inspected only confirms what we already know – that he did not expect new insights into calculus methods from others working in this field.

Leibniz' aim in making these and many other extracts was quite different. His method so far had been to examine the results of others for their basic insights, to penetrate and assimilate their very essence, and to extract from them the general viewpoints that were methodologically of value to him. Whoever studies his sharply outlined summaries of the content of what he has read, and ponders them will realize that his concern is to bring out its underlying concepts, formal structure and method of presentation. Turned into the mathematical: that his intention is not to reproduce the full text of the original in its chronological sequence of composition, but to delineate in its full purity the characteristic idea which is there described. In this sense Leibniz, enthusiastically and unreservedly gathering and absorbing all the knowledge in any way accessible to him and then forming it in a grand new synthesis into a unified whole, is the first true historian of science. From him has been borrowed the method used in this present study: namely, of reporting all essential details as faithfully as possible and connecting the links which clarify the origin and growth of central ideas, their structure, their effectiveness and their final aim – a laborious task since it is not sufficient to confine one's treatment of a topic to sketching its principal developments in broad outline, but also a rewarding one for anybody who is prepared to devote adequate attention to the smaller details and to elaborate their meaning and significance in the total pattern.

22

ABBREVIATIONS

AE   Acta Eruditorum, Leipzig (from 1682).

AS   Académie Royale des Sciences, Paris.

BH   Th. Birch, *History of the Royal Society of London*, London 1756, 1757.

BJC   Johann Bernoulli, *Opera Omnia* I (ed. G. Cramer), Lausanne/Geneva 1742.

BJS   Johann Bernoulli, *Briefwechsel* I (ed. O. Spiess), Basel 1955.

BKC   Jakob Bernoulli, *Opera* I (ed. G. Cramer), Geneva 1744.

C   *Catalogue critique des manuscrits de Leibniz* II (ed. A. Rivaud), Poitiers 1914–24.

CE   *Commercium epistolicum D. Joannis Collins et aliorum de analysi promota*, London 1712 (1713), ₂1722 (1725) (Leibniz' annotations in his personal copy are listed under 'Leibniz' in the index of names).

CR   *Correspondence of scientific men of the seventeenth century* I, II (ed. S. P. Rigaud), Oxford 1841.

DC   R. Descartes, *Lettres* I, II, III (ed. C. Clerselier), Paris 1657, 1659, 1667.

DO   R. Descartes, *Œuvres* I–XII + supplément (ed. C. Adam and P. Tannery), Paris 1897–1913.

FO   P. de Fermat, *Œuvres* I–V (ed. P. Tannery, C. Henry and C. de Waard), Paris 1891–1922.

GT   *J. Gregory Tercentenary Memorial Volume* (ed. H. W. Turnbull), London 1939.

HO   Chr. Huygens, *Œuvres complètes* I–XXII, The Hague 1888–1950.

JS   *Journal des Sçavans*, Amsterdam edition (from 1665).

LBG   G. W. Leibniz, *Der Briefwechsel mit Mathematikern* I (ed. C. I. Gerhardt), Berlin 1899.

LBr   Leibniz Briefe (*manuscripts*), Niedersächsische Landes bibliothek, Hanover.

LD G. W. Leibniz, *Opera omnia* I–VI (ed. L. Dutens), Geneva 1768.

LH Leibniz Handschriften (*manuscripts*), Niedersächsische Landesbibliothek, Hanover.

LMG G. W. Leibniz, *Mathematische Schriften* I–VII (ed. C. I. Gerhardt), Berlin/Halle 1849–63.

LPG G. W. Leibniz, *Die philosophischen Schriften* I–VII (ed. C. I. Gerhardt), Berlin 1875–90.

LSB G. W. Leibniz, *Sämtliche Schriften und Briefe*, Darmstadt/Leipzig/Berlin 1923–.

LWG G. W. Leibniz, *Briefwechsel zwischen Leibniz und Chr. Wolf* (ed. C. I. Gerhardt), Halle 1860.

MC M. Mersenne, *Correspondance* (ed. P. Tannery and C. de Waard), Paris 1932–.

NC I. Newton, *Correspondence* (ed. H. W. Turnbull, J. F. Scott, A. R. Hall), Cambridge 1959–.

NP I. Newton, *Mathematical Papers* (ed. D. T. Whiteside), Cambridge 1967–.

OC H. Oldenburg, *Correspondence* (ed. A. R. and M. B. Hall), Madison/Milwaukee/London 1965–.

PO B. Pascal, *Œuvres* I–XIV (ed. L. Brunschwicg, P. Boutroux and F. Gazier), Paris 1904–14.

PT *Philosophical Transactions*, London (from 1665).

RS Royal Society, London.

SL R. F. de Sluse, *Correspondance* (ed. C. Le Paige), in *Bollettino di bibliografia e di storia delle scienze matematiche e fisiche* **17**, Rome 1884.

SO B. de Spinoza, *Opera quotquot reperta sunt* (ed. J. van Vloten and J. P. N. Land), vol. III, The Hague 1895.

TO E. Torricelli, *Opere* (ed. G. Loria and G. Vassura), Faënza 1919–44.

WCE J. Wallis, *Commercium epistolicum de quæstionibus quibusdam mathematicis*, Oxford 1658.

WO J. Wallis, *Opera mathematica* I, II, III, Oxford 1695, 1693, 1699.

23
CHRONOLOGICAL INDEX

This index consists of five parts. The main section (1) gives a chronological register of all letters or original publications in contemporary periodicals that have been adduced in the text or in the footnotes; plain figures denote pages of text, pairs refer to the notes, thus **7**: 23 signifies note 23 of chapter 7; bracketed figures indicate an indirect or implied reference. The second portion (2) lists hitherto unpublished Hanover manuscripts according to Rivaud's *Catalogue Critique II*; here dates printed in italics are those actually found on the documents, other dates had to be inferred and so are to be taken as provisional. Locations for other manuscripts are briefly given in section (3). We add (4) a calendar of relevant meetings of the Royal Society of London and the Académie Royale des Sciences of Paris, followed by a chronological checklist (5) of publications in the periodical literature of the time.

1. CORRESPONDENCE

1638

18(?) Jan. Descartes–   *MC* **7**: 17–18
     Mersenne
     14: 20

Jan./Feb. Mersenne–   (*MC* **7**: 52)
     Fermat **7**: 35

Feb.   Fermat–   *MC* **7**: 49–52
     Mersenne **7**:
     35

1 June   Roberval–   *MC* **7**: 247–
     Fermat   50
     7: 35

13 July   Descartes–   *MC* **7**: 340–6
     Mersenne
     14: 37

20 July   Fermat–   *MC* **7**: 377–
     Mersenne   80
     7: 35

1638 (*cont.*)

27 July   Fermat–   *MC* **7**: 397–
     Mersenne   402
     7: 35

12 Sept. Descartes–   *MC* **8**: 70–9
     Mersenne
     8: 7

16 Oct.   De Beaune   *MC* **8**: 142–3
     for Roberval
     17: 45

22 Oct.   Fermat–   *MC* **8**: 153–
     Mersenne   62
     8: 8

1639

20 Feb.   Descartes–   *DC* III:
     De Beaune   409–16
     17: 42, 45

1640

1 Feb.	Descartes–Waessenaer 13: 61	*DO* III: 21–33
8 Feb.	Descartes–Hogelande 15: 66	*MC* 9: 308–9
Feb./Mar.	Frénicle–Mersenne 11: 8	*MC* 9: 154–8
18 Oct.	Fermat–Frénicle 13: 61	*FO* II: 206–12

1641

1 May	Mersenne–Pell 3: 20	*MC* 10: 610–1
15 June	Fermat–Mersenne for Frénicle 13: 62	*MC* 10: 655–60
17 Dec.	Cavalieri–Torricelli 13: 51	*TO* III: 65–6

1642

25 Mar.	Descartes–Dotzen 17: 37	*DO* III: 553–6

1643

June	Torricelli–Nicéron 13: 51	*TO* III: 125–30
Summer	Roberval–Mersenne for Torricelli 13: 51	*TO* III: 133–8

1644

27 Aug.	Torricelli–Ricci 13: 49	*TO* III: 222–3
10 Sept.	Ricci–Torricelli 13: 49	*TO* III: 225–6

1645

17 Jan.	Torricelli–Ricci 8: 13–14	Tannery (1933): 133–8
28 Jan.	Ricci–Torricelli 13: 48	*TO* III: 258–61

1645 (*cont.*)

4 Feb.	Mersenne–Torricelli 8: 13, 16	*TO* III: 268–70
6 Feb.	Torricelli–Ricci 13: 50	*TO* III: 271–2
Feb.	Torricelli–Carcavy 8: 14	*TO* III: 279–81
June	Descartes–(?) 7: 44	*DO* III: 443–50

1646

1 Jan.	Roberval–Torricelli 5: 44, 7: 35	*TO* III: 349–56
17 Mar.	Torricelli–Ricci 8: 14	*TO* III: 360–2
23 Mar.	Torricelli–Cavalieri 8: 14	*TO* III: 363–5
7 Apr.	Torricelli–Ricci 8: 18	*TO* III: 368–9
5 May	Torricelli–Cavalieri 13: 52	*TO* III: 373–4
7 July	Torricelli–Roberval 8: 14, 19	*TO* III: 389–92
8 July	Torricelli–Carcavy 8: 19	*TO* III: 405–7
26 Aug.	Mersenne–Torricelli 8: 17	*TO* III: 410–12
late 1646	Fermat–Torricelli 13: 53	(*FO* II: 338)

1647

15 Aug.	Torricelli–Cavalieri 8: 14	*TO* III: 466–71
24 Aug.	Torricelli–Ricci 8: 19	*TO* III: 473–5
31 Aug.	Torricelli–Cavalieri 8: 19	*TO* III: 475–7

CORRESPONDENCE

1657 (*cont.*)

Date	Correspondents	Reference
21 Nov. (1 Dec.)	Wallis–Digby 4: 29, 17: 74–5	WCE: 33–51
18 Dec.	Huygens–Schooten 8: 32–3	HO 2: 110–13
20 Dec.	Huygens–Sluse 8: 32	HO 2: 104
22 Dec.	Schooten–Huygens 8: 35	HO 2: 105
28 Dec.	Huygens–Schooten 8: 32–3	HO 2: 110–13

1658

Date	Correspondents	Reference
18 Jan.	Heuraet–Schooten 5: 22	HO 2: 131
20 Feb.	Frénicle–Digby 4: 22	WCE: 116–21
24 Feb.	Heuraet–Huygens 8: 34	HO 2: 138–9
4 Mar.	Sluse–Huygens 7: 21	HO 2: 144–5
14 Mar.	Sluse–Huygens 7: 22	HO 2: 150–2
5 Apr.	Huygens–Sluse 7: 24	HO 2: 163–4
7 Apr.	Fermat–Digby 4: 30	WCE: 158–161
28 May	Huygens–Sluse 7: 24	HO 2: 178–80
20 (30) June	Wallis–Digby 4: 31, 15: 12	WCE: 180–1
6 Sept.	Huygens–Wallis 7: 17, 25; 11: 30	HO 2: 210–14
19 Sept.	Schooten–Huygens 11: 32	HO 2: 221–2

1658 (*cont.*)

Date	Correspondents	Reference
4 Oct.	Huygens–Schooten 11: 31	HO 2: 235–6
11 Oct.	Sluse–Huygens 7: 64	HO 2: 248–9
19 Oct.	Sluse–Huygens 7: 65	HO 2: 259
1658	Pascal–A.D.D.S. 8: 55	Pascal (1659)

1659

Date	Correspondents	Reference
22 Dec. 1658 (1 Jan. 1659)	Wallis–Huygens 7: 26	HO 2: 296–308
6 Jan.	Pascal–Huygens 8: 45	HO 2:309–10
13 Jan.	Heuraet–Schooten 8: 36, 48	Descartes (1659): 517–20
14 Jan.	Huygens–Sluse 8: 46	HO 2: 312–13
16 Jan.	Huygens–Carcavy 8: 47, 49	HO 2: 315–17
31 Jan.	Huygens–Wallis 7: 29	HO 2: 329–31
31 Jan.	Mylon–Huygens 8: 43, 50, 54	HO 2: 332–5
7 Feb.	Huygens–Schooten 8: 49	HO 2: 343–5
	Carcavy–Huygens 8: 55	HO 2: 346
13 Feb.	Schooten–Huygens 8: 48, 51	HO 2: 352–4
18 (28) Feb.	Wallis–Huygens 7: 28	HO 2: 357–60
8 May	Huygens–Boulliau 8: 56	HO 2: 402

1668		
Mar.	Collins– Gregory **20**: 18	*GT*: 45–8
26 Mar.	Gregory– Collins **6**: 29, **7**: 66, **20**: 19	*GT*: 49–51
13 (23) Apr.	Brouncker **7**: 67	*PT* **3** No. 34: 645–9
	Wallis **11**: 51	654–5
2 July	Huygens for Gallois **7**: 36	*JS* No. 5: 361–8
2 (12) July	Wallis for Oldenburg **11**: 51	*OC* **4**: 489– 92
8 (18) July	Wallis– Brouncker **5**: 47	*PT* **3** No. 38: 753–6
14 (24) July	Collins– Wallis **10**: 46	(*CR* **II**: 490)
Summer	Collins– Barrow **10**: 46	(*LBG*: 114)
5 (15) Aug.	Wallis– Brouncker **5**: 47	*PT* **3** No. 38: 756–9
17 (27) Aug.	Wallis **11**: 53	*PT* **3** No. 38: 744–7
	Wallis **11**: 51	748–50
	Wallis– Brouncker **5**: 47	753–6
	Wallis– Brouncker **5**: 47	756–9
	Mercator **3**: 68	759–64
21 Sept. (1 Oct.)	Wallis **11**: 53	*PT* **3** No. 39: 775–9
26 Sept. (6 Oct.)	Wallis– Collins **20**: 61	*CR* **II**: 500–5
16 (26) Nov.	Wallis **11**: 53	*PT* **3** No. 41: 825–38
15 (25) Dec.	Gregory– Oldenburg **10**: 41	*HO* **6**: 306– 11

1668 (*cont.*)		
late 1668	Huygens v. Gregory **7**: 38	*HO* **6**: 321–3
1669		
30 Dec. 1668 (9 Jan. 1669)	Collins– Gregory **7**: 70, **10**: 54	*GT*: 54–8
mid-Jan.	Collins– Moray **6**: 16, **9**: 55, **10**: 56	*HO* **6**: 372–6
7 (17) Jan.	Collins– Gregory **7**: 70, **10**: 54	*GT*: 59–60
19 (29) Jan.	Wallis– Collins **20**: 61	*CR* **II**: 507– 10
Jan./Feb.	Wallis– Collins **20**: 61	*CR* **II**: 510– 12
2 (12) Feb.	Collins– Gregory **7**: 70, **10**: 54	*GT*: 65–7
15 (25) Feb.	Gregory– Collins **12**: 33, **15**: 35	*GT*: 68–70
15 (25) Feb.	Gregory– Oldenburg **10**: 41	*PT* **3** No. 44: 882–6
15 (25) Mar.	Collins– Gregory **7**: 70, **10**: 54	*GT*: 70–2
Spring	Wallis– Collins **10**: 68	*CR* **II**: 601–4
25 Mar. (4 Apr.)	Review of Sluse (1668) **14**: 13	*PT* **4** No. 45: 903–9
12 (22) Apr.	Collins **4**: 14	*PT* **4** No. 46: 929–34
20 (30) July	Collins– Barrow **10**: 47	*NC* **1**: 13– 14
31 July (10 Aug.)	Barrow– Collins **10**: 48	*NC* **1**: 14

1670 (*cont.*)		
19 (29) Dec.	Gregory–Collins	*GT*: 148–50
	4: 56, 7: 75, 9: 55, 10: 51, 55, 15: 52, 20: 34, 39	
1671		
24 Dec. 1670 (3 Jan. 1671)	Collins–Gregory 10: 49, 61, 11: 5, 50, 16: 12, 18	*GT*: 153–9
15 (25) Feb.	Gregory–Collins 7: 75, 95, 10: 55, 58, 60, 11: 55, 15: 50, 16: 13, 15, 18: 8, 19: 47, 20: 12, 29, 36, 38, 45, 53	*GT*: 168–72
21 Feb. (3 Mar.)	Collins–Bertet 4: 54	*NP* 3: 21–2; *CE* (*1712*): 26–7
9 Mar.	Sluse–Oldenburg 6: 41	*OC* 7: 477–81
11 Mar.	Leibniz–Oldenburg 1: 40, 3: 4	*LSB* II. 1: 88–91
23 Mar. (2 Apr.)	RS meeting 3: 4	*BH* II: 475
25 Mar. (4 Apr.)	Collins–Gregory 10: 64, 11: 67	*GT*: 178–81
7 (17) Apr.	Wallis–Oldenburg 3: 4–5	*PT* 6 No. 74 (for 14 (24) Aug.): 2227–31
20 (30) Apr.	RS meeting 3: 4	*BH* II: 477
early May	Leibniz–Velthuysen 1: 40	*LSB* II. 1: 97–9
9 May	Leibniz–Oldenburg 3: 4	*LSB* II. 1: 102–5
4 (14) May	RS meeting 3: 4	*BH* II: 479

1671 (*cont.*)		
11 (21) May	RS meeting 3: 4	*BH* II: 481
17 (27) May	Gregory–Collins 10: 52, 55, 11: 56, 15: 56, 16: 13–14, 19: 50, 20: 7, 37, 39, 53	*GT*: 187–91
18 (28) May	RS meeting 3: 4	*BH* II: 482
25 May (4 June)	RS meeting 3: 4	*BH* II: 482
2 (12) June	Wallis–Oldenburg 3: 4–5	*PT* 6 No. 74 (for 14 (24) Aug.): 2230–1
22 June	Leibniz–Carcavy 1: 40	*LSB* II. 1: 125–9
12 (22) June	Oldenburg–Leibniz 3: 5	*LSB* II. 1: 131–4
5 (15) July	Collins–Newton 19: 13	*NC* 1: 65–6
17 (27) July	Review of Leibniz (Mainz, 1671) 3: 4	*PT* 6 No. 73: 2213–14
20 (30) July	Newton–Collins 3: 49	*NC* 1: 67–9
early Aug.	Oldenburg–Leibniz 3: 7	(*LSB* II. 1: 142)
5 (15) Aug.	Oldenburg–Leibniz 3: 7	*LSB* II. 1: 142–3
17 Aug.	Leibniz–Carcavy 1: 40	*LSB* II. 1: 143–4
14 (24) Aug.	Wallis–Oldenburg Wallis–Oldenburg 3: 4	*PT* 6 No. 74: 2227–30 2230–1
5 Oct.	Leibniz–Spinoza 8: 66, 12: 51	*LSB* II. 1: 155

317

1673		
early 1673	Huet– Leibniz **3**: 9	*LSB* II. 1: 229
26 Dec. 1672 (5 Jan. 1673)	Collins– Gregory **15**: 43	*GT*: 248–9
14 Jan.	Huygens– Oldenburg **3**: 1	*HO* **7**: 242– 4
14 Jan.	Meyersperg– Schröter **3**: 36	*C* 294
17 Jan.	Sluse– Oldenburg **3**: 13, **19**: 21	*SL*: 673–7
20 (30) Jan.	Sluse– Oldenburg **3**: 13, 41, **4**: 49, **6**, 50	*PT* **7** No. 90: 5143–7
	Notice of Kersey (1673–4) **4**: 16, **7**: 61	5152–3
Jan.	Sinold– Leibniz **5**: 4	*LSB* I. 1: 311
22 Jan. (1 Feb.)	RS meeting **3**: 10	*BH* III: 72–3
23 Jan. (2 Feb.)	Schröter– Hocher **3**: 36	*C* 313
29 Jan. (8 Feb.)	RS meeting **3**: 12	*BH* III: 73–4
29 Jan. (8 Feb.)	Oldenburg– Sluse **19**: 18	*LBG*: 232–3
30 Jan. (9 Feb.)	Oldenburg– Leibniz **3**: 14	*LBG*: 73–4
1 (11) Feb.	Schröter– Hocher **3**: 36	*C* 319
2 (12) Feb.	Leibniz meets Boyle **3**: 18, **7**: 17	(*LBG*: 74)
3 (13) Feb.	Leibniz for the RS **3**: 18, 23–4, 28	*LBG*: 74–8

1673 (*cont.*)		
mid-Feb.	Schröter– Hocher **3**: 36	*C* 334
9 (19) Feb.	Oldenburg– Huygens **3**: 40, **6**: 60	*HO* **7**: 256–7
9 (19) Feb.	Oldenburg– Leibniz **3**: 39	*LBG*: 74
10 (20) Feb.	Leibniz–RS **3**: 32	*LBG*: 80
10 (20) Feb.	Leibniz– Oldenburg **3**: 32	(*LBG*: 80)
early Mar.	Ozanam for Leibniz **4**: 4, **7**: 41	*C* 781
early Mar.	Ozanam for Leibniz **4**: 1	*C* 782
8 Mar.	Leibniz– Oldenburg **3**: 11, 19, 32, 42–3, 62, **4**: 1, 5–6, 51	*LBG*: 81–4
10 Mar.	Leibniz– Schönborn **3**: 63	*LSB* I. 1: 313–20
5 (15) Mar.	RS meeting **3**: 64	*BH* III: 77–8
mid-Mar.	Leibniz meets Huygens **6**: 60	(*LBG*: 92)
6 (16) Mar.	Oldenburg– Leibniz **3**: 43	(*LBG*: 84)
14 (24) Mar.	Briegel– Leibniz **3**: 19	*LSB* I. 1: 326–8
15 (25) Mar.	Habbeus– Leibniz **10**: 3	(*LSB* I. 1: 415)
26 Mar.	Leibniz– Johann Friedrich **3**: 63	*LSB* I. 1: 487–90
Mar.	Leibniz for Ph. W. Boineburg **5**: 4	*LSB* I. 1: 332–3

320

1673 (cont.)

23 June (3 July)	Sluse–Oldenburg 19: 19	*PT* 8 No. 95: 6059
	Review of Kersey (1673) 7: 61	6073–4
23 June (3 July)	Newton–Oldenburg 8: 67, 19: 24	*NC* 1: 291–4
23 June (3 July)	Wallis–Oldenburg 8: 39	*HO* 7: 324–5 (extract)
27 June (7 July)	Oldenburg–Huygens 8: 39	*HO* 7: 324–5
10 July	Huygens–Oldenburg 8: 68	*HO* 7: 336–8
10 July	Huygens–Wallis 8: 64, 68	*HO* 7: 339–40
10 (20) July	Oldenburg–Sluse 19: 24	*CE* (*1712*): 31–2
July	Schönborn–Leibniz 3: 37	*LSB* I. 1: 360–1
5 Aug.	Sluse–Oldenburg 19: 25	*SL*: 679
4 (14) Aug.	Oldenburg–Huygens 8: 69	*HO* 7: 353–4
Summer	Oldenburg–Sluse 18: 28	*SL*: 683
Autumn	Leibniz–Münch 5: 6	*LSB* I. 1: 369–73
4 (14) Oct.	Wallis–Oldenburg 8: 39	*PT* 8 No. 98 (for 17 (27) Nov.): 6146–9
8 (18) Oct.	Brouncker–Oldenburg 8: 41	*PT* 8 No. 98 (for 17 (27) Nov.): 6149–50

1673 (cont.)

Oct.	Wren–Oldenburg 8: 41	*PT* 8 No. 98 (for 17 (27) Nov.): 6150
3 (13) Nov.	Oldenburg–Huygens 8: 70	*HO* 7: 360
22 Nov.	Sluse–Oldenburg 18: 28	*SL*: 680–3
17 (27) Nov.	Wallis–Oldenburg 8: 39	*PT* 8 No. 98: 6146–9
	Brouncker–Oldenburg 8: 40	6149–50
	Wren–Oldenburg 8: 41	6150
8 (18) Dec.	Oldenburg–Huygens 8: 70	*HO* 7: 364–5

1674

Feb.	Leibniz–Lincker 10: 2	*LSB* I. 1: 387–94
2 (12) Mar.	Oldenburg–Huygens 8: 70	*HO* 7: 379
30 Mar. (9 Apr.)	Oldenburg–Huygens 8: 70	*HO* 7: 380
2 (12) Apr.	Gregory–Collins 20: 51	*GT*: 278–9
15 May	Huygens–Oldenburg 8: 71	*HO* 7: 382–3
early June	É. Périer–Leibniz 7: 2	*C* 1351
25 May (4 June)	Oldenburg–Huygens 8: 72	*HO* 7: 385–6
20 June	Toinard–Leibniz 6: 4	*C* 683
7 July	Hautefeuille for the AS 9: 14	*HO* 7: 458–60

1675 (*cont.*)		
mid-Mar.	Leibniz–Oldenburg **6**: 6, 15	*LBG*: 112–13
mid-Mar.	Leibniz for La Roque **9**: 21	*C* 913 A–D
25 Mar.	Leibniz **9**: 23, 25	*JS* No. 7: 96–101
29 Mar.	Constantijn Huygens (father)–Oldenburg **9**: 50	*HO* **7**: 431–2
30 Mar.	Leibniz–Oldenburg **3**: 58, **5**: 29, **6**: 6, 8, 15, **9**: 25	*LBG*: 110–12
late Mar.	Leibniz–Huet? **9**: 24	LBr 265, fol. 2
late Mar.	Leibniz for La Roque **9**: 22	*C* 913 G
Spring	Leibniz for Huygens **17**: 32	*C* 911
Spring	Malebranche–Vaughan **10**: 66	(*LBG*: 116)
25 Mar. (4 Apr.)	Huygens **9**: 19	*PT* **10** No. 112: 272–3
27 Mar. (6 Apr.)	Oldenburg–Huygens **9**: 29	*HO* **7**: 433
28 Mar. (7 Apr.)	Gregory–Collins **11**: 78	(*GT*: 31)
10 (20) Apr.	Collins–Oldenburg for Leibniz **10**: 36	*C* 926
12 (22) Apr.	Oldenburg–Leibniz **4**: 45–6, **10**: 37, **11**: 33, 64, **12**: 78, 96, **15**: 4, 50, **18**: 29, **19**: 50, **20**: 36, 52	*LBG*: 113–22

1675 (*cont.*)		
24 Apr.	AS meeting **9**: 26	*C* 930
15 (25) Apr.	RS meeting **9**: 25	*BH* III: 216–17
'15 (25) Apr.'	Oldenburg–Leibniz	*CE* (*1712*): 39–41
12 (22) Apr.	**10**: 38	
19 (29) Apr.	Oldenburg–Huygens **9**: 31	*HO* **7**: 454–6
5 May	Huygens–Contesse **9**: 15	*HO* **7**: 457–8
26 Apr. (6 May)	Leibniz **9**: 23	*PT* **10** No. 113: 285–8
1 (11) May	Collins–Gregory **9**: 30, 58, **10**: 69, **11**: 68	*GT*: 298–301
5 (15) May	Oldenburg–Huygens **9**: 31	*HO* **7**: 462–3
8 (18) May	Newton–John Smith **10**: 70	*NC* **1**: 342–4
20 May	Leibniz–Oldenburg **9**: 27, **10**: 87, **11**: 18, 25, **12**: 77, 93, **15**: 7	*LBG*: 122–4
10 (20) May	Oldenburg–Huygens **9**: 31	*HO* **7**: 463–4
late May	Leibniz–Linsingen **10**: 3, **11**: 91, 99	*LSB* I. 1: 498–9
May	Spinoza–Oldenburg **12**: 7	(*SO*: 201–2)
late May '1675' [1676]	Collins–Oldenburg for Tschirnhaus **12**: 31 *see* 1676	*CE* (*1712*): 43
early June	Oldenburg–Leibniz **11**: 34	*LBG*: 124–5

1675 (*cont.*)

4 June	Leibniz for the brothers Périer **12: 92**	*C* 978
26 May (5 June)	Gregory–Collins **11: 78, 18: 33, 20: 46, 50**	*GT*: 302–4
12 June	Leibniz–Oldenburg **10: 27, 35**	*LBG*: 125–7
mid-June '1675' [1676]	Tschirnhaus–Oldenburg **12: 31** *see* 1676	*CE* (*1712*): 43
7 (17) June	Oldenburg–Huygens **9: 39**	*HO* 7: 469–70
8 (18) June	Gregory–Oldenburg **20: 43**	*GT*: 306–7
8 (18) June	Oldenburg–Spinoza **12: 11**	*SO*: 201–2
21 June	Huygens–Oldenburg **9: 32**	*HO* 7: 471–2
12 (22) June	Linsingen–Leibniz **11: 99**	*LSB* i. 1: 500
15 (25) June	Collins–Oldenburg for Leibniz **11: 62**	BM MS Birch 4398: 139r–40r
June	Frazer–Gregory **7: 46**	(*GT*: 311)
Summer	Leibniz for Huygens **11: 7–8, 17, 83**	*C* 1032
22 June (1 July)	Oldenburg–Huygens **9: 39**	*HO* 7: 472–3
24 June (4 July)	Oldenburg–Leibniz **11: 33, 36, 38, 41, 58, 71, 73, 75, 80, 12: 86–7, 20: 6**	*LBG*: 127–30

1675 (*cont.*)

5 July	Huygens–Constantijn Huygens (brother) **9: 34**	*HO* 7: 474–5
6 July	AS meeting **17: 86**	(*HO* **19**: 182)
28 June (8 July)	Oldenburgh–Huygens **9: 28**	*HO* 7: 475–6
29 June (9 July)	Collins–Gregory **9: 58, 10: 69, 11: 69**	*GT*: 308–11
30 June (10 July)	Smethwick–Huygens **12: 39, 41**	*HO* 7: 487–8
11 July	Huygens–Oldenburg **9: 40**	*HO* 7: 477–8
12 July	Leibniz–Oldenburg **10: 27, 11: 7–9, 24, 37, 39, 59, 76, 81, 12: 88, 97, 13: 1, 14: 32, 20: 6**	*LBG*: 131–2
mid-July	Oldenburg–Wallis **12: 8**	(*WO* ii: 471)
6 (16) July	Tschirnhaus for Oldenburg **12: 60**	Amsterdam, Wisk. Gen. 49a
13 (23) July	Gregory–Frazer **7: 47**	*GT*: 311–12
15 (25) July	Oldenburg–Huygens **9: 28, 33, 38**	*HO* 7: 481–2
25 July	Schuller–Spinoza **12: 11**	*SO*: 203–4
early Aug.	Tschirnhaus–Gent **12: 15–6, 23, 25, 29, 58**	Amsterdam, Wisk. Gen. 49d
22 July (1 Aug.)	Oldenburg–Huygens **9: 38**	*HO* 7: 482–3

324

1675 (cont.)		
23 July (2 Aug.)	Gregory–Collins 20: 23, 33	GT: 312–14
24 July (3 Aug.)	Newton–John Smith 10: 70	NC 1: 348–9
28 July (7 Aug.)	Strode–Collins 18: 31	CR II: 452–3
9 Aug.	Huygens–Constantijn Huygens (brother) 9: 11	HO 7: 483–5
30 July (9 Aug.)	Oldenburg–Huygens 9: 38, 12: 39–40	HO 7: 486
30 July (9 Aug.)	Oldenburg–Leibniz 12: 42	(LBG: 143)
10 Aug.	Huygens–Oldenburg 9: 36	HO 7: 488–90
10 Aug.	Papin–Huygens 9: 28, 37, 12: 39, 41	HO 7: 490–1
12 Aug.	Tschirnhaus–Spinoza 12: 39	SO: 206–7
3 (13) Aug.	Collins–Gregory 12: 6, 13, 21, 36, 14: 38	GT: 314–17
3 (13) Aug.	Collins for Gregory 12: 18, 18: 13	GT: 317–19
10 (20) Aug.	Collins–Gregory 12: 12, 23, 25, 39	GT: 320–2
10 (20) Aug.	Frazer–Gregory 7: 47	GT: 323
28 Aug.	de Nyert–Huygens 9: 35	HO 7: 493

1675 (cont.)		
20 (30) Aug.	Gregory–Collins 11: 78, 12: 17, 19, 22, 24, 26, 34, 14: 26, 15: 57, 20: 46–7	GT: 324–6
4 Sept.	Gallois–Leibniz 11: 93	C 1049
27 Aug. (6 Sept.)	Newton–John Smith 10: 70	NC 1: 350–1
10 Sept.	Thuret–Huygens 9: 12	HO 7: 498
4 (14) Sept.	Collins–Gregory 12: 28, 39	GT: 327–8
mid-Sept.	Leibniz–Huygens 10: 27, 11: 7, 22, 24, 84	HO 7: 500–4
11 (21) Sept.	Gregory–Collins 11: 78, 15: 44, 58, 20: 48	GT: 328–9
13 (23) Sept.	Oldenburg–Huygens 9: 28, 38, 41	HO 7: 499
16 (26) Sept.	Mohr–Collins 11: 48	C 1066A
30 Sept.	Huygens–Leibniz 10: 27, 11: 22, 87	HO 7: 504–6
21 Sept. (1 Oct.)	Collins–Gregory 11: 50, 12: 10, 27, 15: 44	GT: 330–3
3 Oct.	Gallois–Leibniz 11: 93	C 1057
3 Oct.	Huygens–Leibniz 11: 88, 12: 44	C 1104
24 Sept. (4 Oct.)	Oldenburg–Huygens 9: 28, 38	HO 7: 506–7

1675 (cont.)		
28 Dec.	Leibniz– Oldenburg 10: 27, 11: 19, 49, 72, 79, 89, 96, 12: 47, 90, 97, 104, 13: 34, 15: 2	LBG: 143–7
20 (30) Dec.	Oldenburg– Leibniz 12: 109	LBG: 143
Dec.	Leibniz– Kahm 11: 99	LSB I. 1: 503–4
late 1675	Damian v.d. Leyen– Leibniz 11: 98	C 1276
late 1675	Leibniz– La Roque 5: 38, 13: 26, 43	C 1228
late 1675	Leibniz for Gallois (?) 13: 45	C 1227, 1230–1
1675	Dary–Collins 20: 62	CR I: 220–1

1676

11 Jan.	Leibniz– Colbert 11: 97	LSB I. 1: 500
11 Jan.	Leibniz– Johann Friedrich 11: 99	LSB I. 1: 504–5
11 Jan.	Leibniz– Kahm 11: 99	LSB I. 1: 505–6
11 Jan.	AS meeting 17: 87	(HO 19: 185)
mid-Jan.	Leibniz– Johann Friedrich 11: 99, 12: 110	LSB I. 1: 506–7
mid-Jan.	Leibniz– Kahm 11: 99, 12: 110	LSB I. 1: 507–8

1676 (cont.)		
18 Jan.	Leibniz– Damian v.d. Leyen 11: 100	LSB I. 1: 398–9
17 (27) Jan.	Kahm– Leibniz 11: 99	LSB I. 1: 508
1 Feb.	AS meeting 17: 87	(HO 19: 185)
8 Feb.	AS meeting 17: 87	(HO 19: 185)
9 Feb.	Bertet for Leibniz 13: 47	C 1113, 1304
11 Feb.	Schönborn– Leibniz 5: 7, 11: 101	LSB I. 1: 400–1
14 Feb.	Leibniz– Habbeus 12: 49	LSB I. 1: 444–6
14 Feb.	Leibniz– Kahm 12: 49	LSB I. 1: 509–10
15 Feb.	AS meeting 17: 87	(HO 19: 185)
22 Feb.	AS meeting 17: 87	(HO 19 185)
18 (28) Feb.	Kahm– Leibniz 11: 102	LSB I. 1: 510–11
Feb.	Mariotte for Leibniz 13: 62	C 1306
7 Mar.	AS meeting 17: 87	(HO 19: 185)
9 (19) Mar.	Kahm– Leibniz 11: 102	LSB I. 1: 513
22 Mar.	Leibniz– Habbeus 12: 49	LSB I. 1: 447–8
2 (12) Apr.	Kahm– Leibniz 11: 105, 15: 10	LSB I. 1: 515
late Apr.	Walter– Leibniz 12: 3	C 1410
27 Apr. (7 May)	RS meeting 15: 19	BH III: 313– 14

1676 (*cont.*)

26 July (5 Aug.)	Oldenburg's postscript to Newton's 'Epistola prior' **15**: 33, **16**: 1	*LBG*: 192
mid-Aug.	Collins–Baker **16**: 21	(*CR* II: 4)
11 (21) Aug.	Collins–David Gregory **15**: 46	*GT*: 344–5
24 Aug.	drafts for Leibniz–Oldenburg, 27 Aug. **17**: 1, 2, 29	*C* 1505 A 1505 B
27 Aug.	Leibniz–Oldenburg **6**: 6, **10**: 90, **11**: 17, 22, **15**: 37, **17**: 2, **19**: 1, 4, 8, 52, 54	*LBG*: 193–200
19 (29) Aug.	Collins–Baker **16**: 21, **18**: 35, **20**: 60	*CR* II: 4–10
30 Aug.	Leibniz–É. Périer **12**: 101, **20**: 72	*LBG*: 133–5
late Aug.	Leibniz' annotation to Newton's 'Epistola prior' **17**: 16, 29	*LBG*: 180
Aug.	Leibniz–Mariotte **17**: 39	*LSB* II. 1: 269–71
22 Aug. (1 Sept.)	Tschirnhaus–Oldenburg **13**: 27, **17**: 3, 67, **19**: 1, 4	RS MS LXXXI, 33
31 Aug. (10 Sept.)	Collins–Newton **18**: 37	*NC* 2: 88–90
1 (11) Sept.	Wallis–Collins **20**: 60	*CR* II: 589–90

1676 (*cont.*)

13 Sept.	Brosseau–Leibniz **11**: 110	*LSB* I. 1: 516–17
5 (15) Sept.	Newton–Collins **15**: 32, **17**: 36	*NC* 2: 95–6
mid-Sept.	Collins–Newton **17**: 2, **18**: 17, **19**: 8	(*NC* 2: 99)
9 (19) Sept.	Collins–Newton **18**: 17	*NC* 2: 99–100
9 (19) Sept.	Collins–Wallis **18**: 3, 18	(*CR* II: 591)
11 (21) Sept.	Wallis–Collins **12**: 9, **18**: 4, 18–20, 38	*CR* II: 591–7
22 Sept.	Walter–Leibniz **12**: 3	*C* 1525
14 (24) Sept.	Collins–Wallis **11**: 22, **17**: 2	(*CR* II: 598)
26 Sept.	Brosseau–Leibniz **11**: 111	*LSB* I. 1: 517
16 (26) Sept.	Wallis–Collins **11**: 22, **18**: 19, 26	*CR* II: 598–600
21 Sept. (1 Oct.)	Collins–Baker **16**: 21, **17**: 2, **19**: 1	(*CR* II: 10)
30 Sept. (10 Oct.)	Collins–Oldenburg for Tschirnhaus **18**: 20, 23	*CR* I: 211–20
10 (20) Oct.	Oldenburg–Tschirnhaus **18**: 24	LBr 695, 66r
18–29 Oct.	Leibniz in London **11**: 7, 17, **20**: 1	*LSB* I. 2: 3

1677 (cont.)			1678 (cont.)		
Apr.–	Tschirnhaus–	LBG: 337–8	28 Feb.	Leibniz	JS No. 7:
May	Leibniz			12: 80	78–9
	12: 126		Feb.	Leibniz–	(LBG: 370)
2 (12)	Oldenburg–	LBG: 238–9		Tschirnhaus	
May	Leibniz			12: 119	
	15: 21, 19: 8,				
	20: 71		early	Leibniz–	LBG: 352–4
'11 June'	Leibniz–	NC 2: 212	Mar.	Tschirnhaus	
[21 June	Oldenburg			5: 26, 12: 117,	
(1 July)]	19: 73			127	
			10 Apr.	Tschirnhaus–	LBG: 354–
21 June	Leibniz–	NC 2: 212–		Leibniz	60; 370–1
(1 July)	Oldenburg	19		6: 61, 12: 117,	
	6: 55, 13: 5,			119, 125, 128	
	17: 4, 19: 8,		10 Apr.	Tschirnhaus	LBG: 360–70
	66, 73, 77			for Leibniz	
12 (22)	Leibniz–	NC 2: 231–2		12: 114	
July	Oldenburg		mid-May	draft for	LBG: 520–4
	13: 5, 19: 74			Leibniz–	
12 (22)	Oldenburg–	LBG: 249–52		Tschirnhaus,	
July	Leibniz			late May	
	20: 71			12: 113, 115,	
9 (19)	Oldenburg–	LBG: 253–5		120, 17: 37	
Aug.	Leibniz		23 May	Leibniz	JS No. 18:
	19: 75			5: 26	119–20
20 (30)	Collins–	(CR II: 608)	late May	Leibniz–	LBG: 372–
Sept.	Wallis			Tschirnhaus	82
	14: 28			5: 26, 12: 118,	
Sept.	Leibniz–	LSB II. 1:		120, 122, 124,	
	Gallois	385–9		129, 17: 55	
	17: 78		**1679**		
8 (18)	Wallis–	CR II: 608–9	Feb.	Leibniz–	LSB I. 2:
Oct.	Collins			Johann	120–6
	14: 29			Friedrich	
late Nov.	Tschirnhaus–	LBG: 399–		11: 103	
	Leibniz	401	Mar.	Tschirnhaus–	LBG: 382–
	12: 123			Leibniz	98
1677	Collins–	NC 2: 241–4		6: 61, 12: 38,	
	Wallis			118, 121	
	19: 8		early	Leibniz–	LSB II. 1:
1677	Leibniz–	LSB II. 1:	July	Placcius	421–2
	Bertet	382–5		17: 78	
	13: 5		4 (14)	Leibniz–	LSB II. 1:
1678			Aug.	Malebranche	482–4
27 Jan.	Tschirnhaus–	LBG: 339–52		11: 7	
	Leibniz		8 (18)	Leibniz–	HO 8: 214–
	10: 32, 12: 85,		Sept.	Huygens	19
	128, 18: 44			12: 73, 13: 58,	
				17: 78	

1691 (*cont.*)

31 Dec.	Foucher–Leibniz 17: 78	*LPG* i: 400–2

1692

Jan.	Leibniz–Foucher 6: 38, 10: 82	*LPG* i: 402–6
26 July	L'Hospital–Huygens 20: 39	*HO* **10**: 304–5
27 Aug. (6 Sept.)	Newton–Wallis 19: 20	*WO* ii: 391–6
25 Sept. (5 Oct.)	Leibniz–Bodenhausen 1: 29	*LMG* vii: 374–5
23 Nov.	L'Hospital–Huygens 20: 39	*HO* **10**: 342–8
14 Dec.	L'Hospital–Leibniz 20: 39	*LMG* ii: 216–18

1693

Feb.	Huygens–Basnage 19: 63, 20: 39	*HO* **10**: 407–17
Dec. 1692–Feb. 1693	Huygens–Basnage 19: 63, 20: 39	*Histoire des ouvrages des sçavans*: 244–57
Summer	Leibniz–Foucher 3: 8	*LPG* i: 415–16
Sept.	Leibniz 19: 63	*AE*: 385–92
16 (26) Oct.	Newton–Leibniz 13: 5	*NC* **3**: 285–6

1694

29 Dec. 1693 (8 Jan. 1694)	Leibniz–Tschirnhaus 10: 27	*LBG*: 483–4

1694 (*cont.*)

21 (31) Mar.	Leibniz–Johann Bernoulli 1: 44	*LMG* iii: 135–7
21 (31) Mar.	Leibniz–Tschirnhaus 10: 27	*LBG*: 491–5
7 (17) June	Leibniz–Johann Bernoulli 13: 5	*LMG* iii: 141–3
12 (22) June	Leibniz–Huygens 1: 44	*HO* **10**: 639–46
6 (16) Aug.	Leibniz–L'Hospital 1: 44	(*LMG* ii: 252)
2 (12) Sept.	Johann Bernoulli–Leibniz 13: 5	*LMG* iii: 143–52
Oct.	Johann Bernoulli 1: 44	*AE*: 393–8
6 (16) Dec.	Leibniz–Johann Bernoulli 2: 16	*LMG* iii: 152–7

1695

27 Dec. 1694 (6 Jan. 1695)	Leibniz–L'Hospital 5: 12, 26	*LMG* ii: 255–62
14 (24) Mar.	Leibniz–Bodenhausen 11: 8	*LMG* vii: 379–80
Mar.	Leibniz–L'Hospital 12: 83	*LMG* ii: 274–7
25 Apr.	L'Hospital–Leibniz 12: 83	*LMG* ii: 277–81
May	Newton–Wallis 16: 1, 19: 8, 61	(*NC* **4**: 129)

1698 (cont.)

24 Mar. (3 Apr.)	Leibniz– Wallis 17: 2	LMG iv: 44–5
15 (25) May	Leibniz– Johann Bernoulli 11: 103	LMG iii: 487–9
May	De Moivre 11: 17, 16: 5	PT 20 No. 240: 190–3
22 July (1 Aug.)	Wallis– Leibniz 17: 39	LMG iv: 45–51
29 July (8 Aug.)	Leibniz– Johann Bernoulli 17: 78	LMG iii: 521–7
16 (26) Aug.	Johann Bernoulli– Leibniz 17: 78	LMG iii: 528–33
22 Aug. (1 Sept.)	Leibniz– Johann Bernoulli 17: 78	LMG iii: 534–8
29 Dec. 1698 (8 Jan. 1699)	Leibniz– Wallis 17: 37	LMG iv: 52–6

1699

16 (26) Jan.	Wallis– Leibniz 17: 37	LMG iv: 56–61
30 Mar. (9 Apr.)	Leibniz– Wallis 17: 37	LMG iv: 62–5
20 (30) Oct.	Leibniz– Magliabecchi 11: 8	LMG vii: 303–16

1700

20 (31) Jan.	Leibniz– Römer 17: 88	LD iv. 2: 115–17
14 May	Leibniz– Pinsson 12: 103	LD v: 469– 70

1700 (cont.)

May	Review of WO iii 17: 56, 65 Leibniz 11: 17, 17: 91	AE: 193–8 198–208

1701

26 Sept.	Leibniz– L'Hospital 3: 16	LMG ii: 341–3

1702

2 Feb.	Leibniz– Varignon 11: 22	LMG iv: 91–5
15 Nov.	Jakob Bernoulli– Leibniz 3: 61	LMG iii: 62–4

1703

Apr.	Leibniz– Jakob Bernoulli 3: 16, 6, 61 (draft) Postscript 1: 2, 26, 29, 30, 41, 2: 19, 5: 11, 16, 6: 61	LMG iii: 66–71 71–3
3 July	Leibniz– Römer 17: 90	LD iv. 2· 123–4

1704

(1702) 1704	Jakob Bernoulli 16: 5	Mémoires de l'AS 4: 281–8

1705

Jan.	Review of Newton (1704) 5: 26	AE: 30
28 July	Leibniz– Johann Bernoulli 5: 16, 13: 19	LMG iii: 770–2
17 Sept.	Leibniz– Hermann 12: 4	LMG iv: 398–402

2. 'CATALOGUE CRITIQUE'

279	c. May 1673	3: 59	724	summer 1674	7: 52
294	14 Jan. 1675	3: 36	727	January 1676	13: 60
313	23 Jan. (2 Feb.) 1673	3: 36	742	10 Sept. 1674	7: 55
			743	September 1674	7: 55
319	1 (11) Feb. 1673	3: 36	744	September 1674	7: 55
326₁	16 Feb. 1673	3: 38	745	late summer 1674	11: 8
334	mid-February 1673	3: 36	746	late summer 1674	11: 8
339	mid-February 1673	6: 17	747B	mid-September 1674	11: 9
366	mid-April 1673	3: 19	757	September 1674	7: 55
409A	6 (16) April 1673	3: 46, 4: 10, 7: 61, 11: 33	758	September 1674	7: 55
			759	September 1674	7: 55
			762	early September 1674	12: 4
500	early 1674	5: 26, 6: 46	764	September 1674	11: 27
503	early 1674	6: 5	769	autumn 1674	5: 19
504	early 1674	6: 5	772	September 1674	10: 26
505	early 1674	6: 5	773	3 Oct. 1674	5: 16, 10: 25
510B	1672/3	2: 20			
515	winter 1672–3	3: 27	773₁	October 1674	7: 14, 13: 44
541	spring 1673	5: 36			
544	spring 1673	5: 36	774₂	20 Oct. 1674	12: 4
545	spring 1673	5: 36, 45	775	October 1674	10: 20
546	spring 1673	5: 36	779	October 1674	17: 78
547	summer 1673	13: 10	781	early March 1673	4: 4
548	spring 1673	5: 36	782	March 1673	4: 1
549	spring 1673	5: 36	786	October 1674	6: 36
551	May 1673	13: 10	787	October 1674	10: 28
560	summer 1673	5: 48, 13: 26	789	October 1674	11: 3
			790	October 1674	6: 36, 11: 3–4
561	summer 1673	5: 48, 13: 26			
562	summer 1673	10: 23	791	October 1674	10: 24
564	May 1673	13: 10	794	October 1674	2: 20
575	August 1673	5: 36, 54, 13: 32	796	autumn 1674	7: 53
			797	October 1674	7: 14, 39, 13: 44
612	early 1674	5: 26			
616	spring 1673	6: 50	815	December 1674	10: 27
617	May 1673	5: 36	816	December 1674	10: 27
635	early 1674	5: 26, 28	820	December 1674	2: 20, 13: 20
683	20 June 1674	7: 4			
687	15 July 1674	7: 5, 56	821	December 1674	11: 1
691	summer 1674	7: 13	823	December 1674	13: 26
696	summer 1673	5: 36	826	December 1674	10: 22
697	summer 1673	5: 36	827	December 1674	10: 20, 20: 65
703A	17 Feb. 1674	7: 49			
704	spring 1674	7: 41	828	24 Dec. 1674	10: 22
705	spring 1674	7: 42	829	December 1674	10: 22
706	spring 1674	7: 43	830	December 1674	10: 22
712	summer 1674	7: 52	831	December 1674	13: 8
714	summer 1674	7: 52	832	December 1674	10: 21, 17: 52
715	summer 1674	7: 52			
719	spring 1674	7: 43	833	December 1674	10: 21, 17: 52
721	spring 1674	7: 43			
723	spring 1674	7: 43	834	December 1674	5: 20, 11: 2

1061	*10 Oct. 1675*	**11**: 27		1141	autumn 1675	**17**: 59
1063	*10 Oct. 1675*	**13**: 38		1142	autumn 1675	**17**: 59
1066A	*30 Sept. (10 Oct.)*	**11**: 48		1144	*8 Dec. 1675*	**12**: 66, **17**:
	1675					80, **20**: 68
1067	*30 Sept. (10 Oct.)*	**11**: 44		1150	*December 1675*	**13**: 58
	1675			1151	December 1675	**13**: 58
1071	*15 Oct. 1675*	**13**: 8		1157	*14 Dec. 1675*	**17**: 81
1072	October 1675	**13**: 8		1161	*17 Dec. 1675*	**12**: 69
1073	October 1675	**13**: 8		1164	*21 Dec. 1675*	**17**: 70
1074	mid-October 1675	**13**: 22		1165	*22 Dec. 1675*	**13**: 3, 15, 35
1076	*19 Oct. 1675*	**12**: 58, **13**: 6		1166	22(?) Dec. 1675	**13**: 3, 15, 35
1078	October 1675	**13**: 9		1167	22(?) Dec. 1675	**13**: 15
1079	*20 Oct. 1675*	**13**: 9		1168	December 1675	**5**: 19
1085	*24 Oct. 1675*	**13**: 3		1170A	*December 1675*	**11**: 72
1086	*24 Oct. 1675*	**13**: 11		1170B	late 1675	**11**: 72
1087	*24 Oct. 1675*	**2**: 20, **13**: 7		1172	December 1675	**12**: 106
1089	*25 Oct. 1675*	**13**: 3, 12		1173	December 1675	**12**: 107
1090	*25 Oct. 1675*	**12**: 126, **13**:		1175	December 1675	**12**: 107
		13, 21		1176	28 Dec. 1675	**17**: 82
1091	*26 Oct. 1675*	**13**: 3		1181	*December 1675*	**12**: 81
1092	*29 Oct. 1675*	**13**: 24, 28		1187	*December 1675*	**17**: 89
1095	*30 Oct. 1675*	**17**: 68		1188	December 1675	**17**: 89
1097	31 Oct. 1675	**13**: 9		1189	late autumn 1675	**17**: 63
1098	31 Oct. 1675	**13**: 9		1197	December 1675	**17**: 83
1102	early 1675	**6**: 8		1198	December 1675	**12**: 70
1103	*October 1675*	**17**: 80		1199	December 1675	**12**: 70
1104	3(?) Oct. 1675	**11**: 88, **12**:		1200	December 1675	**17**: 72
		44		1201	April 1676	**13**: 63–4
1104₂	October 1675	**11**: 16		1202	December 1675	**12**: 70
1106	*29 Oct. 1675*	**11**: 94		1208	1675	**12**: 58, **17**:
1106₁	*1 Nov. 1675*	**6**: 61, **13**:				81
		14–15		1209	1675	**6**: 50
1107	*2 Nov. 1675*	**11**: 72, **13**:		1210	1675	**12**: 83
		33		1211	April 1676	**13**: 64
1110	*3 Nov. 1675*	**13**: 40		1216	1675	**12**: 108
1111	3(?) Nov. 1675	**13**: 39		1217	1675	**12**: 108
1112	early November	**13**: 40–1		1218	1675	**12**: 108
	1675			1227	late 1675	**13**: 45
1113	9(?) Feb. 1676	**13**: 47		1228	late 1675	**5**: 38, **13**:
1119	November 1675	**15**: 13				26, 43
1120	*11 Nov. 1675*	**6**: 55, 61,		1230	late 1675	**13**: 45
		13: 31		1231	late 1675	**13**: 45
1125	*22 Nov. 1675*	**5**: 21, **17**:		1232	autumn 1676	**12**: 85
		81		1233	autumn 1676	**6**: 46, **12**:
1127	November 1675	**12**: 55, 58				85, **17**: 20,
1131	*27 Nov. 1675*	**12**: 63, **13**:				29, 78
		35, **17**: 81		1237	early 1676	**13**: 26
1132	November 1675	**12**: 61		1238	early 1676	**13**: 26
1133	*28 Nov. 1675*	**17**: 59		1242	early 1676	**13**: 26, **17**:
1134	*29 Nov. 1675*	**17**: 36				82
1135	November 1675	**12**: 64		1244	early 1676	**13**: 26, **17**:
1136	November 1675	**12**: 64				82
1137	November 1675	**12**: 64		1245	early 1676	**17**: 69
1138	*November 1675*	**11**: 17, **12**:		1246	early 1676	**17**: 69
		75, **19**: 68		1247	early 1676	**17**: 69
1139	November 1675	**6**: 46		1253	*3 Jan. 1676*	**13**: 60

3. OTHER MANUSCRIPTS

Copenhagen
Kongelige Bibliotek (Royal Library)
MS Boll. Ud 1 **12**: 102, **20**: 72
Hanover
Niedersächsische Landesbibliothek
(Lower Saxony State Library)
LBr 265 **9**: 24
LBr 695 **18**: 24, **20**: 9
LBr 870 **3**: 19
LBr 943 (Tschirnhaus)
LH 35 III A32 ('Accessio')
LH 35 III B8 **13**: 57
LH 35 IV 17 **12**: 80
LH 35 VIII 19 **20**: 5
LH 35 VIII 23 **20**: 9
LH 35 XV 4 **20**: 58
London
British Museum
MS Birch 4398 **11**: 62
Royal Society
Letter Book VI 86 **3**: 44
Register Book IV 148 (**9**: 54)
Register Book IV 197 **3**: 65
MS LXXXI 25 **12**: 31, **14**: 1, 2, 23,
36, **15**: 26, **18**: 6
MS LXXXI 30 **15**: 27, 36

MS LXXXI 31 **10**: 75, **15**: 25, 61,
20: 11
MS LXXXI 33 **13**: 27, **17**: 3, 67,
19: 1, 4
Oxford
Bodleian Library
MS Savile 33   pp. *38, 166* (Merry)
MS Savile 43–4 **20**: 20
MS Savile 62 (**4**: 42)
Paris
Bibliothèque Nationale
MS fds. fr. 9119–20 **12**: 90
MS suppl. graec. 883 **3**: 9
Rome
Bibliotheca Vaticana
MS Vat. Lat. 6901, 6905
p. *205* (Ricci)
Shirburn Castle
(Macclesfield Collection)
N 4. 7.   p. *291* (Barrow)
Wroclaw
Biblioteka Uniwersytecka (Breslau
Univ. Library)
Akc. 1948/562
p. *166* (Tschirnhaus)

4. MEETINGS OF LEARNED SOCIETIES

Académie Royale des Sciences, Paris

9 Jan.	1675	**10**: 1
23 Jan.	1675	**9**: 6
24 Apr.	1675	**9**: 26
6 July	1675	**17**: 86
11 Jan.	1676	**17**: 87
1 Feb.	1676	**17**: 87
8 Feb.	1676	**17**: 87
15 Feb.	1676	**17**: 87
22 Feb.	1676	**17**: 87
7 Mar.	1676	**17**: 87

Royal Society, London

20 (30) Jan.	1669/70	**20**: 42
27 Jan. (6 Feb.)	1669/70	**20**: 42
23 Mar. (2 Apr.)	1670/1	**3**: 4
20 (30) Apr.	1671	**3**: 4
4 (14) May	1671	**3**: 4
11 (21) May	1671	**3**: 4

18 (28) May	1671	**3**: 4
25 May (4 June)	1671	**3**: 4
22 Jan. (1 Feb.)	1672/3	**3**: 10
29 Jan. (8 Feb.)	1672/3	**3**: 12
5 (15) Feb.	1672/3	p. 29
5 (15) Mar.	1672/3	**3**: 64
9 (19) Apr.	1673	**3**: 67
7 (17) May	1673	**3**: 65
18 (28) June	1673	**19**: 23
28 Jan. (7 Feb.)	1674/5	**9**: 17
18 (28) Feb.	1674/5	**9**: 20
15 (25) Apr.	1675	**9**: 25
16 (26) Dec.	1675	**15**: 17
27 Apr. (7 May)	1675	**15**: 19
18 (28) May	1676	**15**: 15
15 (25) June	1676	**15**: 30

Council of the RS

12 (22) Oct.	1676	**9**: 49, 52
2 (12) Nov.	1676	**9**: 49, 52
20 (30) Nov.	1676	**9**: 49, 52 (53)

5. PERIODICALS

24

INDEX OF NAMES AND WORKS

INDEX OF NAMES AND WORKS

Collins, John (1625–83) (*cont.*)

'Account concerning the resolution of equations in numbers', *PT* **4** No. 46 (of 12 (22) Apr. 1669): 929–34 *37*

see also 'Historiola', RS MS LXXXI, 31 = Collins to Oldenburg, May/June 1676; 'Abridgment', RS MS LXXXI, 30 = Collins to Oldenburg, 14 (24) June 1676

Comiers, Claude (d. 1693) *245n*

Commandino, Federigo (1509–75) *190*

Commercium epistolicum D. Joannis Collins, et aliorum de analysi promota, London 1712/13: CE (*1712*) *12, 18n, 28–9n, 42n, 51n, 48–100n, 127, 130n–3, 135–6, 140, 142–6n, 171, 191n, 195, 202n, 205, 211, 214n, 228–9n, 231–2n, 235–9n, 241, 243n, 250, 261n, 263n–4n, 265, 267–8, 274–5, 280, 305*

Second edition: CE (*1722*) *18n, 51n, 74n, 171n, 231, 264–5, 274n–5, 280*

Contesse (Paris attorney) *119n*

Conti, Antonio Schinella (1677–1749) *5n, 51n, 211n, 241n–2n, 275–6n*

Costabel, Pierre

'Traduction française de notes de Leibniz sur les Coniques de Pascal', *Revue d'Histoire des Sciences* **15** (1962): 253–68 *180n*

'Le traité de l'angle solide de Florimond de Beaune', *XIᵉ Congrès International d'Histoire des Sciences* (*Varsovie–Cracovie, 1965*), Actes III, Wroclaw 1968: 189–94 *38n*

Couturat, Louis (1868–1914)

La logique de Leibniz, Paris 1901 *7n, 13n–15n, 21n, 24n*

Opuscules et fragments inédits de Leibniz, Paris 1903 *13n–14n*

Dalencé, Joachim (d. 1707) *160n*

Dary, Michael (1613–79) *43, 124n, 149–50, 255, 284n, 290*

Miscellanies, London 1669 *43n*

Interest epitomized, both compound and simple... A short appendix for the solution of adfected equations in numbers by approachment: performed by logarithms, London 1677 *291*

Dati, Carlo (1619–76)

Lettera della vera storia della cicloide, Florence 1663 *50n*

Davenant, Edward (d. 1680) *138, 255, 259n, 289*

De Beaune, Florimond (1601–52) *37–8, 241n–2n*

'In Geometriam R. des Cartes notæ breves', *in* Descartes, *Geometria* I, ed. Schooten, Amsterdam 1659: 107–42 *94n, 245, 175n*

'De natura et constitutione æquationum', ed. E. Bartholinus, *in* Descartes, *Geometria* II, ed. Schooten, Amsterdam 1659/61: 49–116 *38*

De angulo solido = La doctrine de l'angle solide contenu sous trois angles planes (manuscript) *see* Costabel (1968) *37, 138*

Dechales, Claude François Milliet (1621–78) *178, 195*

Cursus mathematicus, Lyons 1674 *195*

Dechales' pupil *177–8*

Dehn, Max (1878–1952)

(with E. Hellinger), 'On James Gregory's Vera quadratura', *GT* (1939): 468–78 *64n*

Desargues, Girard (1591–1661) *138, 197*

Brouillon project d'une atteinte aux evenemens des rencontres d'un cone avec un plan, Paris 1639 *37*

Leçons des tenebres (manuscript, now lost) *178, 253*

347

INDEX OF NAMES AND WORKS

Descartes, René (1596–1650)   *9, 31, 42, 47, 49–50, 82, 89, 94n–5, 102–3, 108, 111,
    116, 129, 136, 139n, 144, 150–1n, 155, 165–7, 170, 181–3, 186, 197, 202–10, 214,
    223, 225, 237, 240n–2, 256n, 300–2, 305*
    Geometrie, third appendix to *Discours de la methode*, Leyden 1637   *3, 38, 94n,
        102–3n, 139n, 145, 156, 158, 172, 203, 205, 240n, 256n, 296*
    Geometria, ed. Fr. van Schooten, Leyden 1649   *139n*
    second edition, Amsterdam 1659/61   *3–42n, 49–50, 69–70, 103n, 108n,
        111n–12n, 116, 139, 143–5, 148, 151, 167n–8n, 175n, 202n–3n, 239n–40n, 256n*
    Principia pilosophiæ, Amsterdam 1644   *179n*
    'Recherche de la verite', first pubished in *DO* x: 498–527   *183n*
    Lettres, ed. C. Clerselier (3 vols), Paris 1657, 1659, 1667 (*DC*)   *151, 253*
    Literary remains   *182–3, 301, 305*
    Œuvres, ed. C. Adam and P. Tannery (12 vols+supplément), Paris 1897–1913
        (*DO*)
Desprez, Guillaume (Paris publisher)   *37, 180*
Dettonville, Amos, *see* Pascal, Blaise
De Witt, Jan, *see* Witt
Digby, Kenelm (1603–65)   *38n–40n, 53n, 96, 212n, 244n–5n*
Dijksterhuis, Eduard Jan (1892–1965)
    'James Gregory and Christiaan Huygens', *GT* (1939): 478–86   *63n*
Diophantus of Alexandria (c. A.D. 270)   *39*
    Arithmetica, ed. Cl. G. Bachet, Paris 1621   *38n–9n, 95*
    re-issued by S. Fermat with the marginal notes of P. Fermat, Toulouse 1670
        38n–9n, 148n
Dotzen, Roderich (1618–70)   *240n*
Dulaurens, François (d. before 1670)   *38, 42, 152, 154*
    Specimina mathematica, Paris 1667   *42n, 152*
    Responsio ad epistolam D. Wallisii, Paris, June 1668   *152*
Duyck, Adriaen van (b. 1628)   *273n*

Ebert (Ramus professor, Paris 1676)   *162, 212*
Eckhard, Arnold (d. 1685)   *144n, 201n*
Elsevier, (publishers) Daniel (1626–80) and Lodewijk (1604–70)   *245n*
Emperor *see* Leopold
Encyclopédie ou dictionaire raisonné des sciences, des arts et des métiers, Paris/
    Amsterdam 1751–80   *130n*
'Epistola prior': Newton to Oldenburg for Leibniz and Tschirnhaus, 13 (23) June
    1676   *225–31, 292n, 302*
'Episotla posterior': Newton to Oldenburg for Leibniz, 24 Oct. (3 Nov.) 1676
    226n, 259–76, 291n–2n
D'Espagnet, Jean   *198*
Euclid (c. 325 B.C.)   *2, 7, 12–13n, 21, 41, 76, 102, 147, 150, 200–1, 284*
    see Proclus (1533, 1560), Leibniz (c. 1696)
    Elementa, Latin: ed. Clavius, Rome 1574 (frequently reprinted)   *2n*
    ed. Barrow, Cambridge 1655   *3*
Euler, Leonhard (1707–83)   *34n, 53*
    Opuscula analytica II, St Petersburg 1785   *34n*
    'De transformatione seriei divergentis...in fractionem continuam', *Nova acta
        Acad. Scient. Petropol.* 2 (1784) 1788: 46–79 (*Opera omnia* (1) XVI, Zürich 1933:
        34–46)   *53n*

INDEX OF NAMES AND WORKS

INDEX OF NAMES AND WORKS

351

INDEX OF NAMES AND WORKS

Malebranche, Nicolas (1638–1715) *94, 136–7n, 144n*
Manolessi, Carlo *42n, 190n*
  see Pappus (1660)
Mariotte, Edmé (1620–84) *93, 118, 181, 201, 240n, 245n, 248*
  see Fréniole de Bessy (1676)
Mathion, Oded Louis (1620–1700) *46n, 174*
Maurolico, Francesco (1494–1575)
  Emendatio et restitutio conicorum Apollonii Pergæi, ed. G. A. Borelli, Messina 1654
    281
Mecklenburg-Schwerin, Duke of, see Christian Ludwig I
Meibom, Marcus (1630–1711) *96*
Mengoli, Pietro (1625–86) *26, 30–3, 89, 188, 244n, 281*
  Novæ quadraturæ arithmeticæ, Bologna 1650 *26, 32–3n, 281*
  *Via regia ad mathematicas per arithmeticam, algebram speciosam et plani-
    metriam...*, Bologna 1655 *44*
  Geometria speciosa, Bologna 1659 *281*
  Speculazioni di musica, Bologna 1670 *44*
  Circolo, Bologna 1672 *244*
  Theorema arithmeticum, Bologna 1674 *89*
Mercator (Kauffman), Nicolaus (1620–87) *74, 89, 95–6, 98, 128, 134–5, 234–6,
    273, 282, 285–6*
  see Ricci (1668)
  Logarithmotechnia, London 1668 *26, 42, 60, 68, 75n, 96, 128n, 133, 262, 281,
    290n*
  'An Illustration of the Logarithmotechnia', *PT* 3 No. 38 of 17 (27) Aug. 1668:
    759–64 *97n*
  'The apogeum of the way of planetes', *PT* 5 No. 57 of 25 March (4 Apr.) 1670:
    1168–75 *285n*
  Latin translation of Kinckhuysen, *Algebra ofte stelkonst* (1661), 1670 (manu-
    script, Oxford, Bodleian Library MS Savile 62) *NP* 2: 295–364 *41n, 256n*
Merry, Thomas
  'Invention and Demonstration of Hudden's Rules for Reducing Equations...'
    (manuscript, Oxford, Bodleian Libary, MS Savile 33) *38, 291*
Mersenne, Marin (1588–1648) *26n, 69, 86n, 103–4, 144n, 149n, 154n, 198n, 205n,
    208n*
  Cogitata physico-mathematica, Paris 1644 *37, 103*
  Correspondence, Paris 1932– (*MC*) *154n*
Metternich, Lothar Friedrich von (1617–75, Elector of Mainz 1673) *46, 161*
Meyer, Ludwig (physician in Amsterdam) *165*
Meyersperg von Meyern, August Freiherr von *30n*
Möbius, August Ferdinand (1790–1868)
  Über eine neue Behandlungsweise der analytischen Sphaerik, Leipzig 1846 (*Gesam-
    melte Werke* II, Leipzig 1885) *13n*
Møller-Pedersen, Kirsti
  'Roberval's comparison of the arclength of a spiral and a parabola', *Centaurus* 15
    (1970): 26–43 *103n*
Mohr, Georg (1640–97) *140, 151–2, 171, 211, 249*
Moivre, Abraham de (1667–1754) *239n, 273n*
  'A method of extracting the root of an infinite equation', *PT* 20 No. 240 (for
    May 1698) 1699: 190–93 *146n, 227n*

INDEX OF NAMES AND WORKS

INDEX OF NAMES AND WORKS

INDEX OF NAMES AND WORKS

Raey, Jan de (d. 1702) *165*
Rahn, Johann Heinrich (1622–76)
  Algebra, Zurich 1659 *38, 41*
Ramée (Ramus), Pierre de la (1515–72) *86n, 161*
Regiomontanus (Müller), Johannes (1436–76) *102*
  see Archimedes (1464, printed 1544)
  De triangulis omnimodis libri quinque, ed. J. Schöner, Nuremberg 1533 (quoted
    as 'Trigonometry') *102*
Regnauld, François *27, 31, 190*
Remond, Nicolas (d. 1716) *180n*
Renaldini, Carlo (1651–98)
  Ars analytica mathematica, Florence 1665, Padua 1668 *69*
Ricci, Michelangelo (1619–82) *69, 71, 72, 74, 103–4n, 198n*
  Exercitatio geometrica de maximis et minimis, Rome 1666, reprinted with Mer-
    cator, *Logarithmotechnia*, London 1668, *69, 205*
  Manuscript on algebra (Rome, Vat. Lat. 6901, 6905) *205*
Rivaud, Albert (1876–1956) *73–4*
Roannes, Artus Gouffier, Duc de (d. 1696) *247*
Roberval, Gilles Personne de (1602–75) *44, 49, 58, 69–70n, 75, 86, 103–5n, 112, 137,
    139, 160, 174, 178–9, 196, 204, 244, 248, 301, 305*
  see Auger (1962), Møller-Pedersen (1970)
  Note on Descartes' concept of space (summer 1648), *DO* xi: 687–91 *204n*
  'Elementa geometrica' (manuscript) *179*
  'De locis planis' (by Fermat; wrongly attributed to R.) *137n*
  'Traité des indivisibles' (*c.* 1640), *Divers ouvrages*: 190–245 *86n*
  'Observations sur la composition des mouvemens et sur le moyen de trouver les
    touchantes des lignes courbes', *Divers ouvrages*: 69–111 *196*
  *Divers ouvrages de mathematique et de physique par Messieurs de l'Académie Royale
    des Sciences*, ed. Ph. de La Hire, Paris 1693 *179n*
  Manuscripts (Paris, Bibl. Nat.) *178–9, 181, 305*
Römer, Ole (1644–1710) *160, 247–9*
  Manuscript on epicycloidal gearwheels (?) *247*
Rosenfeld, Louis
  'La théorie des couleurs de Newton et ses adversaires', *Isis* 9 (1927): 44–65
    213n
  'R. Fr. de Sluse et le problème des tangentes', *Isis* 10 (1928): 416–34
    74n
Royal Society, London (1660–) (RS) *23, 24, 29, 71, 80, 99–101, 117, 121, 123, 131,
    141, 212–13, 231, 257, 277, 299*
Rudolff, Christoph (1500?–45?), see Stifel (1553) *151*

Sacrobosco, Johannes de (d. 1256?) *3*
St Petersburg, Academy, see Academia Petropolitana
Saint-Vincent, Grégoire de, see Grégoire
Santini, Antonio (1577–1662) *198*
Sauveur, Joseph (1653–1716) *245n*
Schelhammer, Günther Christoph (1649–1716) *165*
  De auditu liber unus, Leipzig 1684 *165n*
  'Novæ institutiones medicæ' (manuscript, present location uncertain) *165n*
Schloer, Friedrich (1647?–1727) *25*

362

INDEX OF NAMES AND WORKS

Torricelli, Evangelista (1608–47) (*cont.*)
  'De infinitis spiralibus', *TO* i. 2: 349–73; 381–92   *103n–4n*
  Opera geometrica, Florence 1644   *51, 104, 198n*
  Opere, ed. G. Loria and G. Vassura, Faënza 1919 (*TO*)   *51n*
Tschirnhaus, Ehrenfried Walter von (1651–1708)   *7n, 31, 47n–9n, 51n, 64n, 74n,*
    76n–7, 93n, 95n, 130n, 136n, 145n–6n, 150, 156, 159, 161, 165–86, 190n–1n, 195,
    202, 205–10, 212–15, 221–2, 226, 233, 239–40n, 242n–3, 246n, 249–60, 266,
    270n, 272, 285, 292, 300–4
  'Methodus exhibendi omnium æquationum radices, si sint in proportione arith-
    metica' (*GT*: 317–19)   *167*
  'Nova methodus tangentes curvarum expedite determinandi', *AE* December
    1682: 391–3   *175n*
  Manuscript on algebra and geometrical loci (Wroclaw, Bibl. Uniw. Akc 1948/
    562)   (*166*)
  'Methodus auferendi omnes terminos intermedios ex æquatione', *AE* May 1683:
    204–7   *176*
  'Methodus datæ figuræ rectis lineis et curva geometrica terminatæ aut quad-
    raturam aut impossibilitatem ejusdem quadraturæ determinandi', *AE* October
    1683: 433–7   *191n*
  Medicina mentis, Amsterdam 1687, Leipzig ₂1695   *185n*
Turnbull, Herbert Westren (1885–1961)   *11n, 124n*
  see Gregory, James (*GT*)
  ed. Newton, *Correspondence* 1–3 (1959–61) (*NC* 1–3)

Vagetius, Johannes (1633–91)   *180n*
Valens, Vettius (*c.* A.D. 300)   *24*
Valerio, Luca (1552–1618)   *69*
Van Duyck, Adriaen, see Duyck
Van Gent, Pieter, see Gent
Varignon, Pierre (1654–1722)   *71n, 95n, 139n, 147n, 244n*
Vaughan, William (1648?–1712)   *137n*
Velthuysen, Lambert van (1622–85)   *8n*
Venatorius (Gechauff), Thomas, see Archimedes 1544
Vernon, Francis (1637?–77)   *42n, 76n, 180n, 206n*
Viète, François (1540–1603)   *9, 39, 89, 102–3, 143–5, 177n, 197, 223, 227n, 252n,*
    256
  Variorum de rebus mathematicis responsorum liber octvaus, Tours 1593 (*Opera*:
    347–435)   *102n, 177n*
  Supplementum geometriæ, Tours 1593 (*Opera*: 240–57)   *102n*
  Zeteticorum libri quinque, Tours 1595 (*Opera*: 42–81)   *39n*
  *Ad problema, quod omnibus mathematicis totius orbis construendum proposuit
    Adrianus Romanus*, Paris 1595 (*Opera*: 305–24)   *146*
  Appollonius Gallus, Paris 1600 (*Opera*: 325–46)   *102n*
  De numerosa potestatum purarum et adfectarum ad exegesin resolutione, ed. M.
    Ghetaldi, Paris 1600 (*Opera*: 162–228)   *89n, 226, 256n*
  De æquationum recognitione et emendatione (1591) ed. A. Anderson, Paris 1615
    (*Opera*: 82–161)   *94n*
  Ad angularium sectionum analyticen theoremata, ed. A. Anderson, Paris 1615
    (*Opera*: 286–304)   *227n*
  Opera mathematica, ed. F. van Schooten, Leyden 1646

INDEX OF NAMES AND WORKS

25

SUBJECT INDEX

This index was compiled by Herbert Mehrtens.